Active Flow Control: Recent Advances in Fundamentals and Applications — Volume II

Active Flow Control: Recent Advances in Fundamentals and Applications — Volume II

Guest Editors

Feng Ren
Xin Wen
Hui Tang

MDPI

Basel • Beijing • Wuhan • Barcelona • Belgrade • Novi Sad • Cluj • Manchester

Guest Editors

Feng Ren
School of Marine Science and Technology
Northwestern Polytechnical University
Xi'an
China

Xin Wen
School of Mechanical Engineering
Shanghai Jiao Tong University
Shanghai
China

Hui Tang
Department of Mechanical Engineering
The Hong Kong Polytechnic University
Hong Kong
China

Editorial Office
MDPI AG
Grosspeteranlage 5
4052 Basel, Switzerland

This is a reprint of the Special Issue, published open access by the journal *Actuators* (ISSN 2076-0825), freely accessible at: https://www.mdpi.com/journal/actuators/special_issues/G0TT6V72YC.

For citation purposes, cite each article independently as indicated on the article page online and as indicated below:

Lastname, A.A.; Lastname, B.B. Article Title. *Journal Name* **Year**, *Volume Number*, Page Range.

ISBN 978-3-7258-3481-5 (Hbk)
ISBN 978-3-7258-3482-2 (PDF)
https://doi.org/10.3390/books978-3-7258-3482-2

Contents

actuators

MDPI

Review

Model-Free Closed-Loop Control of Flow Past a Bluff Body: Methods, Applications, and Emerging Trends

Feng Ren [1], Xin Wen [2,*] and Hui Tang [3]

1 School of Marine Science and Technology, Northwestern Polytechnical University, Xi'an 710072, China; renfeng@nwpu.edu.cn
2 School of Mechanical Engineering, Shanghai Jiaotong University, Shanghai 200241, China
3 Department of Mechanical Engineering, The Hong Kong Polytechnic University, Hong Kong, China; h.tang@polyu.edu.hk
* Correspondence: wenxin84@sjtu.edu.cn

Abstract: Flow past one or multiple bluff bodies is almost ubiquitous in nature and industrial applications, and its rich underlying physics has made it one of the most typical problems in fluid mechanics and related disciplines. The search for ways to control such problems has attracted extensive attention from both the scientific and engineering fields, as this could potentially bring about benefits such as reduced drag, mitigated noise, suppressed vibration, and enhanced heat transfer. Flow control can be generally categorized into passive and active approaches, depending on whether there is an external energy input to the flow system. Active control is further divided into open-loop approaches and closed-loop approaches, depending on whether the controller depends on feedback signals extracted from the flow system. Unlike in many other applications of passive flow control and open-loop active flow control, theoretically advantageous closed-loop controls are quite rare in this area, due to the complicated features of flow systems. In this article, we review the recent progress in and future perspectives of flow past a single or multiple bluff bodies using model-free closed-loop control so as to outline the state-of-the-art research, determine the physical rationale, and point to some future research directions in this field.

Keywords: active flow control; flow past a bluff body; machine learning

Citation: Ren, F.; Wen, X.; Tang, H. Model-Free Closed-Loop Control of Flow Past a Bluff Body: Methods, Applications, and Emerging Trends. *Actuators* **2024**, *13*, 488. https://doi.org/10.3390/act13120488

Academic Editor: Luigi de Luca

Received: 30 September 2024
Revised: 16 November 2024
Accepted: 27 November 2024
Published: 29 November 2024

1. Background

Flow past a single or multiple bluff bodies is among the most typical flow types of fluid mechanics and is almost ubiquitous in nature and engineering applications [1,2]. In general, bluff bodies produce a stronger separated flow, have a broader wake with low pressure and complicated vortical structures, and have higher drag compared with streamlined bodies. Extensive studies have been carried out, using theoretical, computational, and experimental approaches to explore the very rich physics of such flows, including flow transition, flow instability, flow separation, and diverse vortical structures, as well as to provide engineering solutions [1–4]. Taking the flow past a finite-length circular cylinder with one free end and one fixed end as an example, as shown in Figure 1, very complicated flow features can be observed, such as the horseshoe vortex that appears in proximity to the fixed end, tip vortices, shear layers, an alternately shedding Kármán vortex street, and time-mean vortices [5,6]. These flow features exert complicated forces on the bluff body, which play a crucial role in engineering applications. For instance, in the case of an underwater vehicle, reducing hydrodynamic drag not only saves energy but also improves cruising speed and range [7,8].

In addition to these pure flow systems, flow past a bluff body is also accompanied by flow-induced vibrations [9–12], flow-induced noise [13–15], heat transfer [16–18], etc., making the problem more complicated and challenging. For example, vortex-induced vibration (VIV) is one of the most typical flow-induced vibrations that occurs when asymmetrical vortex pairs shed from a bluff body that is immersed in a uniform flow and is elastically mounted [9,11]. For different combinations of vibration amplitude and wavelength, the phase diagram (Figure 2) shows diverse wake patterns, where vortex shedding and structural motion are synchronized; these are categorized as the "mS + nP" mode. In this mode, during one vibration cycle, m single vortices and n vortex pairs are shed from the vibrating body and convect downstream [9]. In many engineering scenarios, VIV and other flow-induced vibrations are often unwanted, as they can induce mechanical fatigue or even structural damage [12,19]. Conversely, the mechanical energy that is generated can also be harnessed and exploited as green energy.

The noise generation and radiation accompanied by flow past a bluff body are also a key issue [13,15,20]. As reviewed and summarized by [13], fluctuations at the wall, viscous shear stresses at the wall, and turbulent velocity fluctuations are identified as the apparent sources of boundary-layer noise, which are closely related to the near wake. Figure 3 demonstrates the far-field dilatation of flow past a circular cylinder at $Re = 3900$ and $Ma = 0.4$, as well as the near-field wake [20]. It is straightforward to note that sound pressure waves are closely related to vortex shedding. In laminar cases, noise is primarily generated by vortex formation and shedding, where a negative pressure pulse is produced on one side of the cylinder surface from which a vortex is shed, while a positive pressure pulse is produced on the opposite side [21]. In turbulent cases, the mechanism becomes more complex, involving the effects of turbulent fluctuations [15,20]. The generation and propagation of noise play a significant role in the field of aeronautics, where environmental noise is a key factor influencing passengers' experience [15]. Flow-induced noise is particularly important for underwater vehicles, as it reduces the signal-to-noise ratio of communication and navigation devices that primarily rely on sound, thereby increasing the likelihood of detection.

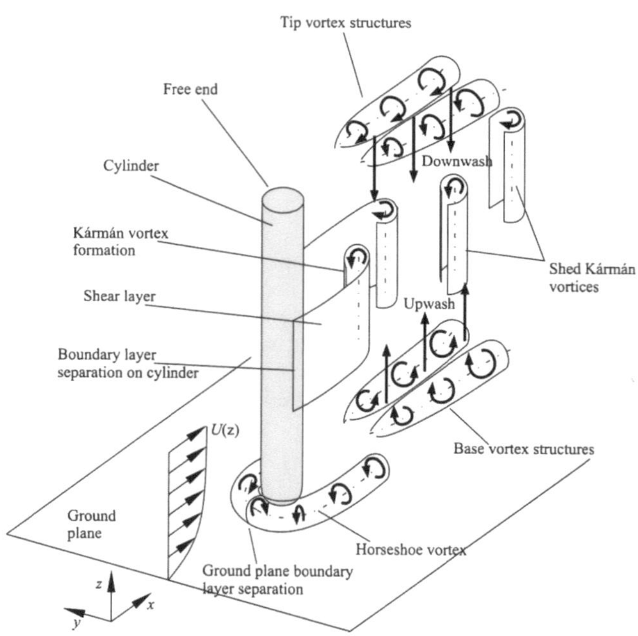

Figure 1. Schematics of flow features past a circular cylinder with one free end and the other end mounted on the wall. This figure is reproduced from the work of Sumner [5], with permission.

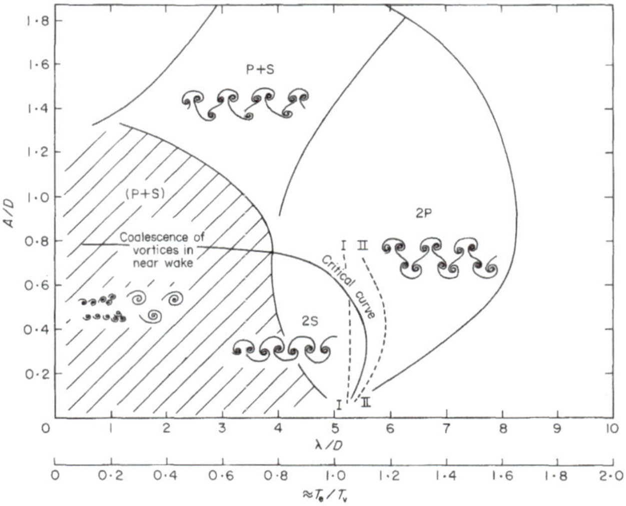

Figure 2. Map of vortex synchronization patterns of vortex-induced lateral (cross-flow) vibrations of a circular cylinder in a uniform stream. This figure is reproduced from the work of Williamson and Roshko [9], with permission.

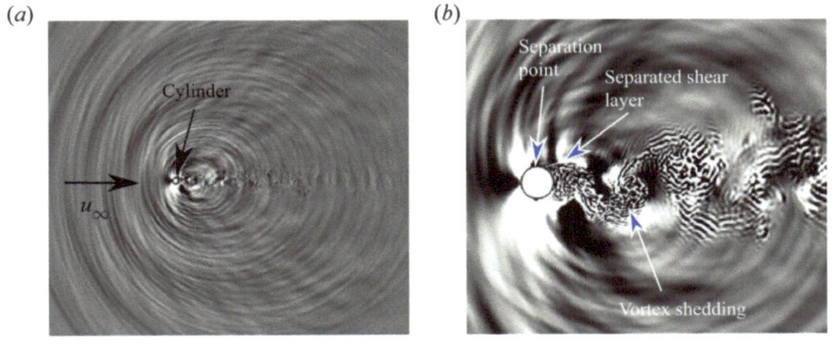

Figure 3. Instantaneous snapshots of flow past a circular cylinder in the turbulent regime: (**a**) far-field dilatation field, and (**b**) near-field wake and dilatation field. This figure is reproduced from the work of Li et al. [20], with permission.

Controlling flow past a bluff body can provide many benefits, such as drag reduction, noise mitigation, and vibration suppression [22,23]. Control strategies can generally be categorized into two main types: passive approaches and active approaches [3,12,22,24]. In recent years, as technologies such as micro-electro-mechanical systems (MEMS) have been developed [25], many advanced actuators have been invented [26], including blowing/suction jets, synthetic jets, sweeping jets, pulse jets, plasma actuators, and rotating rods. A properly chosen actuator can yield significant control effects. For example, Dong et al. [27] used the windward-suction–leeward-blowing (WSLB) actuator and successfully eliminated the Kármán vortex street. The pulse jet modulated by Abdolahipour et al. [28] can use both high and low frequencies [29], which can then control flow past an airfoil in a highly deflected flap and at high angles of attack [30], as well as controlling the vortex shedding [31]. Mirzaei et al. [32] also proved that the plasma actuator is efficient in controlling the vortex shedding. However, even with well-designed and manufactured actuators, it is often challenging to exert the precise amount of forcing needed to mediate the flow dynamics and achieve optimal performance [26].

Closed-loop flow control exerts a certain amount of energy on the flow system in a self-adaptive manner, according to the feedback signals sampled from the dynamically varying flow system and the adopted controller [33]. In scenarios where the flow system is linear, weakly linear, or moderately linear, a linear control design can usually be utilized [33,34], such as optimal control methods including the linear–quadratic regulator (LQR) and linear–quadratic–Gaussian (LQG) method. Researchers have also developed methods that rely on a model reduction in the flow systems [35,36], such as flow control methods based on proper orthogonal decomposition (POD), dynamic mode decomposition (DMD), resolvent analysis, etc. In contrast, there are also model-free methods of designing the controller that are independent of the mathematical modeling of the flow system, instead relying on methods such as parameter tuning or optimization or data-driven dynamic optimization. In the engineering field, flow systems always share nonlinear and high-dimensional features; for example, a turbulent flow and separation flow. Therefore, even though there are many successful examples of closed-loop flow control, designed using rigorous mathematical models, there is still an urgent demand for a general solution for closed-loop control that is effective, efficient, and robust, representing an ongoing research challenge [23,33].

It is thus worth reviewing previous studies in which researchers apply model-free closed-loop control strategies to address specific problems related to flow past a bluff body, especially the key elements involved therein, such as the control method, the control objective, the sensors, and the actuators. It would be interesting to lead readers to further consider what can one expect from closed-loop AFC, such as an excellent control performance or physical insights. It is also necessary to mention that, with the rapid development of machine learning, two quite promising ideas and effective tools, such as genetic programming (GP) and reinforcement learning (RL), were provided to guide the closed-loop AFC in a data-driven and model-free manner. These have attracted extensive attention in the field of fluid mechanics, especially from researchers working on flow control [37–40]. Reviewing the research cases and emerging trends can help to identify existing problems in this specific field and future perspectives. In the remaining sections of this manuscript, we categorize the current model-free closed-loop control into four types, i.e., model-free PID control, GP-based control, RL-guided control, and cluster-based control. In Table 1, some examples of model-free and closed-loop control strategies are summarized, including their methods, objectives, and the sensors and actuators they utilize.

Table 1. An overview of the methods and typical applications of closed-loop control of flow past a bluff body.

Method	Objective	Sensor	Actuation	Reference
PID control	Stabilize wake	Transverse velocity	Synthetic jets	Park et al. [41]
	Reduce vortex strength and suppress vibration	Velocity and vibration displacement	Piezoelectric ceramic actuators	Zhang et al. [42]
	Mitigate lift fluctuations	Lift	Self-rotation	Lu et al. [43]
	Reduce noise	Lift	Self-rotation	Du and Sun [44]
	Stabilize wake	Spanwise-averaged velocity	Blowing/suction jets	Yun and Lee [45]
	Suppress vibration	Transverse velocity	WSLB jets	Wang et al. [46]
	Mediate vibration	Vibration displacement/velocity	Self-rotation	Vicente-Ludlam et al. [47]
	Suppress vibration	Vibration velocity	Damping force	Song et al. [48]
	Suppress vibration	Lift	Self-rotation	Hasheminejad et al. [49]
	Suppress vibration	Vibration displacement	Control force	Rabiee and Esmaeili [50]

Table 1. *Cont.*

Method	Objective	Sensor	Actuation	Reference
GP-based control	Mitigate separation	Skin friction	Active vortex generator	Debien et al. [51]
	Reduce drag	Wall pressure	Pulsed jets	Li et al. [52]
	Stabilize wake and reduce drag	velocity	Self-rotation	Raibaudo et al. [53]
	Stabilize wake and reduce drag	Velocity	Self-rotation	Raibaudo and Martinuzzi [54]
	Suppress vibration	Vibration displacement	Blowing/suction jets	Ren et al. [55]
	Reduce drag	Velocity	Blowing/suction jets	Castellanos et al. [56]
RL-guided control	Reduce drag	Velocity array	Blowing/suction jets	Rabault et al. [57]
	Reduce drag	Velocity array	Blowing/suction jets	Ren et al. [58]
	Reduce drag	Velocity array	Blowing/suction jets	Paris et al. [59]
	Suppress vortex shedding	Velocity array	Blowing/suction jets	Li and Zhang [60]
	Reduce drag	Velocity array	Blowing/suction jets	Varela et al. [61]
	Mitigate lift	Velocity array	Self-rotation	Zhao et al. [62]
	Reduce drag	Wall pressure	Blowing/suction jets	Wang et al. [63]
	Mitigate lift	Velocity array	Self-rotation	Ren et al. [64]
	Reduce drag	Wall-pressure	Blowing/suction jets	Chen et al. [65]
	Achieve hydrodynamic stealth	Velocity array	WSLB jets	Ren et al. [66]
	Reduce drag	Force	Rotary rod	Fan et al. [67]
	Reduce drag	Force	Rotary rod	Wang et al. [68]
	Suppress vibration	Velocity array	Blowing/suction jets	Zheng et al. [69]
	Suppress vibration	Velocity array	Self-rotation	Ren [70]
	Reduce vibration	Velocity array	Blowing/suction jets	Chen et al. [71]
	Suppress vibration	Sensorymotor cues	Self-rotation	Ren et al. [72]
	Enhance vibration	Velocity array	Blowing/suction jets	Mei et al. [73]
	Enhance heat transfer	Sensorymotor cues	Blowing/suction jets	Ren et al. [74]
	Suppress vibration	Vibration displacement and velocity	Blowing/suction jets	Zheng et al. [75]
	Reduce drag	Velocity	Blowing/suction jets	Castellanos et al. [56]
Cluster-based control	Control flow separation	Force	Reduce power consumption	Nair et al. [76]
	Reduce drag	Velocity field	Self-rotation	Wang et al. [77]

2. Model-Free PID Control

In classical control theory, controllers are typically designed based on linear assumptions. The AFC also follows this principle [34]. PID control is perhaps the most classical control approach and is extensively utilized in engineering fields. In the closed-loop AFC of flow systems, the interpretation of PID control differs slightly from that in classical control theory. In our view, PID control is primarily considered a model-free approach. By tuning the control parameters, certain control performance can be achieved.

The PID control has been proven to be effective and efficient in controlling flow past a bluff body. For example, Park et al. [41] applied a proportional control strategy to stabilize the Kármán vortex street at Reynolds numbers lower than 100, where a pair of synthetic jets is deployed near the separation point, and the transverse velocity at a selected downstream position is used to provide feedback signals. With this relatively simple strategy, vortex shedding is completely suppressed at $Re = 60$. Zhang et al. [42] performed an experimental study of PID control for the vortex shedding and vibration of an elastically supported square cylinder in the turbulent flow regime. They investigated different feedback signals, including turbulent velocity, structural vibration displacement, and combinations thereof, showing that feedback from both flow and structural vibrations can remarkably reduce vortex strength as well as vibration amplitude. Lu et al. [43] applied proportional control to suppress the transverse lift of a circular cylinder at low Reynolds numbers, where the rotational velocity of the cylinder is proportional to the instantaneous lift coefficient exerted on the cylinder. It was found that the control performance is closely related to the proportional control parameters, and the optimal control parameters vary with the Reynolds number. Using the same method but different flow conditions and a different control objective, Du and Sun [44] applied proportional control to reduce noise in a circular cylinder at $Re = 1000$, showing that the noise generated by the cylinder can be reduced by 10 dB due to the suppression of vortex shedding. Figure 4 shows the instantaneous vortical structures with and without control. Based on Curle's acoustic analogy [78], the noise generated by the solid wall is primarily attributed to force fluctuations. Thus, the objectives of lift suppression and noise reduction would lead to similar physical attributes. Yun and Lee [45] applied proportional control to turbulent flow over a circular cylinder with a similar blowing/suction jet configuration to that in Ref. [41], while the feedback transverse velocity was averaged along the spanwise direction.

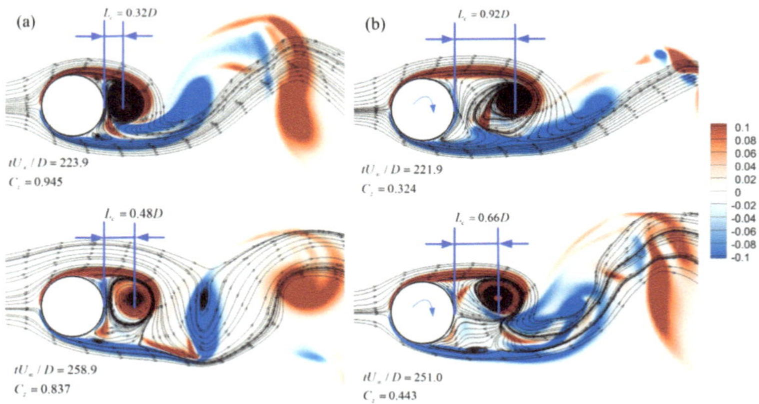

Figure 4. Instantaneous vortical structures of flow past a circular cylinder at $Re = 1000$: (**a**) uncontrolled case, and (**b**) a case with proportional control. This figure is reproduced from the work of Du and Sun [44], with permission.

Controlling the VIV using the PID controller is of great interest. Wang et al. [46] carried out a systematic parametric study on the effects of proportional and PI control, where two groups of WSLB actuators are used to mediate the flow dynamics near the cylinder, and a transverse velocity feedback signal in the near wake is utilized. Within certain control parameter ranges, VIV can be effectively suppressed. Vicente-Ludlam et al. [47] studied rotary control to manage the VIV of a circular cylinder in an experimental environment within a Reynolds number range of 1500 to 10,000. The rotation rate was made proportional to either the cylinder's vibration displacement or its vibration velocity, similar to the concepts of proportional control and proportional–derivative control. Both control laws resulted in reduced or enhanced oscillations. A recent study by Song et al. [48] applied PID control to suppress VIV with two degrees of freedom and demonstrated that PI or PID control can significantly reduce the vibration range in both the streamwise and transverse directions by 68.4% to 97.1%.

In PID control, determining the control variables is usually a difficult task, especially when the flow system involves turbulence or fluid–structure interaction (FSI), or when the controller has multiple inputs and/or multiple outputs (MIMO). In the field of control theory, there are also problems when the system to be controlled is regarded as a black box whose transfer function cannot be directly obtained. Feasible solutions include adaptive control methods, fuzzy control, sliding mode control, and others. Researchers working on closed-loop AFC have also used this idea. For example, Hasheminejad et al. [49] applied fuzzy proportional control for the VIV of a circular cylinder under laminar flow conditions. Three different proportional controllers were studied, all of which outperformed the representative open-loop control with the prescribed rotational oscillations. Rabiee and Esmaeili [50] further applied fuzzy PID control to control the flow-induced vibrations of two circular cylinders in tandem configuration with the aid of a fuzzy logic system. A schematic diagram of the coupling framework of the controller and the numerical environment is demonstrated in Figure 5, where the desired control force is determined using the feedback signals from vibration displacement.

Because the flow system is nonlinear in nature, while PID control theory is developed based on linear approximations, the effectiveness and robustness of PID control may be limited to flow systems with weak nonlinearity, primarily in the laminar regime. In turbulent conditions, special treatments must be implemented, such as using the spanwise-averaged sensing signal Yun and Lee [45]. Due to the gap between linear control theory and nonlinear flow systems, PID control must be systematically studied before being applied in practical scenarios.

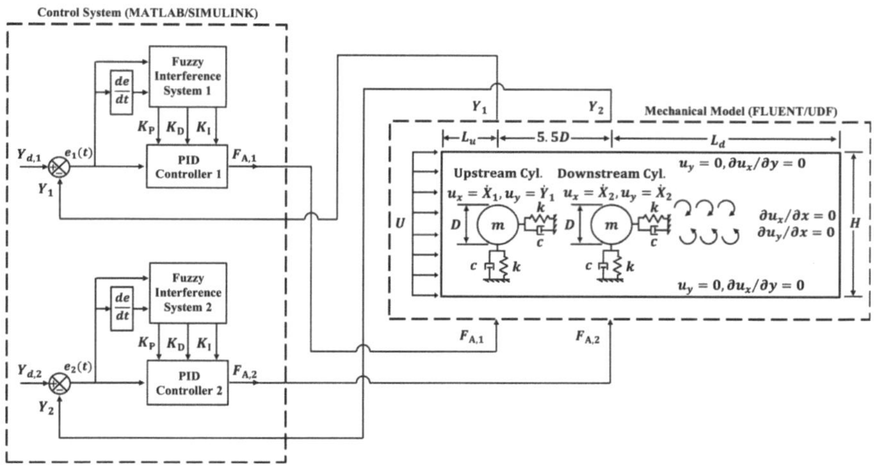

Figure 5. Schematics of the coupling framework between the fuzzy PID controllers and the CFD model. This figure is reproduced from the work of Rabiee and Esmaeili [50], with permission.

3. Genetic Programming-Based Control

Compared to the aforementioned PID control, control methods that are independent of prior knowledge of the flow system are attractive, as they can provide a general solution for closed-loop flow control, especially when addressing the challenging task of modeling complicated flows and determining appropriate parameters to adjust control performance. Among these model-free methods, machine learning has attracted increasing attention from the flow control community. Most machine learning approaches share data-driven features, making their application in problems in fluid mechanics feasible due to the large amount of data available from experiments and numerical simulations. The applications of machine learning in fluid mechanics were reviewed previously [39,79–81].

In this section, GP-based control is reviewed. GP stems from the genetic algorithm (GA), which is usually used to optimize one or multiple values. In GP, symbolic expressions are involved in the evolution process, with the aid of the locator/identifier separation protocol (LISP) language. Because these symbolic expressions can naturally function as explicit control laws, GP can be used to generate and select the optimal control law, similarly to the way in which optimal parameters are determined by the GA. Both GP and GA share an evolutionary algorithm. Here, a group of individuals participate in the natural selection, and only a few individuals are selected to generate the next generation through crossover, replication, mutation, etc. Figure 6 shows the mutation of an individual and the crossover between two individuals Gautier et al. [82].

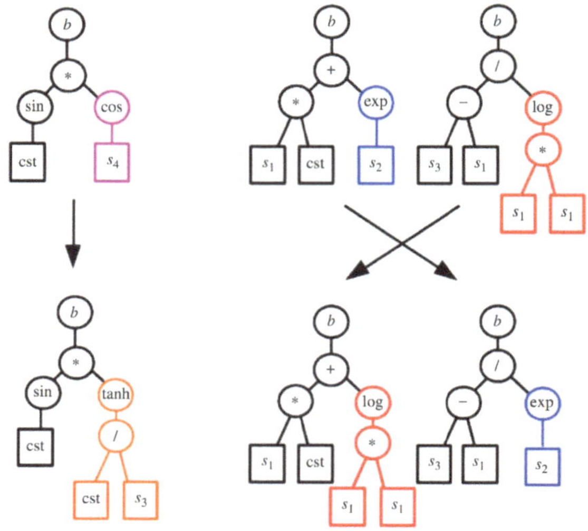

Figure 6. An example of mutation of an individual (**left subfigure**) and an example of crossover between two individuals (**right subfigure**). This figure is reproduced from the work of Gautier et al. [82], with permission.

Gautier et al. [82] were perhaps the first to apply GP to closed-loop AFC, where the backward-facing step is controlled by blowing/suction jets deployed near the step corner. Using a similar method and idea as those of Gautier et al. [82], Debien et al. [51] conducted a closed-loop control of flow over a sharp edge ramp, targeting separation mitigation and early re-attachment. Active vortex generators were used to mediate the flow, and downstream skin-friction sensors provided feedback information. Comparisons with the open-loop control experiments show that the open-loop system utilizes the lock-on effect, while the GP-based control accelerates shear layer growth with a lower energy. Li et al. [52] carried out a closed-loop control of flow past the blunt-edged Ahmed car model based on linear GP. Multiple pulsed jets at all trailing edges were utilized as actuators, which were also combined with a deflection surface to exploit the Coanda effect. A group of wall pressure sensors deployed at the rear side was then used to provide feedback signals. In an unsupervised manner, the linear GP-based control achieved an approximately 22% drag reduction rate.

Raibaudo et al. [53] performed experiments in which linear GP was utilized to stabilize the wake of a fluidic pinball and reduce drag. The velocities from three hot-wire sensors were used to provide feedback signals and the constant rotational velocities of three cylinders were used to mediate the flow dynamics. The results show that the GP-based strategies reveal unanticipated solutions or parameter relationships. To account for the effects of unsteady actuation, Raibaudo and Martinuzzi [54] conducted further experimental studies in which each cylinder of the fluidic pinball was controlled independently and could vary periodically. The results indicate that the GP-based control was more efficient than traditional methods at optimizing within a large parametric space and was also more robust than open-loop control. Ren et al. [55] used the GP to actively control the VIV of a circular cylinder under lock-in conditions. Figure 7 shows the schematics of the GP-based control, where the upper module represents the dynamic FSI system and the lower module represents the GP agent. The converged control law suppresses the VIV amplitude by 94.2% and achieved a 21.4% better overall performance than the best proportional control at a Reynolds number of $Re = 100$. In robustness tests across a Reynolds number range of 100 to 400, the GP-based control law remains highly effective, while the proportional control, which is suitable for $Re = 100$, fails the test.

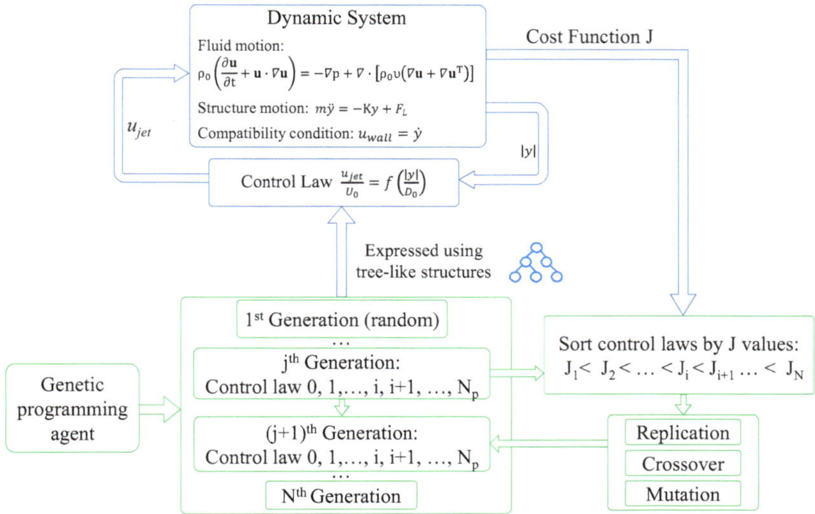

Figure 7. Framework of the GP-based control for vortex induced vibration control in the numerical environment based on the LBM. This figure is reproduced from the work of Ren et al. [55], with permission.

In a numerical environment, the evolutionary algorithm typically utilizes hundreds or thousands of individuals until the control performance converges. Therefore, it is crucial to employ a computationally efficient simulation tool. For example, in the aforementioned study by Ren et al. [55], a lattice Boltzmann solver accelerated on a graphics processing unit device [83,84] is used, allowing for a single case initialized with a fully developed flow field to be completed in 20 min or less.

In an experimental environment, an individual trial can usually be completed within just a few minutes, which is more advantageous than the numerical approach. The hardware costs in terms of real-time signal recording and a fast actuation response, as well as the uncertainties during the experimental study, are other crucial issues that need to be addressed.

4. Reinforcement Learning-Guided Control

RL is an important branch of machine learning. In RL, an agent interacts with a dynamic environment, perceives the environment's state, and learns actions through trial and error in order to gain a high reward [85]. RL is usually combined with deep neural networks for decision-making and performance evaluation, and is then referred to as DRL. RL is applied to environment-interactive tasks such as the game of Go [86], where the RL agent learns from scratch rather than relying on the human knowledge that AlphaGo adopts [87]. RL has also been applied to train a simplified glider model to exploit upward plumes in sunny weather to soar higher [88]. Verma et al. [89] applied RL to train a fish to swim efficiently. These successes have inspired researchers in the field of fluid mechanics to tackle the closed-loop AFC problem. Thorough reviews of the use of RL in AFC can be found in the previous literature [79–81,90,91].

Unlike conventional approaches to constructing a controller, which rely heavily on prior knowledge of the system, the RL agent does not have any prior knowledge of fluid dynamics, and is thus regarded as a model-free approach. Diverse RL algorithms and their associated framework can be applied to AFC [92]. Herein, we use the popular proximal policy optimization (PPO) [93] as an example. In the interactive framework, effective control strategies are learned through interactions between the RL agent and the flow-related environment. Initially, the RL agent interacts with the environment using randomized actions. Through trial and error, it learns how to exert a specific action when the system is in a particular state. Simultaneously, it learns how to predict the long-term reward based on this state information. The state of the environment can be represented by velocity probes, wall normal and/or shear stresses, sensory–motor cues in flow-induced vibration problems, etc. The model's actions can include adjustable jet velocity, rotary motion, and body force. During the training, the reward is evaluated and fed back to the RL agent, providing a baseline for the agent to learn how to assess control performance and encouraging it to achieve a greater reward. Figure 8 shows the schematics of RL-guided AFC for enhancing thermal convection from an elastically mounted circular cylinder, which involves the basic interactive relationship between the RL agent and the flow environment [74]. In this setup, the agent receives the state of the flow system, determines the action, and judges whether the current action earns a good reward. The control strategy is updated after sampling a state–action–reward chain within one episode.

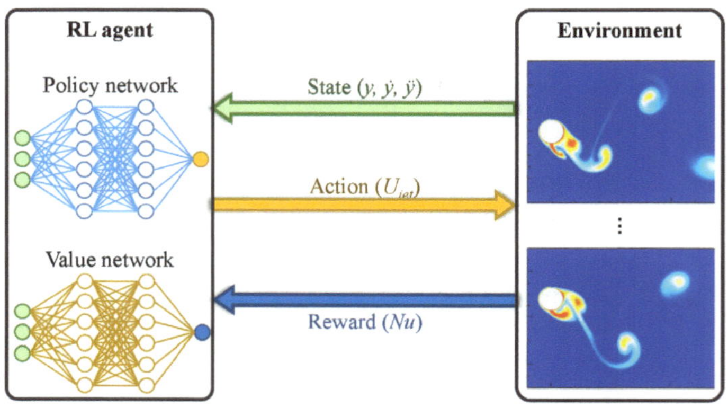

Figure 8. Schematics of RL-guided AFC for enhancing the thermal convection of a circular cylinder undergoing VIV. This figure is reproduced from the work of [74].

Before being applied to AFC, RL is utilized to maneuver the motion of a gliding bird to achieve higher altitudes by taking advantage of rising plumes in sunny weather [88], as well as to guide fish-swimming to exploit wake vortices and save energy [89]. In the more straightforward applications of RL in AFC, Rabault et al. [57] were perhaps the first to apply RL to obtain a closed-loop AFC strategy for drag reduction in a circular cylinder confined to a channel. In this pioneering work, two synthetic jet actuators were deployed at the top and bottom stagnation points of the cylinder and operated in a one-blowing–one-suction mode. A group of sensors was placed both around the cylinder and in the wake. The final converged strategy achieves an approximately 8% drag reduction rate at a fixed Reynolds number of 100. Ren et al. [58] extended intelligent flow control in laminar situations to weakly turbulent conditions where the Reynolds number was 1000, and the baseline flow exhibited many chaotic features. With a similar control setup to that of Rabault et al. [57], an approximately 30% drag reduction rate was achieved, along with mitigated lift fluctuations. Moreover, through a sensitivity analysis, it was noted that only a subset of the 151 velocity sensors play key roles in the converged control strategy. This finding provides a simple yet promising solution for reducing the number of sensors and optimizing the sensor layout. Paris et al. [59] applied RL to reduce the drag of a cylinder under laminar conditions while focusing on the efficiency and robustness of the identified control strategy, and proposed an improved RL algorithm to optimize the sensor layout. The obtained control strategy was shown to be robust within a Reynolds number range of [100, 216] and to measurement noise. Li and Zhang [60] applied RL to suppress vortex shedding from a wall-confined cylinder by embedding the physical information of the flow into the RL-based control. Global linear stability and sensitivity analyses based on the time-mean flow and steady flow were conducted across a range of blockage ratios and Reynolds numbers, which were then used to design RL-based control policies that can outperform the gradient-based optimization method. Varela et al. [61] explored the use of an intelligent AFC of a circular cylinder over a wide range of Reynolds numbers and identified different control strategies using DRL as the Reynolds number increased. For $Re \leq 1000$, the control strategy was mainly based on classical opposition control relative to wake oscillation. For $Re = 2000$, the agent applied a high-frequency actuation that energized the boundary layers and the separation zone, further modulating flow separation and reducing drag in a manner similar to that of the drag crisis. Zhao et al. [62] used DRL-guided AFC to mitigate the lift fluctuations in a circular cylinder placed in the wake of an upstream, equal-size cylinder in a tandem configuration. Through an analysis of the flow structures, it was revealed that the learned policy accelerates the shear layer development in the rear cylinder, subsequently adjusting its interaction with the wake of the front cylinder. Wang et al. [63] proposed a self-learning algorithm for reducing drag and mitigating lift fluctuations in a cylinder based on DRL. By transforming the sensor signals into dynamic features that can predict future flow states, the performance of RL-guided control can be significantly improved. Ren et al. [64] carried out intelligent controls aiming to mitigate the fluctuations in a single circular cylinder and a wake-interfered circular cylinder, respectively. With six feedback velocity signals placed near the wake, a lift mitigation rate of larger than 90% was achieved for both scenarios. In addition to the velocity sensors, many researchers used this method to study feedback; Chen et al. [65] used wall pressure, monitored at the surface of a circular cylinder, and achieved both drag reduction and lift mitigation for $Re \in [100, 400]$.

The RL can also be used to realize unconventional AFC objectives. Ren et al. [66] proposed the concept of hydrodynamic stealth and provided an RL-based solution. In this concept, five pairs of WSLB jets are used to eliminate the velocity deficit, while a velocity rake consisting of 33 velocity sensors placed in the near wake provides feedback signals. Figure 9 shows the learning process, which consists of 1000 episodes. As the learning proceeds, the wake signature gradually vanishes. Ultimately, the wake signature disappears, and the downstream velocity profile at a distance of two diameters from the cylinder becomes almost identical to the clean stream in front of the cylinder.

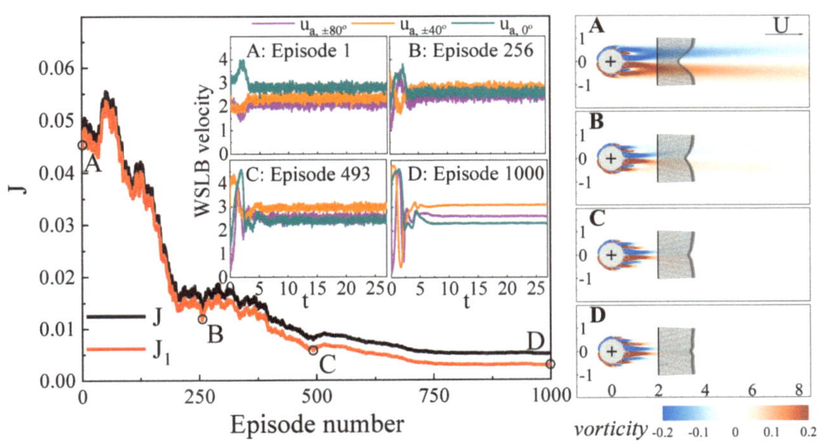

Figure 9. Training process of the reinforcement learning guided closed-loop control to achieve hydrodynamic stealth. Four subplots denote the instantaneous wake patterns and measured velocity profiles at four selected stages. This figure is reproduced from the work of Ren et al. [66], with permission.

The applications of RL-based AFC in complicated environments, such as turbulent flows, have attracted considerable attention. Fan et al. [67] demonstrated the effectiveness of RL in bluff body control through experiments, aiming to maximize power gain efficiency by adjusting the rotational speed of two small cylinders located parallel to and downstream of the main cylinder. By properly defining rewards and designing noise reduction techniques, the RL agent discovered a control strategy that is comparable to the optimal strategy found through systematically planned control experiments. Wang et al. [68] considered the canonical flow past a circular cylinder, whose wake is controlled by two small rotating cylinders. The trained DRL agent, developed from inexpensive simulations at low Reynolds numbers, is transferred to cases at Reynolds numbers of up to 1.4×10^5, leading to successful control strategies that are feasible for high-Re conditions.

In the aforementioned work [66], the hydrodynamic stealth in the VIV scenario is also tested, which clearly shows that the VIV is almost fully suppressed alongside an almost vanished wake signature. It is thus quite natural to apply the RL-guided AFC to control classical flow-induced vibration problems. Zheng et al. [69] applied RL to control the VIV of a circular cylinder and made comparisons with an active learning approach, where a Gaussian process regression surrogate model is used to predict the VIV amplitude, combined with a Bayesian optimization algorithm for specified control actions. The RL agent was found to be able to suppress the VIV by 82.7%, outperforming the active learning approach, which reduced the VIV amplitude by only 28.3%. Ren [70] studied VIV control using feedback signals consisting of an array of velocity sensors. The final VIV suppression rate was 89%, achieving a slightly better control performance than that shown in Ref. Zheng et al. [69] while using far fewer sensors. Chen et al. [71] used DRL-based AFC to mitigate the VIV of a square cylinder, reducing the vibration amplitude by 86%, 79%, and 96% for three jet positions deployed at the front, middle, and back sides of the square cylinder, respectively. Using a state space that consists solely of sensory–motor cues, Ren et al. [72] employed DRL-guided AFC for VIV control, highlighting the interpretability of the exploration path. It was found that different state spaces yield different trajectories, as illustrated in Figure 10. Furthermore, by analyzing the phase diagram recorded during training, three distinct stages were identified, which can be interpreted from physical perspectives.

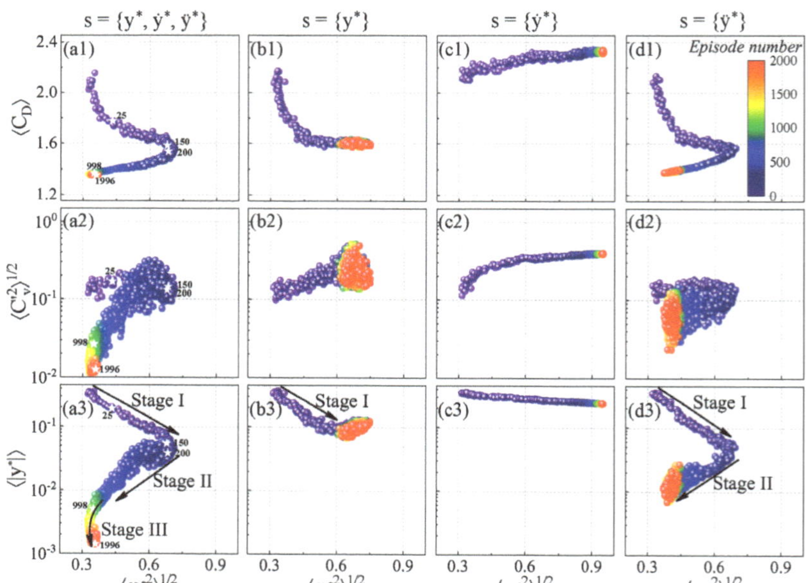

Figure 10. Trajectories of the (**a**) mean drag, (**b**) root-mean-square (RMS) of the vortex force, and (**c**) absolute value of the transverse displacement against the AFC forcing strength during training. Four training processes with different combinations of sensory–motor cues are shown in the four columns. The scattered points are colored according to the episode number. This figure is reproduced from the work of Ren et al. [72], with permission.

In contrast, increasing the VIV amplitude is sometimes advantageous; for example, Mei et al. [73] applied RL-guided jet control to enhance the energy-harnessing efficiency of a circular cylinder undergoing VIV. The jets on the cylinder were found to destabilize the vortex shedding and enhance the VIV, leading to a net energy output increase of 357.63%. Ren et al. [74] applied RL to enhance the heat transfer from a circular cylinder undergoing VIV. Unlike the typically adopted state space, which consists of high-dimensional state representations using velocity feedback signals, this study utilizes the sensory–motor cues of the cylinder, specifically the cross-flow displacement, velocity, and acceleration. The learned strategy ultimately achieved a remarkable improvement in the heat transfer rate of 76.7%.

A reliable flow environment is the basis of successful RL-guided AFC, whether numerical or experimental. The low efficiency of data acquisition and data sampling has attracted a lot of attention. Until now, most applications using numerical simulations were based on the finite element method [57] combined with the Python TensorForce package. Developing the DRL framework with open-source software [94] or commercial software [95] is also quite constructive for the community. Other approaches, such as LBM [58,66,72] or its variants [74], coupled with the Python TensorFlow library, also provide powerful solutions for RL-guided AFC due to the high-fidelity features of the LBM and its superior parallel efficiency with GPU devices. More recent studies [96,97] provide even more effective and efficient numerical tools for flow past a bluff body and are capable of simulating the flow field, structural vibrations, and acoustic field simultaneously. This is believed to provide more solutions, enabling efficient data acquisition to be achieved using numerical approaches. Rabault and Kuhnle [98] further proposed an accelerating method utilizing multiple environments to learn in a distributed configuration. Because the environments can be simulated in a parallel manner, the learning efficiency can be greatly improved. Other approaches, such as expert demonstrations that rely on a simplified mathematical modeling of the system and model optimization [75], also provide very promising directions to explore.

Comparisons of RL-based control with other existing approaches have also attracted attention. Pino et al. [99] performed a comparison of GP-based control, RL-guided control, and global optimization methods, including Bayesian optimization and Lipshitz optimization, along with an in-depth analysis. The difference in exploration versus exploitation, as well as in model capacity and required complexity, were comprehensively illustrated. Furthermore, the authors point out that the hybridization of these methods for flow control is a promising field. In the study of Castellanos et al. [56], both the DRL and the linear GP were applied to control the laminar flow past a circular cylinder, with a few velocity sensors providing feedback. Straightforward comparisons were provided using this well-designed setup, which implied that DRL is more robust with respect to its initial conditions and observation noise while linear GP is more efficient, requiring less sensor data.

Finally, with the emergence of the successful applications of RL-guided AFC in complicated flow systems and the elucidation of more rationales for these successes, it is anticipated that RL-guided control will become a general and reliable solution for the closed-loop control of flow past a bluff body and beyond.

5. Cluster-Based Control

Based on the methods from statistical mathematics, as well as a physical understanding of the flow system, other data-driven approaches are emerging, with one of the most representative being cluster-based control. Cluster-based control stems from cluster analysis, where the feature space is partitioned into groups sharing similar attributes (clusters), which correspond to coarse-grained characteristic phases in a low-dimensional (typically two-dimensional or three-dimensional) feature space [76,100]. The system dynamics are then represented as linear and probabilistic Markov chains, and transitions between clusters are considered transitions between flow states [76,100]. Active control is then established on the low-dimensional representations of the flow system in a model-free and relatively low-cost manner. The framework of cluster-based AFC for flow past a post-stall airfoil is shown in Figure 11.

Nair et al. [76] proposed a cluster-based control strategy for the feedback control of post-stall separated flows past an NACA 0012 airfoil in the turbulent regime, utilizing a limited number of force sensor measurements to reduce power consumption during aerodynamic flight. Wang et al. [77] presented a cluster-based control strategy for model-free feedback drag reduction with multiple actuators and full-state feedback, which was demonstrated in the case of flow past a fluidic pinball using three flow regimes: symmetric periodic, asymmetric periodic, and chaotic vortex shedding. The net drag reductions for these cases amounted to 33.06%, 24.15%, and 12.23%, respectively, highlighting the distinct advantages of robust control under varying flow conditions.

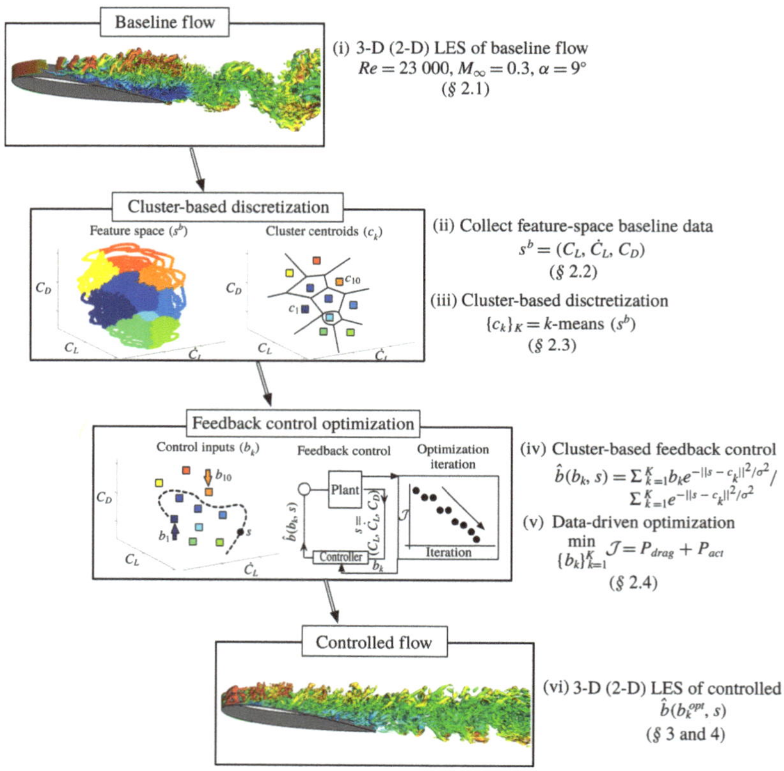

Figure 11. Overview of the cluster-based control framework. This figure is reproduced from the work of Nair et al. [76], with permission.

6. Future Prospects and Challenges

In this article, we reviewed progress in the use of model-free closed-loop control for flow past a bluff body. From the presented applications, closed-loop control is shown to be quite promising in effectively and efficiently managing flow past a bluff body. Although it has attracted attention from multiple disciplines, challenges in closed-loop active flow control also exist, such as the following:

- In numerical approaches, the problem of low-efficiency data acquisition and sampling still needs to be addressed. This area of research is quite promising given the rapid development of computational algorithms and hardware. For experimental approaches, due to the uncertainty in perturbed flows and measurement devices, the need for reliable data acquisition still needs to be addressed. Regarding these efforts, it is quite promising to see the increase in the applications of closed-loop control in realistic scenarios.

- The effectiveness of the control performance is always the first priority. With the aid of advanced control methods, such as machine learning-based approaches, further improving effectiveness in tasks such as flow past a circular cylinder in the laminar regime is no longer challenging. However, in complicated situations involving turbulent flow, multiphysics fields, chemical reactions, etc., more efforts are essential. In our view, machine learning, especially reinforcement learning, is the most promising approach to provide a general solution to these complicated problems.
- Efficiency is a significant requirement due to the demands in the practical engineering field. Net energy savings are essential when viewing the system as a whole. For instance, it is usually not a good choice to reduce the drag of an object at the cost of increased energy consumption.
- Robustness is a special and vital topic. For the AFC, robustness involves the control responses when certain levels of perturbations are applied or when the system conditions are altered. In our view, to achieve better robustness, the process through which the final control strategy is derived should involve sufficient randomness and/or take into account a wide range of realistic perturbations.
- For data-driven, machine learning-based approaches, the interpretability of the evolution or learning path, as well as the physical rationale behind the final converged strategy, can be significant obstacles. Elucidating both issues requires knowledge of statistical mathematics, stability analyses, sensitivity analyses, resolvent analyses, etc.

Author Contributions: Conceptualization, F.R., X.W. and H.T.; investigation, F.R., X.W. and H.T.; resources, F.R.; writing—original draft preparation, F.R., X.W. and H.T.; writing—review and editing, F.R., X.W. and H.T.; visualization, F.R.; supervision, H.T.; project administration, F.R., X.W. and H.T.; funding acquisition, F.R., X.W. and H.T. All authors have read and agreed to the published version of the manuscript.

Funding: This research was funded by the National Natural Science Foundation of China (12102357), the Key Research and Development Program of Shaanxi (2023-YBGY-389), the Young Talent Fund of Xi'an Association for Science and Technology (959202413092), and the NPU Proof of Concept for Future Aerospace and Propulsion Industry (24GNYZ0001-03).

Data Availability Statement: The data used in this work are available upon request to the corresponding author.

Conflicts of Interest: The authors declare no conflicts of interest.

Abbreviations

The following abbreviations are used in this manuscript:

DMD	Dynamic mode decomposition
FSI	Fluid–structure interaction
GA	Genetic algorithm
GP	Genetic programming
GPU	Graphic processing unit
LBM	Lattice–Boltzmann method
PID	Proportional Integral and Differential
POD	Proper orthogonal decomposition
RL	Reinforcement learning
DRL	Deep reinforcement learning
WSLB	windward–suction–leeward blowing

References

1. Derakhshandeh, J.; Alam, M.M. A review of bluff body wakes. *Ocean. Eng.* **2019**, *182*, 475–488. [CrossRef]
2. Lekkala, M.R.; Latheef, M.; Jung, J.H.; Coraddu, A.; Zhu, H.; Srinil, N.; Lee, B.H.; Kim, D.K. Recent advances in understanding the flow over bluff bodies with different geometries at moderate Reynolds numbers. *Ocean. Eng.* **2022**, *261*, 111611. [CrossRef]

3. Rashidi, S.; Hayatdavoodi, M.; Esfahani, J.A. Vortex shedding suppression and wake control: A review. *Ocean. Eng.* **2016**, *126*, 57–80. [CrossRef]
4. Mondal, R.; Alam, M.M. Blockage effect on wakes of various bluff bodies: A review of confined flow. *Ocean. Eng.* **2023**, *286*, 115592. [CrossRef]
5. Sumner, D. Flow above the free end of a surface-mounted finite-height circular cylinder: A review. *J. Fluids Struct.* **2013**, *43*, 41–63. [CrossRef]
6. Zhang, D.; Cheng, L.; An, H.; Draper, S. Flow around a surface-mounted finite circular cylinder completely submerged within the bottom boundary layer. *Eur. J. Mech.-B/Fluids* **2021**, *86*, 169–197. [CrossRef]
7. Hasan, K.; Ahmad, S.; Liaf, A.F.; Karimi, M.; Ahmed, T.; Shawon, M.A.; Mekhilef, S. Oceanic Challenges to Technological Solutions: A Review of Autonomous Underwater Vehicle Path Technologies in Biomimicry, Control, Navigation, and Sensing. *IEEE Access* **2024**, *12*, 46202–46231. [CrossRef]
8. Wynn, R.B.; Huvenne, V.A.; Le Bas, T.P.; Murton, B.J.; Connelly, D.P.; Bett, B.J.; Ruhl, H.A.; Morris, K.J.; Peakall, J.; Parsons, D.R.; et al. Autonomous Underwater Vehicles (AUVs): Their past, present and future contributions to the advancement of marine geoscience. *Mar. Geol.* **2014**, *352*, 451–468. [CrossRef]
9. Williamson, C.; Roshko, A. Vortex formation in the wake of an oscillating cylinder. *J. Fluids Struct.* **1988**, *2*, 355–381. [CrossRef]
10. Gabbai, R.; Benaroya, H. An overview of modeling and experiments of vortex-induced vibration of circular cylinders. *J. Sound Vib.* **2005**, *282*, 575–616. [CrossRef]
11. Zhao, M. A review of recent studies on the control of vortex-induced vibration of circular cylinders. *Ocean. Eng.* **2023**, *285*, 115389. [CrossRef]
12. Hong, K.S.; Shah, U.H. Vortex-induced vibrations and control of marine risers: A review. *Ocean. Eng.* **2018**, *152*, 300–315. [CrossRef]
13. Devenport, W.; Alexander, N.; Glegg, S.; Wang, M. The Sound of Flow Over Rigid Walls. *Annu. Rev. Fluid Mech.* **2018**, *50*, 435–458. [CrossRef]
14. Wang, M.; Freund, J.B.; Lele, S.K. Computational predictions of flow-induced sound. *Annu. Rev. Fluid Mech.* **2006**, *38*, 483–512. [CrossRef]
15. Moreau, S. The third golden age of aeroacoustics. *Phys. Fluids* **2022**, *34*, 031301. [CrossRef]
16. Zhao, F.; Hung, D.L. Applications of machine learning to the analysis of engine in-cylinder flow and thermal process: A review and outlook. *Appl. Therm. Eng.* **2023**, *220*, 119633. [CrossRef]
17. Krishan, G.; Aw, K.C.; Sharma, R.N. Synthetic jet impingement heat transfer enhancement–a review. *Appl. Therm. Eng.* **2019**, *149*, 1305–1323. [CrossRef]
18. Arshad, A.; Jabbal, M.; Yan, Y. Synthetic jet actuators for heat transfer enhancement—A critical review. *Int. J. Heat Mass Transf.* **2020**, *146*, 118815. [CrossRef]
19. Williamson, C.; Govardhan, R. Vortex-induced vibrations. *Annu. Rev. Fluid Mech.* **2004**, *36*, 413–455. [CrossRef]
20. Li, S.; Rival, D.E.; Wu, X. Sound source and pseudo-sound in the near field of a circular cylinder in subsonic conditions. *J. Fluid Mech.* **2021**, *919*, A43. [CrossRef]
21. Inoue, O.; Hatakeyama, N. Sound generation by a two-dimensional circular cylinder in a uniform flow. *J. Fluid Mech.* **2002**, *471*, 285–314. [CrossRef]
22. Collis, S.S.; Joslin, R.D.; Seifert, A.; Theofilis, V. Issues in active flow control: Theory, control, simulation, and experiment. *Progress Aerosp. Sci.* **2004**, *40*, 237–289. [CrossRef]
23. Oruç, V. Strategies for the applications of flow control downstream of a bluff body. *Flow Meas. Instrum.* **2017**, *53*, 204–214. [CrossRef]
24. Choi, H.; Jeon, W.P.; Kim, J. Control of Flow Over a Bluff Body. *Annu. Rev. Fluid Mech.* **2008**, *40*, 113–139. [CrossRef]
25. Kasagi, N.; Suzuki, Y.; Fukagata, K. Microelectromechanical Systems–Based Feedback Control of Turbulence for Skin Friction Reduction. *Annu. Rev. Fluid Mech.* **2009**, *41*, 231–251. [CrossRef]
26. Cattafesta, L.N.; Sheplak, M. Actuators for Active Flow Control. *Annu. Rev. Fluid Mech.* **2011**, *43*, 247–272. [CrossRef]
27. Dong, S.; Triantafyllou, G.S.; Karniadakis, G.E. Elimination of Vortex Streets in Bluff-Body Flows. *Phys. Rev. Lett.* **2008**, *100*, 204501. [CrossRef]
28. Abdolahipour, S.; Mani, M.; Taleghani, A.S. Parametric study of a frequency-modulated pulse jet by measurements of flow characteristics. *Phys. Scr.* **2021**, *96*, 125012. [CrossRef]
29. Abdolahipour, S.; Mani, M.; Shams Taleghani, A. Pressure Improvement on a Supercritical High-Lift Wing Using Simple and Modulated Pulse Jet Vortex Generator. *Flow Turbul. Combust.* **2022**, *109*, 65–100. [CrossRef]
30. Abdolahipour, S.; Mani, M.; Shams Taleghani, A. Experimental Investigation of Flow Control on a High-Lift Wing Using Modulated Pulse Jet Vortex Generator. *J. Aerosp. Eng.* **2022**, *35*, 05022001. [CrossRef]
31. Abdolahipour, S. Effects of low and high frequency actuation on aerodynamic performance of a supercritical airfoil. *Front. Mech. Eng.* **2023**, *9*, 1290074. [CrossRef]
32. Mirzaei, M.; Taleghani, A.S.; Shadaram, A. Experimental Study of Vortex Shedding Control Using Plasma Actuator. *Appl. Mech. Mater.* **2012**, *186*, 75–86. [CrossRef]
33. Brunton, S.L.; Noack, B.R. Closed-Loop Turbulence Control: Progress and Challenges. *Appl. Mech. Rev.* **2015**, *67*, 050801. [CrossRef]

34. Kim, J.; Bewley, T.R. A Linear Systems Approach to Flow Control. *Annu. Rev. Fluid Mech.* **2007**, *39*, 383–417. [CrossRef]
35. Rowley, C.W.; Dawson, S.T. Model Reduction for Flow Analysis and Control. *Annu. Rev. Fluid Mech.* **2017**, *49*, 387–417. [CrossRef]
36. Taira, K.; Hemati, M.S.; Brunton, S.L.; Sun, Y.; Duraisamy, K.; Bagheri, S.; Dawson, S.T.M.; Yeh, C.A. Modal Analysis of Fluid Flows: Applications and Outlook. *AIAA J.* **2020**, *58*, 998–1022. [CrossRef]
37. Brenner, M.P.; Eldredge, J.D.; Freund, J.B. Perspective on machine learning for advancing fluid mechanics. *Phys. Rev. Fluids* **2019**, *4*, 100501. [CrossRef]
38. Vinuesa, R. Perspectives on predicting and controlling turbulent flows through deep learning. *Phys. Fluids* **2024**, *36*, 031401. [CrossRef]
39. Brunton, S.L.; Noack, B.R.; Koumoutsakos, P. Machine learning for fluid mechanics. *Annu. Rev. Fluid Mech.* **2020**, *52*, 477–508. [CrossRef]
40. WU, Z.; FAN, D.W.; ZHOU, Y. Advances in control of turbulence by artificial intelligence: Systems, algorithms, achievements and data analysis methods. *Adv. Mech.* **2023**, *53*, 273.
41. Park, D.; Ladd, D.; Hendricks, E. Feedback control of von Kármán vortex shedding behind a circular cylinder at low Reynolds numbers. *Phys. Fluids* **1994**, *6*, 2390–2405. [CrossRef]
42. Zhang, M.; Cheng, L.; Zhou, Y. Closed-loop-controlled vortex shedding and vibration of a flexibly supported square cylinder under different schemes. *Phys. Fluids* **2004**, *16*, 1439–1448. [CrossRef]
43. Lu, L.; Qin, J.M.; Teng, B.; Li, Y.C. Numerical investigations of lift suppression by feedback rotary oscillation of circular cylinder at low Reynolds number. *Phys. Fluids* **2011**, *23*, 033601. [CrossRef]
44. Du, L.; Sun, X. Noise reduction by feedback rotary oscillation of a three-dimensional circular cylinder. *J. Fluids Struct.* **2019**, *84*, 421–439. [CrossRef]
45. Yun, J.; Lee, J. Active proportional feedback control of turbulent flow over a circular cylinder with averaged velocity sensor. *Phys. Fluids* **2022**, *34*, 095133. [CrossRef]
46. Wang, C.; Tang, H.; Yu, S.C.M.; Duan, F. Active control of vortex-induced vibrations of a circular cylinder using windward-suction-leeward-blowing actuation. *Phys. Fluids* **2016**, *28*, 053601. [CrossRef]
47. Vicente-Ludlam, D.; Barrero-Gil, A.; Velazquez, A. Flow-induced vibration control of a circular cylinder using rotational oscillation feedback. *J. Fluid Mech.* **2018**, *847*, 93–118. [CrossRef]
48. Song, J.; Du, J.; Gao, X. Applying PID control to 2-dof vortex induced vibration of a cylinder. *Ocean. Eng.* **2024**, *300*, 117295. [CrossRef]
49. Hasheminejad, S.M.; Rabiee, A.H.; Bahrami, H. Active closed-loop vortex-induced vibration control of an elastically mounted circular cylinder at low Reynolds number using feedback rotary oscillations. *Acta Mech.* **2018**, *229*, 231–250. [CrossRef]
50. Rabiee, A.H.; Esmaeili, M. Simultaneous vortex- and wake-induced vibration suppression of tandem-arranged circular cylinders using active feedback control system. *J. Sound Vib.* **2020**, *469*, 115131. [CrossRef]
51. Debien, A.; Von Krbek, K.A.; Mazellier, N.; Duriez, T.; Cordier, L.; Noack, B.R.; Abel, M.W.; Kourta, A. Closed-loop separation control over a sharp edge ramp using genetic programming. *Exp. Fluids* **2016**, *57*, 40. [CrossRef]
52. Li, R.; Noack, B.R.; Cordier, L.; Borée, J.; Harambat, F. Drag reduction of a car model by linear genetic programming control. *Exp. Fluids* **2017**, *58*, 103. [CrossRef]
53. Raibaudo, C.; Zhong, P.; Noack, B.R.; Martinuzzi, R.J. Machine learning strategies applied to the control of a fluidic pinball. *Phys. Fluids* **2020**, *32*, 015108. [CrossRef]
54. Raibaudo, C.; Martinuzzi, R.J. Unsteady actuation and feedback control of the experimental fluidic pinball using genetic programming. *Exp. Fluids* **2021**, *62*, 219. [CrossRef]
55. Ren, F.; Wang, C.; Tang, H. Active control of vortex-induced vibration of a circular cylinder using machine learning. *Phys. Fluids* **2019**, *31*, 093601. [CrossRef]
56. Castellanos, R.; Cornejo Maceda, G.Y.; de la Fuente, I.; Noack, B.R.; Ianiro, A.; Discetti, S. Machine-learning flow control with few sensor feedback and measurement noise. *Phys. Fluids* **2022**, *34*, 047118. [CrossRef]
57. Rabault, J.; Kuchta, M.; Jensen, A.; Réglade, U.; Cerardi, N. Artificial neural networks trained through deep reinforcement learning discover control strategies for active flow control. *J. Fluid Mech.* **2019**, *865*, 281–302. [CrossRef]
58. Ren, F.; Rabault, J.; Tang, H. Applying deep reinforcement learning to active flow control in weakly turbulent conditions. *Phys. Fluids* **2021**, *33*, 037121. [CrossRef]
59. Paris, R.; Beneddine, S.; Dandois, J. Robust flow control and optimal sensor placement using deep reinforcement learning. *J. Fluid Mech.* **2021**, *913*, A25. [CrossRef]
60. Li, J.; Zhang, M. Reinforcement-learning-based control of confined cylinder wakes with stability analyses. *J. Fluid Mech.* **2022**, *932*, A44. [CrossRef]
61. Varela, P.; Suárez, P.; Alcántara-Ávila, F.; Miró, A.; Rabault, J.; Font, B.; García-Cuevas, L.M.; Lehmkuhl, O.; Vinuesa, R. Deep Reinforcement Learning for Flow Control Exploits Different Physics for Increasing Reynolds Number Regimes. *Actuators* **2022**, *11*, 359. [CrossRef]
62. Zhao, F.; Zhou, Y.; Ren, F.; Tang, H.; Wang, Z. Mitigating the lift of a circular cylinder in wake flow using deep reinforcement learning guided self-rotation. *Ocean. Eng.* **2024**, *306*, 118138. [CrossRef]
63. Wang, Q.; Yan, L.; Hu, G.; Chen, W.; Rabault, J.; Noack, B.R. Dynamic feature-based deep reinforcement learning for flow control of circular cylinder with sparse surface pressure sensing. *J. Fluid Mech.* **2024**, *988*, A4. [CrossRef]

64. Ren, F.; Du, J.; Li, G. Intelligent self-adaptive control for mitigating lift fluctuations of a circular cylinder. *Chin. J. Theor. Appl. Mech.* **2024**, *56*, 972–979. [CrossRef]
65. Chen, J.; Chen, S.; Ren, F.; Hu, H. Artificiallyintelligent control of drag reduction around a circular cylinder based on wall pressure feedback. *Acta Phys. Sin.* **2022**, *71*, 084701-1. [CrossRef]
66. Ren, F.; Wang, C.; Tang, H. Bluff body uses deep reinforcement learning trained active flow control to achieve hydrodynamic stealth. *Phys. Fluids* **2021**, *33*, 093602. [CrossRef]
67. Fan, D.; Yang, L.; Wang, Z.; Triantafyllou, M.S.; Karniadakis, G.E. Reinforcement learning for bluff body active flow control in experiments and simulations. *Proc. Natl. Acad. Sci.* **2020**, *117*, 26091–26098. [CrossRef]
68. Wang, Z.; Fan, D.; Jiang, X.; Triantafyllou, M.S.; Karniadakis, G.E. Deep reinforcement transfer learning of active control for bluff body flows at high Reynolds number. *J. Fluid Mech.* **2023**, *973*, A32. [CrossRef]
69. Zheng, C.; Ji, T.; Xie, F.; Zhang, X.; Zheng, H.; Zheng, Y. From active learning to deep reinforcement learning: Intelligent active flow control in suppressing vortex-induced vibration. *Phys. Fluids* **2021**, *33*, 063607. [CrossRef]
70. Ren, F. Intelligent flow control for vortex-induced vibration of cylinder. *Chin. J. Hydrodyn.* **2022**, *37*, 757–762. [CrossRef]
71. Chen, W.; Wang, Q.; Yan, L.; Hu, G.; Noack, B.R. Deep reinforcement learning-based active flow control of vortex-induced vibration of a square cylinder. *Phys. Fluids* **2023**, *35*, 053610. [CrossRef]
72. Ren, F.; Wang, C.; Song, J.; Tang, H. Deep reinforcement learning finds a new strategy for vortex-induced vibration control. *J. Fluid Mech.* **2024**, *990*, A7. [CrossRef]
73. Mei, Y.F.; Zheng, C.; Aubry, N.; Li, M.G.; Wu, W.T.; Liu, X. Active control for enhancing vortex induced vibration of a circular cylinder based on deep reinforcement learning. *Phys. Fluids* **2021**, *33*, 103604. [CrossRef]
74. Ren, F.; Zhang, F.; Zhu, Y.; Wang, Z.; Zhao, F. Enhancing heat transfer from a circular cylinder undergoing vortex induced vibration based on reinforcement learning. *Appl. Therm. Eng.* **2024**, *236*, 121919. [CrossRef]
75. Zheng, C.; Xie, F.; Ji, T.; Zhang, X.; Lu, Y.; Zhou, H.; Zheng, Y. Data-efficient deep reinforcement learning with expert demonstration for active flow control. *Phys. Fluids* **2022**, *34*, 113603. [CrossRef]
76. Nair, A.G.; Yeh, C.A.; Kaiser, E.; Noack, B.R.; Brunton, S.L.; Taira, K. Cluster-based feedback control of turbulent post-stall separated flows. *J. Fluid Mech.* **2019**, *875*, 345–375. [CrossRef]
77. Wang, X.; Deng, N.; Cornejo Maceda, G.Y.; Noack, B.R. Cluster-based control for net drag reduction of the fluidic pinball. *Phys. Fluids* **2023**, *35*, 023601. [CrossRef]
78. Curle, N. The Influence of Solid Boundaries Upon Aerodynamic Sound. *Proc. R. Soc. A* **1955**, *231*, 505–514.
79. Rabault, J.; Ren, F.; Zhang, W.; Tang, H.; Xu, H. Deep reinforcement learning in fluid mechanics: A promising method for both active flow control and shape optimization. *J. Hydrodyn.* **2020**, *32*, 234–246. [CrossRef]
80. Ren, F.; Hu, H.B.; Tang, H. Active flow control using machine learning: A brief review. *J. Hydrodyn.* **2020**, *32*, 247–253. [CrossRef]
81. Ren, F.; Gao, C.; Tang, H. Machine learning for flow control: Applications and development trends. *Acta Aeronaut. Astronaut. Sin.* **2021**, *42*, 524686.
82. Gautier, N.; Aider, J.L.; Duriez, T.; Noack, B.; Segond, M.; Abel, M. Closed-loop separation control using machine learning. *J. Fluid Mech.* **2015**, *770*, 442–457. [CrossRef]
83. Ren, F.; Song, B.; Zhang, Y.; Hu, H. A GPU-accelerated solver for turbulent flow and scalar transport based on the Lattice Boltzmann method. *Comput. Fluids* **2018**, *173*, 29–36. [CrossRef]
84. Ren, F.; Song, B.; Hu, H. Lattice Boltzmann simulations of turbulent channel flow and heat transport by incorporating the Vreman model. *Appl. Therm. Eng.* **2018**, *129*, 463–471. [CrossRef]
85. Kaelbling, L.P.; Littman, M.L.; Moore, A.W. Reinforcement learning: A survey. *J. Artif. Intell. Res.* **1996**, *4*, 237–285. [CrossRef]
86. Silver, D.; Schrittwieser, J.; Simonyan, K.; Antonoglou, I.; Hassabis, D. Mastering the game of Go without human knowledge. *Nature* **2017**, *550*, 354–359. [CrossRef]
87. Silver, D.; Huang, A.; Maddison, C.J.; Guez, A.; Sifre, L.; Driessche, G.V.D.; Schrittwieser, J.; Antonoglou, I.; Panneershelvam, V.; Lanctot, M. Mastering the game of Go with deep neural networks and tree search. *Nature* **2016**, *529*, 484–489. [CrossRef]
88. Reddy, G.; Celani, A.; Sejnowski, T.J.; Vergassola, M. Learning to soar in turbulent environments. *Proc. Natl. Acad. Sci. USA* **2016**, *113*, E4877. [CrossRef]
89. Verma, S.; Novati, G.; Koumoutsakos, P. Efficient collective swimming by harnessing vortices through deep reinforcement learning. *Proc. Natl. Acad. Sci. USA* **2018**, *115*, 5849–5854. [CrossRef]
90. Vignon, C.; Rabault, J.; Vinuesa, R. Recent advances in applying deep reinforcement learning for flow control: Perspectives and future directions. *Phys. Fluids* **2023**, *35*, 031301. [CrossRef]
91. Xie, F.; Zheng, C.; Ji, T.; Zhang, X.; Bi, R.; Zhou, H.; Zheng, Y. Deep Reinforcement Learning: A New Beacon for Intelligent Active Flow Control. *Aerosp. Res. Commun.* **2023**, *1*, 11130. [CrossRef]
92. Belus, V.; Rabault, J.; Viquerat, J.; Che, Z.; Hachem, E.; Reglade, U. Exploiting locality and translational invariance to design effective deep reinforcement learning control of the 1-dimensional unstable falling liquid film. *AIP Adv.* **2019**, *9*, 125014. [CrossRef]
93. Schulman, J.; Wolski, F.; Dhariwal, P.; Radford, A.; Klimov, O. Proximal policy optimization algorithms. *arXiv* **2017**, arXiv:1707.06347.
94. Wang, Q.; Yan, L.; Hu, G.; Li, C.; Xiao, Y.; Xiong, H.; Rabault, J.; Noack, B.R. DRLinFluids: An open-source Python platform of coupling deep reinforcement learning and OpenFOAM. *Phys. Fluids* **2022**, *34*, 081801. [CrossRef]
95. Mao, Y.; Zhong, S.; Yin, H. DRLFluent: A distributed co-simulation framework coupling deep reinforcement learning with Ansys-Fluent on high-performance computing systems. *J. Comput. Sci.* **2023**, *74*, 102171. [CrossRef]

96. Song, J.; Zhang, F.; Zhao, Y.; Ren, F.; Hu, H. Direct simulations of external flow and noise radiation using the generalized interpolation-supplemented cascaded lattice Boltzmann method. *Comput. Fluids* **2024**, *280*, 106347. [CrossRef]
97. Song, J.; Hu, H.; Li, G.; Zhang, H.; Ren, F. Generalized interpolation-supplemented cascaded lattice Boltzmann method for noise radiated from a circular cylinder. *J. Comput. Phys.* **2024**, *517*, 113319. [CrossRef]
98. Rabault, J.; Kuhnle, A. Accelerating deep reinforcement learning strategies of flow control through a multi-environment approach. *Phys. Fluids* **2019**, *31*, 094105. [CrossRef]
99. Pino, F.; Schena, L.; Rabault, J.; Mendez, M.A. Comparative analysis of machine learning methods for active flow control. *J. Fluid Mech.* **2023**, *958*, A39. [CrossRef]
100. Hou, C.; Deng, N.; Noack, B.R. Trajectory-optimized cluster-based network model for the sphere wake. *Phys. Fluids* **2022**, *34*, 085110. [CrossRef]

actuators

Article

Dual Synthetic Jets Actuator and Its Applications Part V: Novel Valveless Continuous Micropump Based on Dual Synthetic Jets with a Tesla Structure

Jian-Yuan Zhang, Wen-Qiang Peng, Zhen-Bing Luo *, Zhi-Jie Zhao, Jian-Yu Gong and Zhao-Feng Dong

College of Aerospace Science and Engineering, National University of Defense Technology, Changsha 410073, China; zhangjianyuan1999@163.com (J.-Y.Z.); plxhaz@126.com (W.-Q.P.); ji_yo_100@163.com (Z.-J.Z.); g513666@163.com (J.-Y.G.); dongzhaofeng@nudt.edu.cn (Z.-F.D.)

* Correspondence: luozhenbing@163.com

Abstract: The valveless micropump based on dual synthetic jets is a potential fluid pumping device that has the ability to transport fluid continuously. In order to improve the performance of this device, a novel valveless continuous micropump based on dual synthetic jets with a Tesla structure was proposed by combining a double Tesla symmetrical nozzle and a dual synthetic jets actuator. The mechanism of the novel micropump and its flow field characteristics were analyzed, combined with numerical simulation and a PIV experiment. The performance of the novel micropump was compared with that of a dual synthetic jet micropump based on a traditional shrinking nozzle. The novel micropump achieved continuous flow with a larger and more stable flow rate in one cycle. The maximum pump flow speed reached 12 m/s. Compared with the traditional type, the pump flow rate was increased by 5.27% and the pump flow pulsation was reduced by 214.93%. The backflow and vortex inside the nozzle were prevented and inhibited effectively by the Tesla structure. The velocity and influence range of the pump flow increased with the intensification of driving voltage in a certain range.

Keywords: dual synthetic jets actuators; Tesla structure; valveless micropump; continuous jet flow

Citation: Zhang, J.-Y.; Peng, W.-Q.; Luo, Z.-B.; Zhao, Z.-J.; Gong, J.-Y.; Dong, Z.-F. Dual Synthetic Jets Actuator and Its Applications Part V: Novel Valveless Continuous Micropump Based on Dual Synthetic Jets with a Tesla Structure. *Actuators* **2023**, *12*, 226. https://doi.org/10.3390/act12060226

Academic Editor: Luigi de Luca

Received: 5 May 2023
Revised: 25 May 2023
Accepted: 27 May 2023
Published: 29 May 2023

1. Introduction

The microfluidic system holds immense potential for a wide range of applications, including drug delivery [1–3], biochemical applications [4–8], robotics [9,10], micro-electronics cooling [11–13], and more. As the core of this system, micropump plays a crucial role in its overall functionality. It can be used for energy conversion, directional and quantitative fluid transfer, and even as a power source for micro-robots. Without the micropump, the whole system would not work properly. Micropumps can be classified into two categories: mechanical and non-mechanical [14]. Non-mechanical micropumps can generate constant/steady flows (compared with oscillatory pumping). Their performance, however, is often limited by the properties of the selected fluid. Mechanical micropumps often use the motion of a solid (such as a gear or diaphragm) or liquid to generate the pressure difference needed to move fluid. The motion often causes an unsteady flow. The piezoelectric pump is a type of non-mechanical micropump that utilizes the inverse piezoelectric effect of piezoelectric materials. The application of alternating current excitation can deform the piezoelectric material, thereby driving the fluid. The piezoelectric pump has the advantages of a compact structure, low energy consumption, a flexible control mode, and fast response times. Valveless piezoelectric pumps are particularly noteworthy as they enable the directional transport of fluid by optimizing the pump structure. This design avoids the hysteresis of the mechanical valve under high-frequency conditions and the problem of fatigue damage.

Conventional valveless pumps, such as contract/expansion tube valveless pumps, are not widely applicable in microfluidic systems due to their shortcomings, such as their inability to continuously discharge and large outflow pulsation. In 1998, Glezer, A. et al. [15] developed the first piezoelectric synthetic jet actuator (SJA) with superior working characteristics, which has been extensively studied by researchers and introduced into the field of micropump research [16]. Luo, X.B. et al. [17] first applied synthetic jets to a micropump in 2002. The feasibility of a synthetic jet micropump was verified through numerical simulation, providing a novel approach for valveless micropump design. Building on this foundation, Luo, Z.B. et al. [18] designed a valveless micropump based on synthetic jets with a filter by employing two piezoelectric diaphragms in 2005 that avoided the blockage caused by impurities from the outside and significantly increased the flow rate of the pump. Dau, V.T. et al. [19], Choi, J.P. et al. [20], He, X.H. et al. [21], Munas, F.R. et al. [22], and Van, L.L. et al. [23] developed different structures for valveless piezoelectric micropumps based on synthetic jets, such as the planar [19,20,22,23] and vertical [21] structures, and carried out a large number of numerical simulations and experimental studies on them. Through excellent structure design, these researchers enabled the micropump to have the ability of directional fluid transport and improved the directivity of fluid flow. Some of them achieved a continuous pumping flow of the micropumps [21–23]. However, these micropumps had some drawbacks, such as a less compact structure, insufficient development space for synthetic jets, low pump flow, and large pulsations. Tran, C.D. et al. [24] proposed a micropump based on synthetic jets with a Tesla coupled nozzle in 2020 on the basis of these predecessors. Through the novel Tesla coupled nozzle, the energy contained in the vortices on either side of the synthetic jet was utilized during the blowout phase. During the suction stage of the pump, the main stream stagnated, causing the cavity to only absorb air from the side roads. However, this design still has some drawbacks, such as unstable pump flow and an inability to withstand high ballasts.

In summary, although the valveless micropump based on synthetic jets can generate a continuous jet flow through the structure design, there are some issues such as unsteady pump flow, ballast invalidation, and low energy utilization rate. The fundamental reasons for this are that the synthetic jet actuators cannot pump fluid continuously and the environment on both sides of the diaphragm is inconsistent. In 2006, the piezoelectric dual synthetic jets actuator (DSJA) was invented by Luo, Z.B. et al. [25,26], which solved the above problems by sharing one PZT diaphragm with two cavities. Building on this innovation, Luo's team developed a continuous micropump based on dual synthetic jets [27] and conducted numerical simulations and performance analyses on it. The simulation results showed that the micropump had the characteristics of continuous flow and low pulsation.

In order to further improve the performance of the continuous micropump based on dual synthetic jets, this paper designed a novel valveless continuous micropump based on dual synthetic jets with a Tesla structure. This micropump combined the novel double Tesla symmetrical nozzle with a dual synthetic jets actuator. The mechanism and flow field characteristics of the novel micropump were analyzed through numerical simulation and particle image velocimetry (PIV) experiments, which laid a theoretical foundation for the engineering application of a valveless continuous micropump based on dual synthetic jets.

2. Structure and Numerical Methods

2.1. Structure and Mechanism

The structure and specific parameters of the novel micropump designed in this paper are shown in Figure 1. The micropump was composed of a nozzle and actuator cavity. The actuator cavities featured two shrinking outlets. The interior of the nozzle contained a flow guide baffle, two suction channels, and a Tesla symmetrical structure. The idea of this structure was inspired by the Tesla principle of preventing undesired backflow. The outlet area of the nozzle measured 3 mm × 30 mm.

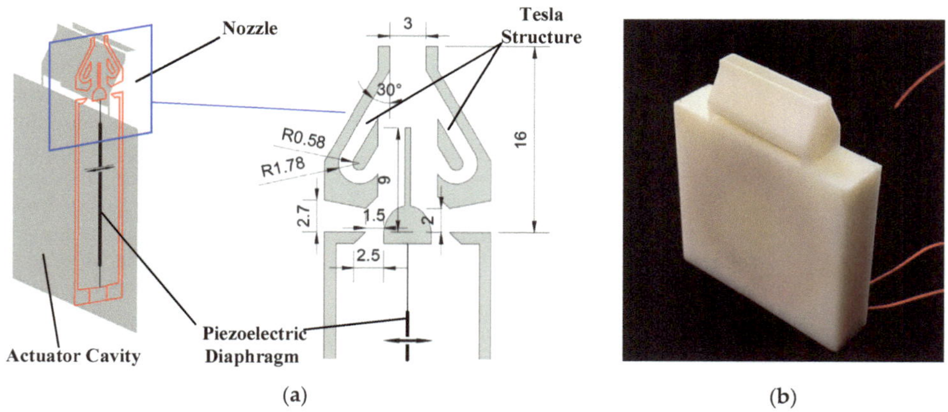

Figure 1. Structure of the novel micropump: (**a**) the interior structure and specific parameters; (**b**) the appearance of the material object.

The working mechanism of the valveless continuous micropump based on dual synthetic jets is illustrated in Figure 2. The piezoelectric diaphragm underwent an inverse piezoelectric effect under the external alternating current excitation, vibrating in the actuator cavity along the normal direction of the diaphragm. In the first half cycle, as shown in Figure 2a, when the piezoelectric diaphragm vibrated to the left, the left cavity was compressed, while the right cavity expanded. The compressed air in the left cavity formed a jet in the left outlet of the actuator cavity. Under the action of the Coanda surface at the bottom of the flow guide baffle, the jet flowed into the main channel along the curved surface. The jet was in the forward flow state and was directly expelled through the nozzle contraction outlet and stabilization section, unaffected by the Tesla backflow channel through which it flowed. At the same time, the right cavity expanded and formed a low-pressure region. Since the Tesla structure only allowed the fluid to pass through in one direction, it could obstruct the reverse flow of the main channel. Therefore, the right cavity could only take in air from the outflow field through the suction channel. In the second half cycle, the diaphragm vibrated to the right; thus, the flow field was contrary to the above description. The left cavity provided the main flow in the first half cycle, while the right cavity did so in the second half cycle, resulting in a continuous flow within one cycle at the nozzle.

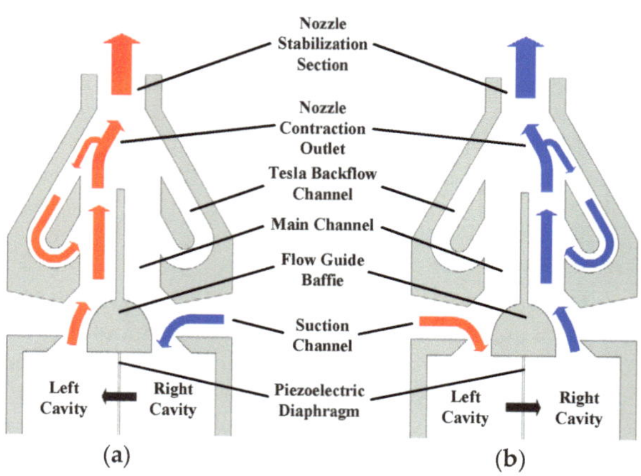

Figure 2. The working mechanism of the micropump: (**a**) first half cycle; (**b**) second half cycle.

2.2. Numerical Model and CFD Method

The established two-dimensional numerical model of the valveless continuous micropump based on dual synthetic jets is shown in Figure 3. The numerical model used was the X-L model [28]. The velocity inlet boundary condition was given directly in terms of its time-dependent displacement. The calculation expression was as follows:

$$u_x = 2\pi f A_m \cdot \sin(2\pi f t + \Phi_0), u_l(l,t) \approx 0 \tag{1}$$

where $u_x(l,t)$ and $u_l(l,t)$ are, respectively, the axial and radial components of the velocity at any point on the surface of the diaphragm (x_1, l), where A_m is the amplitude of the center point of the diaphragm.

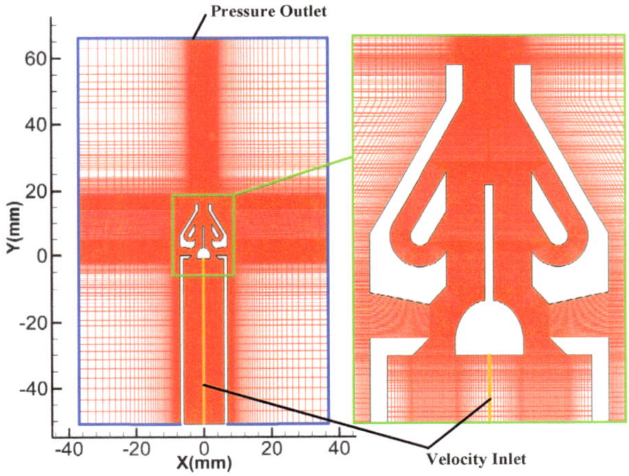

Figure 3. Two−dimensional numerical model.

In this study, the UDF file was imported to define the velocity inlet. The diaphragm vibration amplitude was specified as $A_m = 0.01857$ mm and the frequency was specified as f = 600 Hz. Then, the formulas were as follows:

$$\begin{cases} u_{1x} = 0.7\sin(2\pi \cdot 600 \cdot t) \\ u_{2x} = 0.7\sin(2\pi \cdot 600 \cdot t + \pi) \end{cases} \tag{2}$$

where u_{1x} and u_{2x} are the normal velocities of the left and right diaphragm, respectively.

The simulation software used in this study was Ansys Fluent 2021R2. The SST k-ω turbulence model was selected for numerical calculation. The vibration diaphragm adopted the velocity inlet boundary while the wall adopted the non-slip wall boundary, and the surrounding outlet was set as the pressure outlet. The fluid medium was an incompressible gas. The pressure-based solver was chosen for calculation as it requires a small amount of memory and has a high degree of resilience in solving, making it suitable for a wide range of flow systems, from low-speed incompressible flows to high-speed compressible flows. The Semi-Implicit Method for Pressure Linked Equations (SIMPLE) was used for the pressure and velocity coupling, and the momentum and energy equations were discretized using the second-order upwind scheme. Structured grids were employed throughout the computational domain, which was locally encrypted inside the nozzle and at the exit. The number of grids was 244,666. A cycle of the actuator was divided into 80 steps for calculation, that is, the time step $\Delta t = 1/80\text{T} = 1/(80 \times 600)\text{s} = 2.08333 \times 10^{-5}\text{s}$.

2.3. Grid Independence Verification

The numerical calculation was carried out on the two-dimensional models, with the number of grids being 60,000, 130,000, 240,000, and 400,000 by learning a considerate method [29]. After calculating for 50 cycles, the instantaneous flow rate curve in the nozzle outlet within two cycles was obtained, as shown in Figure 4. The instantaneous flow rate results calculated by the four grid quantities had few differences. The difference and average flow rate are shown in Table 1. The flow rate decreased slightly with the increase in the number of grids and gradually tended to be stable. Compared with the calculation results where the number of grids was 400,000, the errors in the calculations with the other numbers of grids were, respectively, 4.74%, 2.15%, and 0.92%. In order not to compromise the calculation accuracy and efficiency, this paper adopted a model with a grid number of 240,000 for numerical calculation.

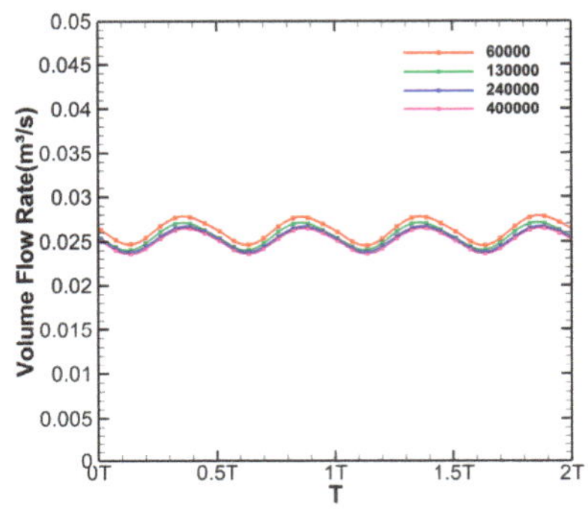

Figure 4. Instantaneous flow rate curves under different mesh quantities.

Table 1. Grid quantity and average flow rate.

Number of Grids	Average Flow Rate (m^3/s)	Difference from 400,000 Results (%)
60,000	0.02632	4.74
130,000	0.02567	2.15
240,000	0.02536	0.92
400,000	0.02513	——

2.4. PIV Experimental Method

The composition of the PIV experimental system is shown in Figure 5. The PIV system mainly included a high-energy dual-channel laser, three-dimensional light guide arm, 24 mm F/2.8 long lens, 532 ± 5 nm narrow-band filter, cross-frame CCD camera (2456 × 2056), synchronous frequency-locking controller, image acquisition card, and MicroVec V3 system PC software. The micropump was placed in a plexiglass chamber (200 mm × 200 mm × 600 mm), which was filled with smoke particles less than 4µm in diameter as PIV tracer particles. Two signals with the same frequency and phase were generated by the signal generator, one of which was provided to the actuator by the piezoelectric ceramic driving power supply, and the other was used as the external trigger signal of the synchronous controller to realize the phase-locked measurement of the flow field. In the experiment, the image magnification was 21.3136 µm/pixel. The operating frequency of the time-averaged flow field image acquisition was 7 Hz. The images taken at this frequency could be traversed through the whole cycle of the valveless duel synthetic jet micropump. A total of 50 pairs of images were taken, and the mean value of the images was taken as the velocity vector of the time-averaged flow field. A cycle of the valveless dual synthetic jet micropump flow field was divided into 16 phase points by using synchronization and phase-locked technology, and 10 pairs of images were taken for the flow field at each phase point. The mean value was taken as the velocity vector of the phase point. The laser light was located at the middle section of the micropump, and the nozzle exit pointed to the laser light.

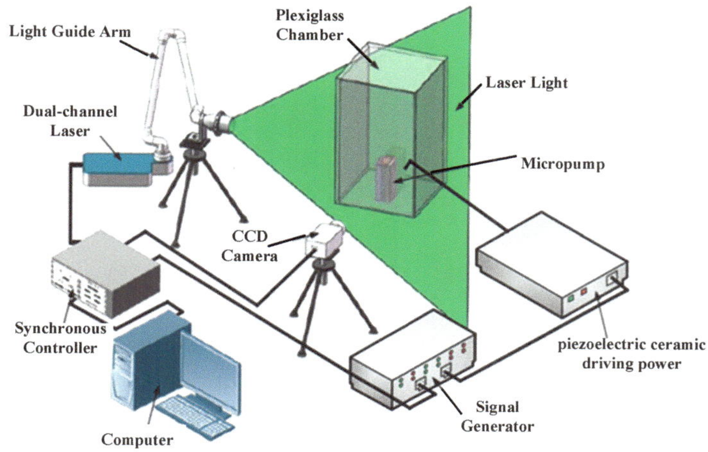

Figure 5. Composition of PIV experimental system.

3. Results and Analysis

3.1. Instantaneous Flow Field

A cycle could be divided into four stages according to the diaphragm vibration regularity (taking the left cavity as the research object): accelerated blowing stage, $0 < t \leq 1/4\text{T}$ (the first stage); deceleration blowing stage, $1/4\text{T} < t \leq 1/2\text{T}$ (the second stage); accelerated suction stage, $1/2\text{T} < t \leq 3/4\text{T}$ (the third stage); and deceleration suction stage, $3/4\text{T} < t \leq \text{T}$ (the fourth stage).

Figure 6 shows the velocity and streamline of the flow field in the simulation and the PIV diagram of the operation of the valveless micropump based on dual synthetic jets in one cycle. When $t = 1/4\text{T}$, as shown in Figure 6a, the left cavity of the micropump was at the end of the accelerated blowing stage. At this stage, the entire micropump was manifested as accelerated blowing in the left cavity and accelerated suction in the right cavity. The air in the left cavity was constantly compressed and ejected from the left cavity outlet. Under the action of the Coanda effect, the jet flowed into the left main channel along the curved surface at the bottom of the flow guide baffle. The jet then continued to develop downstream and was expelled after flowing through the nozzle contraction outlet and stabilization section. During the accelerated blowing stage, the air at the edge of the cavity outlet experienced strong shear action from the jet stream, leading to the formation of a vortex inside the suction channel. The vortex sucked up part of the main stream, resulting in an overflow shunt at the upper end of the channel. The vortex also drove the overflow shunt to move back into the main stream to avoid flow loss. The jet in the contraction outlet deflected towards the center at the end of the flow guide baffle and collided with the wall of the contraction outlet, causing some of the air to separate from the main stream and flow into the Tesla backflow channel. The air in the Tesla backflow channel flowed along the channel and collided with the air in the right main channel, forming a mass of large and small whirlpools that prevented the right cavity from inhaling air in the right main channel. Therefore, the right cavity outlet could only inhale air from the outside through the right suction channel. This entire process also elucidated the working mechanism of the Tesla structure, specifically, how it functioned to impede the reverse flow.

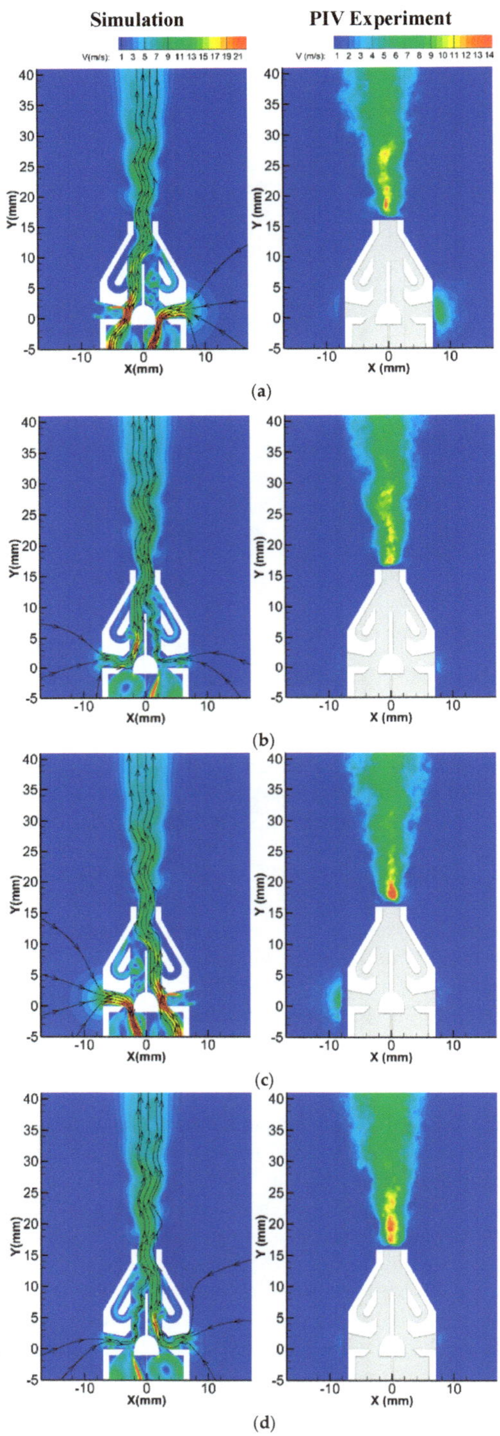

Figure 6. The velocity and streamline of the flow field in the simulation and the PIV diagram of the operation of the valveless micropump based on dual synthetic jets in one cycle: (**a**) 1/4T; (**b**) 1/2T; (**c**) 3/4T; (**d**) 1T.

The PIV diagram shows that there was a clear air inhalation phenomenon around the right suction channel, which was asymmetric in comparison to the left side. This suggested that the right cavity was currently inhaling air from the outside through the right suction channel. In the left suction channel, some air was observed escaping from the channel, confirming the presence of a vortex in the left channel. The flow field around the left and right suction channels was consistent with the simulation results.

When t = 1/2T, as shown in Figure 6b, the left cavity of the micropump was at the end of the deceleration blowing stage. At this stage, the left cavity was still being compressed, but the degree of compression was constantly decreasing, as was the intensity of the jet from the left cavity. The jet ejected in the previous stage developed into the left main channel. The high-speed flow generated low pressure in the left main channel, allowing external air with relatively high pressure to enter the left main stream via the left suction channel due to the pressure difference. There was also a pressure difference between the main jet in the contraction outlet and the relatively static air in the right main channel. Additionally, the main jet had an ejection effect on the air in the right main channel, making the air flow into the main jet. Since it was then at the end of the deceleration blowing stage, the right cavity no longer inhaled air. The air inhaled from the outside during the previous period had no time to stagnate and flowed directly into the right main channel under the guidance of the structure, supplementing the air in the right main channel. These figures illustrate the complete jet flow phenomenon from the exterior into the main jet.

As observed in the PIV diagram, the air speed near the right suction channel at this stage was significantly reduced compared with the previous stage. This indicated that the right cavity was at the end of the deceleration suction stage, and the suction speed was greatly reduced. During this time, the inhaled air was added to the right main channel, and the left suction channel was slightly inhaling air. This action, which was contrary to the previous stage, also confirmed the phenomenon described in the simulation: the outside air was replenished in the left main channel due to the pressure difference.

Figure 6c,d, respectively, show the flow conditions of the micropump when t = 3/4 t and t = T. In these periods, the left cavity was at the end of the accelerated suction stage and the deceleration suction stage, respectively, and the flow conditions were exactly opposite to those when t = 1/4T and t = 1/2T. In the first half cycle, the micropump was characterized by left blowing and right suction, while in the second half cycle, it was characterized by left suction and right blowing. Consequently, the micropump exhibited continuous flow at the jet nozzle within a single cycle.

3.2. TimeMean Flow Field

Figure 7 shows the time-mean flow field of the micropump in one cycle. Within a cycle, the micropump continuously inhaled air from the outside. The air flowed successively through the suction channel, the flow guide baffle, and the main channel, and was finally ejected through the nozzle, showing a shape similar to the steady jet outside the nozzle. This process realized the micropump function of directional transport air. In addition, the time-mean flow field had a clear dividing line at the throat outlet of the cavity, which indicated that the cavity did not eject or inhale air in one cycle, and the total mass flow rate of the cavity in one cycle was still zero, although there was a mass exchange between the cavity and the outside surroundings.

The velocity distribution and maximum velocity magnitude at different altitudes ($0 \leq H_Y \leq 18$ is equal to 16 mm $\leq Y \leq$ 34 mm) are given in Figure 8a, showing the law of velocity distribution in a downstream flow. At position $H_Y = 0$, the velocity presented unimodal symmetry, with lower magnitude on both sides of the center and a large velocity gradient in the X direction. At position $H_Y = 3$, the central maximum velocity magnitude and the velocity on both sides of the center increased. For the velocity gradient in the X direction, the gradient decreased on both sides of the center but increased in the middle. As H_Y continued to grow, the peak magnitude of the unimodal symmetric velocity distribution gradually decreased; this was accompanied by a larger range on both sides of the center and a flatter distribution. Due to the wall boundary layer on both sides of the nozzle stabilization section at $H_Y = 0$, the velocity magnitude here was close to 0. The jet was compressed and ejected at the nozzle contraction section. The closer the jet was to the center, the higher the energy was. The energetic central jet formed strong shear with the stationary air in the boundary layer, thus showing a unimodal symmetric velocity distribution. When H_Y increased, the wall limit disappeared, and the jet gradually expanded to both sides. The expansion became clearer with the increase in altitude. After expansion, the energy concentrated in the middle of the jet was dispersed to both sides, resulting in the development of velocity distribution from unimodal symmetry to flat. The velocity (maximum velocity) of the flow field center at different altitudes was obtained, as shown in Figure 8b. As the distance increased, the maximum velocity magnitude first increased and then decreased, reaching a maximum at approximately 12 m/s at position $H_Y = 3$, and was approximately linear at the velocity decay stage.

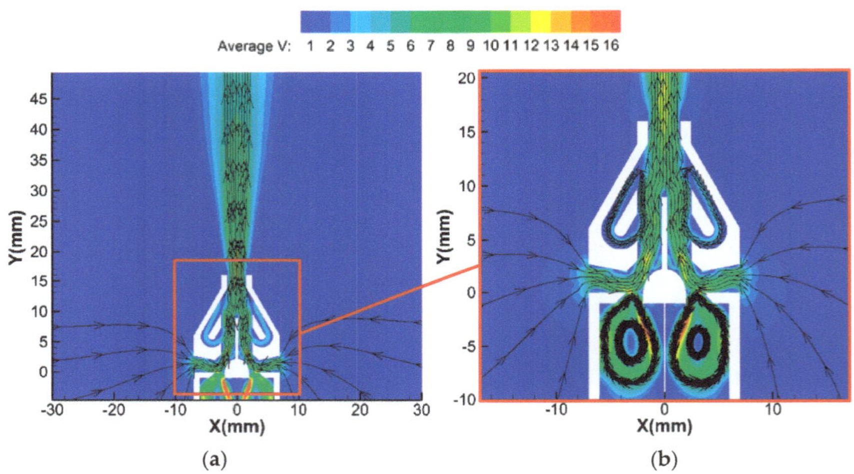

Figure 7. The velocity and streamline time−mean flow field of the micropump in one cycle: (**a**) overview of time−mean flow field; (**b**) partial view of time−mean flow field.

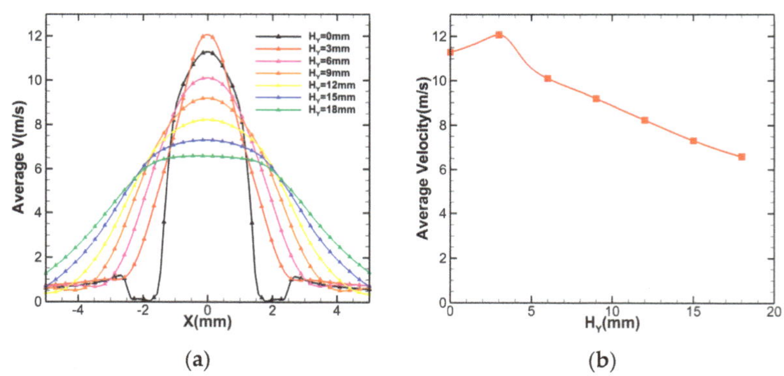

Figure 8. Average velocity in the nozzle downstream: (**a**) velocity distribution of the downstream at different positions ($-5 \leq X \leq 5, 0 \leq H_Y \leq 18$); (**b**) average velocity in the center line ($X = 0, 0 \leq H_Y \leq 18$).

The PIV experimental system was used to photograph the real flow field downstream of the micropump. Time equalization processing was carried out and the simulation flow field was compared, as shown in Figure 9. At this time, the diaphragm driving voltage was 150 V and the driving frequency was 660 Hz. The real flow field diagram shows that the jet flowed out continuously from the nozzle, and there was no backflow at the nozzle exit. The downstream velocity field was similar to that of the steady jet flow and was in agreement with the simulation results. Outside the suction channel, the suction velocity shown in the PIV diagram was not clear. This was because as the momentum during the suction process was smaller than that during the injection process, the disturbance caused by the suction process was weaker. Additionally, after the air had flowed into the suction channel, the camera of the PIV device was unable to capture the particles, resulting in inaccurate shots of the flow field in this region. Figure 10 compares the velocity distribution of the real flow field with the simulation result at $H_Y = 3$. The velocity distribution of the real flow field also presented a form of unimodal symmetry, which was in high agreement with the simulation result.

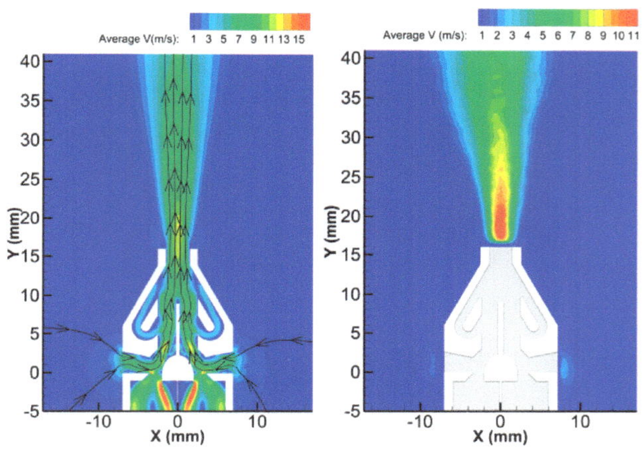

(a) (b)

Figure 9. Comparison of time−mean flow field: (**a**) simulation results with streamline; (**b**) PIV experiment.

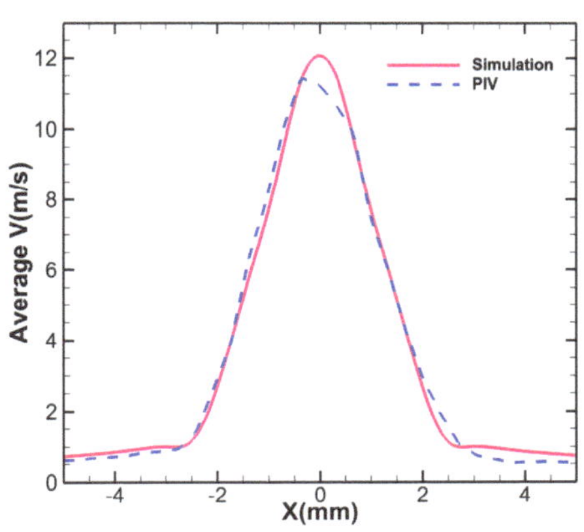

Figure 10. Comparison of velocity distribution between PIV and simulation results ($-5 \leq X \leq 5$, $H_Y = 3$).

3.3. PIV Results Analysis

By changing the driving voltage amplitude, the micropump performance and downstream flow field changes were investigated. The PIV system was used to photograph the time-mean flow field of the micropump under different driving voltage amplitudes, as shown in Figure 11. As shown in the figure, the shape of the downstream jet did not change as the driving voltage amplitude increased, but the influence range of the jet expanded. Here, the angle between the two rays was used to measure the influence range of the jet. Table 2 shows the relationship between voltage amplitude and the angle.

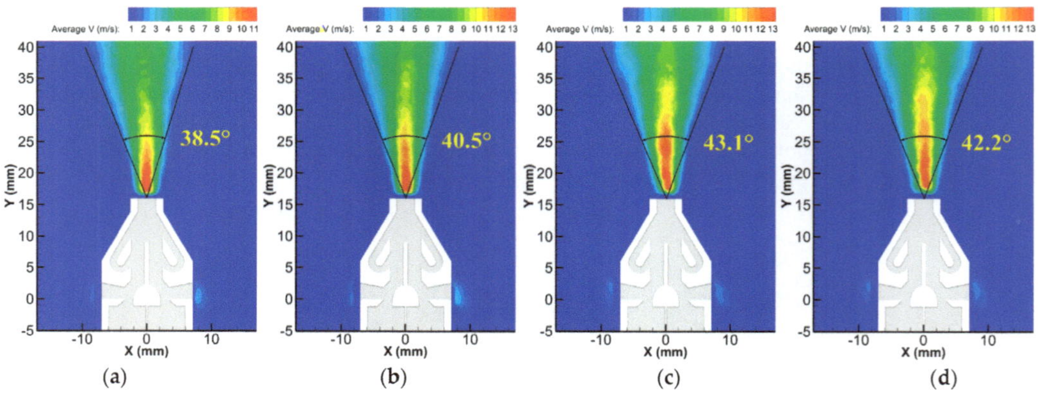

Figure 11. Time−mean flow field of the micropump under different driving voltage amplitudes: (**a**) Case1; (**b**) Case2; (**c**) Case3; (**d**) Case4.

Table 2. Experimental voltage amplitude and angle parameters.

Experimental Serial Number	Driving Voltage Amplitude (±V)	Influence Area Size (°)
Case1	150	38.5
Case2	180	40.5
Case3	210	43.1
Case4	240	42.2

At Y = 40 mm, which was 24 mm downstream of the nozzle, the influence range of the jet increased from 38.5° (in voltage 150 V) to 43.1° (in voltage 210 V), representing an increase of 11.95%. When the driving voltage amplitude was 240 V, the influence range was slightly reduced to 42.2°. Although the range of influence decreased somewhat, it may have been caused by the shooting error. Thus, with the further increase in driving voltage amplitude, the range of influence gradually tended to stabilize.

The velocity magnitude of the flow field center (X = 0 mm) in the range of Y = 17~40 mm was extracted and compared, as shown in Figure 12b. When the voltage was 150 V, the velocity reached its maximum at approximately Y = 20 mm and decayed with the increase in distance. When the voltage was increased to 180 V, the velocity generated by the diaphragm increased, and the velocity of the downstream flow field also increased significantly. The maximum velocity magnitude grew from 11 m/s to approximately 13.5 m/s. The local velocity of the corresponding position also integrally increased by approximately 1–2 m/s, and the penetrability of the jet was enhanced. As the driving voltage continued to rise, the maximum velocity of the jet decreased slightly by approximately 0.5 m/s, but the velocity at the same position still increased. Moreover, the position of the maximum velocity moved downstream constantly. The maximum velocity was near Y = 18 mm at 180 V, Y = 20 mm at 210 V, and Y = 22 mm at 240 V. It is known from previous studies that the influence range of jets increases with voltage amplitude. Since the jet developed downstream in an "S" form, the larger influence range meant that the jet was "getting fatter". Although the velocity generated by the diaphragm increased after the voltage rose, the "fatter" jet led to a dispersion of energy in the jet center. Thus, the maximum velocity of the jet was reduced, which also indicated that the penetrability of the micropump had reached the threshold.

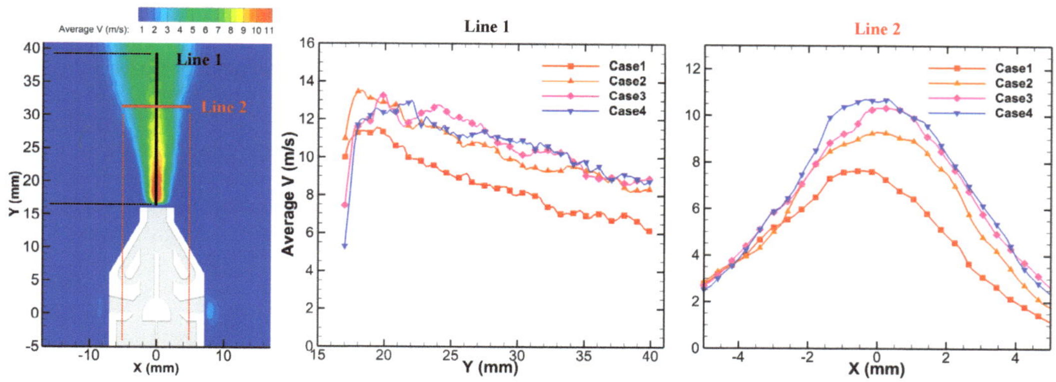

(**a**) (**b**) (**c**)

Figure 12. The velocity magnitude of the flow field under different driving voltage amplitudes: (**a**) schematic diagram of line location; (**b**) velocity of flow field center (X = 0, 17 ≤ Y ≤ 40); (**c**) velocity distribution (−5 ≤ X ≤ 5, Y = 32).

Figure 12c shows the velocity distribution at Y = 32 mm under different driving voltage amplitudes. With the growth in voltage, the velocity magnitude presented an overall upward trend, which was reflected in the increase of the maximum velocity and both sides. The area enclosed by the velocity distribution curve and the coordinate axis became larger, which meant that the larger the flow rate, the more gas could be transported and the stronger the transport capacity. Therefore, although increasing the voltage could not effectively improve the penetrability of the micropump jet, it could enhance the transport capability of the micropump.

3.4. Comparison with the Traditional Shrink Nozzle

In order to reflect the advantages of the novel Tesla symmetrical nozzle over the traditional shrink nozzle, a simulation investigation of the shrink nozzle was carried out. The structure is shown in Figure 13, which adopted the same CFD method. The number of grids was 173,428.

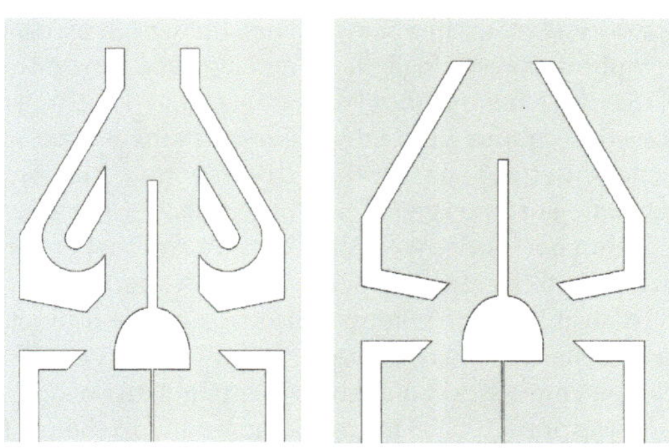

(**a**) (**b**)

Figure 13. Structure of nozzle: (**a**) Tesla symmetrical nozzle; (**b**) shrink nozzle.

Figure 14 shows the flow field inside the shrink nozzle in one cycle. In the first stage, large vortices were formed on both sides inside the nozzle. Part of the main jet flowed into the right vortex under attraction, taking away the energy. At the same time, attracted by the vortex, the main jet deflected to the right at the end of the flow guide baffle and impinged on the wall of the contraction outlet, further consuming the energy of the main jet. The figure also shows that at the bottom of the right vortex, some air flowed out of the vortex and was sucked into the right cavity along with the gas in the suction channel. This phenomenon is called "self-absorption". This phenomenon hinders the jet from developing downstream, affecting the speed of the jet, and can even cause the jet to continuously circulate in the nozzle. Therefore, the occurrence of the "self-absorption" phenomenon should be avoided as far as possible. In the second stage, the two vortices were still in existence. At this time, there was no energy being continuously added to the main jet. The main jet, which had not yet developed downstream, could not resist the attraction of the right vortex and was sucked into it, increasing the energy of the right vortex.

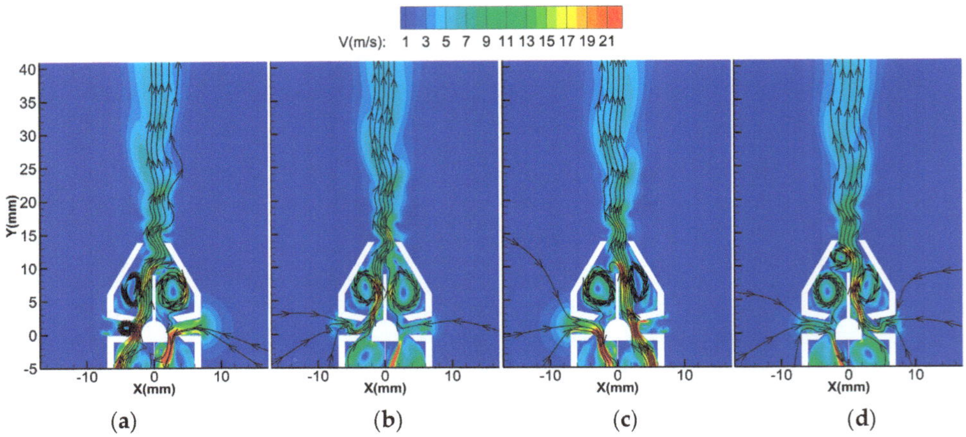

Figure 14. Flow field and streamline of the shrink nozzle in one cycle: (**a**) 1/4T; (**b**) 1/2T; (**c**) 3/4T; (**d**) 1T.

Throughout the whole cycle, vortices always existed on both sides due to the nozzle that had no Tesla structure. This carried several disadvantages. Firstly, the energy of the main jet was weakened by the vortices, which reduced the strength and flow rate of the downstream jet. Secondly, the attraction effect of the vortex made the main jet deflection angle excessive at the end of the flow guide baffle, resulting in the poor directivity of the jet in the downstream direction. Finally, the vortex inside the nozzle made the flow field more complicated, resulting in large and irregular flow rate fluctuations at the nozzle outlet. Figure 15 shows the instantaneous volume flow rate curves of the two structures. The instantaneous volume flow rate of the shrinking nozzle was slightly greater than that of the nozzle with the Tesla structure in some moments, but most times, it was not. From the perspective of average volume flow rate, the average volume flow rate of the shrink nozzle within two cycles was 0.02409 m^3/s and that of the Tesla symmetrical nozzle was 0.02536 m^3/s, an increase of 5.27%. From the perspective of flow rate stability, the flow rate of the shrink nozzle was not periodic, and the difference between the maximum and minimum instantaneous flow rate was as high as 0.00907 m^3/s. Although the flow rate of the Tesla symmetrical nozzle also fluctuated, the maximum difference in instantaneous flow rate was only 0.00288 m^3/s. Compared with the contraction nozzle, the fluctuation degree was greatly reduced by 214.93%. In addition, the flow rate of the Tesla symmetrical nozzle was also periodic, which means that the value of the flow rate at any time could be obtained according to the periodic law.

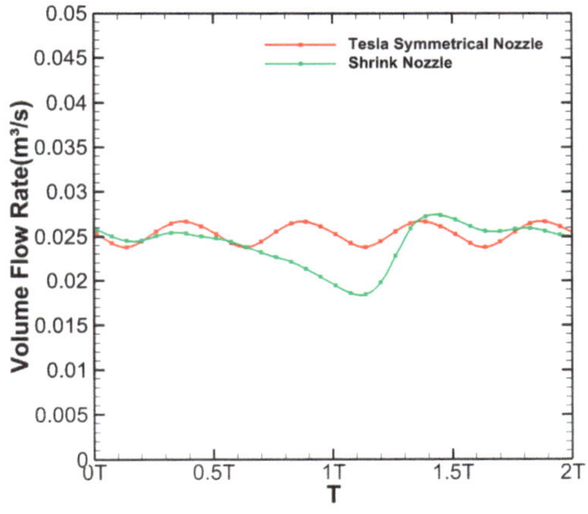

Figure 15. Instantaneous flow curves of the two structures.

This shows that the Tesla symmetrical nozzle was superior to the contraction nozzle in all aspects. First of all, the Tesla structure inside the nozzle could play a directional diversion role on the shunt generated by the main jet impinging on the wall surface of the contraction outlet. This allowed the shunt to be re-injected into the main jet through the Tesla backflow channel, and this energy could be reused to avoid waste. Secondly, the Tesla structure could eliminate the vortex, which improved the flow rate and made the flow fluctuation small and periodic. Finally, the Tesla structure could also effectively stagnate the backflow, and its unique unidirectional conduction function prevented the jet in the nozzle from being sucked into the cavity again; thus, the "self-absorption" phenomenon did not occur.

4. Conclusions

In this paper, a novel valveless continuous micropump based on dual synthetic jets with a Tesla structure was designed. Numerical simulations and PIV tests were carried out. The following conclusions were obtained:

(1) The novel valveless continuous micropump based on dual synthetic jets with a Tesla structure could realize the directional transport of fluid. Through the reciprocating vibration of the diaphragm, the micropump could periodically inhale fluid from both sides and alternately eject the jet from the nozzle. It was able to continuously eject the fluid within one cycle and the pumping flow rate was relatively stable.

(2) The numerical simulation results showed that the downstream time-mean flow field of the micropump presented a unimodal symmetry pattern. The maximum velocity on the center line of the flow field increased first and then decreased as the distance from the nozzle outlet grew. It appeared at a distance of 3 mm from the nozzle outlet and reached 12 m/s. Even at a distance of 24 mm, the velocity reached 6 m/s.

(3) The PIV results showed that the influence range of the micropump jet could be effectively enlarged by increasing the driving voltage amplitude within a certain range ($\pm$150 V~$\pm$210 V), and the maximum influence range was 43.1°. With the increase in driving voltage amplitude, the maximum velocity of the micropump jet first increased and then remained at a stable value. At 180 V driving voltage, the maximum velocity reached 13.5 m/s. Although the increase in the voltage could not effectively improve the maximum velocity of the jet, it could enhance the transport capability of the micropump.

(4) Compared with the traditional shrinking nozzle, the double Tesla symmetrical nozzle could eliminate the internal vortex and enhance the flow rate by 5.27%, making the flow rate more stable and periodic.

Author Contributions: Writing—original draft preparation, J.-Y.Z.; writing—review and editing, W.-Q.P., Z.-B.L., Z.-J.Z., J.-Y.G., Z.-F.D. All authors have read and agreed to the published version of the manuscript.

Funding: This research was funded by the National Natural Science Foundation of China, grant numbers 52075538 and U2141252, and the Provincial Natural Science Foundation of Hunan, grant number 2023JJ30622.

Institutional Review Board Statement: Not applicable.

Informed Consent Statement: Not applicable.

Data Availability Statement: Not applicable.

Conflicts of Interest: The authors declare no conflict of interest.

References

1. Cobo, A.; Sheybani, R.; Tu, H.; Meng, E. A wireless implantable micropump for chronic drug infusion against cancer. *Sens. Actuator A Phys.* **2016**, *239*, 18–25. [CrossRef]
2. Bußmann, A.; Leistner, H.; Zhou, D.; Wackerle, M.; Congar, Y.; Richter, M.; Hubbuch, J. Piezoelectric Silicon Micropump for Drug Delivery Applications. *Appl. Sci.* **2021**, *11*, 8008. [CrossRef]
3. Rubayet, H.; Sevki, C.; Mahmoud, B.; Anthony, P.; Austin, R. A Magnetorheological Duckbill Valve Micropump for Drug Delivery Applications. *Micromachines* **2022**, *13*, 723.
4. Atsushi, T.; Kenichi, K.; Hiroaki, S. Autonomous microfluidic control by chemically actuated micropumps and its application to chemical analyses. *Anal. Chem.* **2010**, *82*, 6870–6876.
5. Do, V.K.; Yamamoto, T.; Ukita, Y.; Takamura, Y. Precise flow control with internal pneumatic micropump for highly sensitive solid-phase extraction liquid electrode plasma. *Sens. Actuators B Chem.* **2015**, *221*, 1561–1569. [CrossRef]
6. Kant, R.; Singh, D.; Bhattacharya, S. Digitally controlled portable micropump for transport of live micro-organisms. *Sens. Actuators A Phys.* **2017**, *265*, 138–151. [CrossRef]
7. Vedran, B.; Denis, D. All Silica Micro-Fluidic Flow Injection Sensor System for Colorimetric Chemical Sensing. *Sensors* **2021**, *21*, 4082.
8. Bußmann, A.; Thalhofer, T.; Hoffmann, S.; Daum, L.; Surendran, N.; Hayden, O.; Hubbuch, J.; Richter, M. Microfluidic Cell Transport with Piezoelectric Micro Diaphragm Pumps. *Micromachines* **2021**, *12*, 1459. [CrossRef]
9. Liu, J.H.; Yan, W.X.; Zhao, Y.Z. A Micropump Sucker Using a Piezo-Driven Flexible Mechanism. *J. Mech. Robot.* **2019**, *11*, 041009. [CrossRef]
10. Chang, X.; Chen, C.; Li, J.; Lu, X.; Liang, Y.; Zhou, D.; Wang, H.; Zhang, G.; Li, T.; Wang, J.; et al. Motile Micropump Based on Synthetic Micromotors for Dynamic Micropatterning. *ACS Appl. Mater.* **2019**, *11*, 28507–28514. [CrossRef]
11. Pramod, K.; Sen, A.K. Flow and Heat Transfer Analysis of an Electro-Osmotic Flow Micropump for Chip Cooling. *J. Electron. Packag.* **2014**, *136*, 031012. [CrossRef]
12. Tang, Y.; Jia, M.; Ding, X.; Li, Z.; Wan, Z.; Lin, Q.; Fu, T. Experimental investigation on thermal management performance of an integrated heat sink with a piezoelectric micropump. *Appl. Therm. Eng.* **2019**, *161*, 114053. [CrossRef]
13. Peng, Y.H.; Wang, D.H.; Li, X.Y.; Zhang, Y. Cooling chip on PCB by embedded active microchannel heat sink. *Int. J. Heat Mass Transf.* **2022**, *196*, 123251. [CrossRef]
14. Abhari, F.; Jaafar, H.; Yunus, N.A. A Comprehensive Study of Micropumps Technologies. *Int. J. Electrochem. Sci.* **2012**, *7*, 9765–9780.
15. Smith, B.L.; Glezer, A. The formation and evolution of synthetic jets. *Phys. Fluids* **1998**, *10*, 2281–2297. [CrossRef]
16. Ikhlaq, M.; Yasir, M.; Demiroğlu, M.; Arik, M. Synthetic Jet Cooling Technology for Electronics Thermal Management—A Critical Review. *IEEE Trans. Compon. Packag. Manuf. Technol.* **2021**, *8*, 1156–1170. [CrossRef]
17. Luo, X.B.; Li, Z.X.; Guo, Z.Y. Principle and Numerical Simulation of A New Valve-less Micropump. *China Mech. Eng.* **2002**, *15*, 1261–1263. (In Chinese)
18. Luo, Z.B.; Xia, Z.X. A novel valve-less synthetic-jet-based micro-pump. *Sens. Actuator A Phys.* **2005**, *122*, 131–140. [CrossRef]
19. Dau, V.T.; Dinh, T.X.; Sugiyama, S. A MEMS-based silicon micropump with intersecting channels and integrated hotwires. *J. Micromech. Microeng.* **2009**, *19*, 125016. [CrossRef]
20. Choi, J.P.; Kim, K.S.; Seo, Y.H.; Kim, B.H. Design and Fabrication of Synthetic Air-Jet Micropump. *Int. J. Precis. Eng. Manuf.* **2011**, *12*, 355–360. [CrossRef]
21. He, X.H.; Zhu, X.B. Performances of Valve-less Synthetic Jet-based Piezoelectric Pumps. *J. Beijing Univ. Technol.* **2015**, *4*, 493–500. (In Chinese)
22. Munas, F.R.; Melroy, G.; Abeynayake, C.B. Development of PZT Actuated Valveless Micropump. *Sensors* **2018**, *18*, 1302. [CrossRef]
23. Le Van, L.; Bui, T.T.; Nhu, C.N.; Ngoc, A.N.; Dinh, T.X.; Dang, L.B.; Tran, C.D.; Duc, T.C. Simulation and Experimental Study of a Synthetic Jet Valveless Pump. *IEEE ASME Trans. Mechatron.* **2019**, *25*, 1162–1170. [CrossRef]
24. Tran, C.D.; Pham, P.H.; Nguyen, T.K.; Phan, H.P.; Dinh, T.; Nguyen, T.V.; Bui, T.T.; Chu, D.T.; Nguyen, N.T.; Dao, D.V. A new structure of Tesla coupled nozzle in synthetic jet micro-pump. *Sens. Actuator A Phys.* **2020**, *315*, 112296. [CrossRef]

25. Luo, Z.B.; Xia, Z.X.; Liu, B. New Generation of Synthetic Jet Actuators. *AIAA J.* **2006**, *44*, 2418–2420. [CrossRef]
26. Luo, Z.B.; Zhao, Z.J.; Deng, X.; Wang, L.; Xia, Z.X. Dual Synthetic Jets Actuator and Its Applications—Part I: PIV Measurements and Comparison to Synthetic Jet Actuator. *Actuators* **2022**, *11*, 205. [CrossRef]
27. Luo, Z.B. Principle of Synthetic Jet and Dual Synthetic Jets, and Their Applications in Jet Vectoring and Micro-Pump. Ph.D. Thesis, National University of Defense Technology, Changsha, China, October 2006.
28. Luo, Z.B.; Xia, Z.X. Luo-Xia Model, A Computing Model of PZT-Driven Synthetic Jet Actuator. In Proceedings of the CCCM 2010, Mianyang, China, 20 August 2010. (In Chinese)
29. Celik, I.B.; Ghia, U.; Roache, P.J.; Freitas, C.J. Procedure for Estimation and Reporting of Uncertainty Due to Discretization in CFD Applications. *J. Fluids Eng.* **2008**, *7*, 078001.

actuators

MDPI

Article

Experimental Investigation on the Combined Blowing Control of a Hybrid Wing Body Aircraft

Jiaxin Pan, Wanbo Wang *, Chen Qin, Xunnian Wang, Qixiang Sun and Xin Zhang

State Key Laboratory of Aerodynamics, China Aerodynamics Research and Development Center, Mianyang 621000, China; bc22005009@mail.ustc.edu.cn (J.P.)
* Correspondence: bowanw@163.com

Abstract: Combined blowing was performed on a Hybrid Wing Body (HWB) aircraft through wind tunnel testing at a Reynolds number of 1.75×10^6. The full cycle of separation and reattachment under the control of combined blowing was implemented using Computational Fluid Dynamics (CFD), and the mechanism of combined blowing inhibiting separation was analyzed. The aerodynamic characteristics of the baseline and the independent effects of the blown deflected trailing edge (TE), blown leading edge (LE), and combined blowing on the TE and LE were investigated. The results clearly show that combined blowing can inhibit the development of cross-flow, reduce the accumulation of a boundary layer at the tip, and inhibit the flow separation effect. The effect of using seamless simple flaps alone to increase the lift is limited; blowing control is required to enhance the lift further. Applying the blown deflected TE can improve the lift linear segment, so that 30° flap achieves the lift gain of 40° flap without control, while the drag coefficient is approximately 0.02 smaller, but the stall gradually advances. Using the blown LE can significantly increase the stall angle from 12° to 18°. However, the lift linear segment remains unaffected. In particular, combined blowing can achieve the control effect of improving the lift linear segment, delaying stall, and decreasing drag. Moreover, the maximum lift coefficient is approximately 0.19, and the lift-to-drag ratio increment in the control state with a 30° flap deflection angle is above 2.2 in the angle of attack range of 4° to 12° compared to the uncontrolled state with a 40° flap deflection angle.

Keywords: Hybrid Wing Body; wind tunnel force test; lift enhancement; blowing flow control

Citation: Pan, J.; Wang, W.; Qin, C.; Wang, X.; Sun, Q.; Zhang, X. Experimental Investigation on the Combined Blowing Control of a Hybrid Wing Body Aircraft. *Actuators* **2023**, *12*, 237. https://doi.org/10.3390/act12060237

Academic Editor: Luigi de Luca

Received: 17 May 2023
Revised: 4 June 2023
Accepted: 5 June 2023
Published: 8 June 2023

1. Introduction

Compared to conventional configurations, the Blended Wing Body (BWB) or Hybrid Wing Body (HWB) layout significantly reduces the wetted area. It has more significant advantages in terms of drag and noise reduction [1–6]. However, limited longitudinal control authority due to the short lever arm of the BWB layout is a critical challenge. This issue is usually solved by lengthening the body, adding a horizontal tail, or using active flow control [3]. The HWB layout developed by the Lock Martin (LM) company keeps the characteristics of the BWB, with a smooth transition between the wings and the body. The remaining part retains the longer cylindrical fuselage of the traditional layout and the "T-shaped" tail to enhance longitudinal control authority [7]. This design not only has the same load distribution advantages as the BWB, but the wing and fuselage transition section also helps to prevent sudden changes in inertial loads, and the rounded fuselage in the rear half supports the internal structural arrangement [1].

The LM-HWB configuration uses the same leading edge slats and Fowler flaps as conventional aircraft [8]. This conventional lift enhancement method requires complex working mechanisms for safe operation, usually covered by fairings and protruding under the wings, creating detrimental drag [9].

Seamless, simple flaps do not involve complex drive mechanisms, but the airflow on the flap surface is easily separated, resulting in limited lift. The concept of a seamless, simple flap based on blowing flow control, which can significantly suppress airflow separation, has been revived [10] and is receiving increasing attention in research [11–14]. Applying internal-blown flaps can significantly increase lift and cause earlier stalls [15]. Therefore, the deployed slat increases the stall angle, as in the case of the HWB STOL (Short Take-Off and Landing) model applied in the LaRC (Langley Research Center) subsonic wind tunnel by Collins et al. The model uses a combination of deployed slats and internal-blown flaps to achieve STOL [16]. The FAST-MAC (Fundamental Aerodynamics Subsonic/Transonic-Modular Active Control) wing-body-composite half-model low-speed high-lift configuration was tested in the NTF (National Transonic Facility) transonic wind tunnel by Milholen et al. using a 10%c deployed slat and an internal-blown flap [17]. In 2020, Lin et al. compared a simple flap lift solution based on a 10% scaled-down CRM (Common Research Model) wind tunnel test model using a combination of swept and constant blowing actuators with a conventional Fowler flap lift solution in the LaRC subsonic wind tunnel, showing that the latter was more efficient [18].

However, when LE slats are deployed, the concave pressure surface has a low-velocity reverse region. The free shear layer developing from the sharp lower TE collides with the slat TE pressure surface, affecting drag and noise during take-off and landing conditions [19]. Research has shown that the combination of LE blowing and internal-blown flaps can effectively increase the stall angle of attack and maximum lift coefficient and reduce the noise of the aircraft [15,20].

The HWB has been proven to have a good high-speed cruise performance [8,21,22], but there are many problems with the application of conventional lift devices to such aircraft with new layouts [7]. Combined blowing lift enhancement solutions can replace leading edge slats and Fuller flaps, but there is little research on the application of this technology to aircraft with new layouts. Based on existing blowing flow control research, this paper uses a half-mode model of an HWB layout with blowing slots on the leading and tailing edges of the main wing, exploring the combined blowing lift enhancement scheme for HWB through CFD methods and wind tunnel tests to investigate the flow characteristics and mechanism of action under blowing control, providing new insights and methods to improve the overall performance of the HWB layout.

2. Test Equipment and Method

2.1. Test Device

The blowing control methods include TE blowing, LE blowing, and combined blowing, as seen in Figure 1.

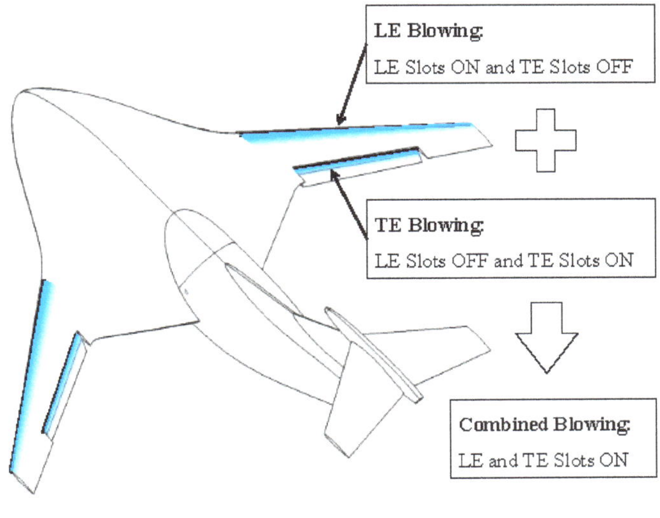

Figure 1. Diagram of the blowing control scheme.

The test was conducted in a recirculating low-speed wind tunnel at the China Aerodynamics Research and Development Center. The cross-section size of the test section was 1.8 m (width) × 1.4 m (height). The stable wind speed range was 10 m/s~105 m/s, the turbulence intensity reached 0.0008 when the wind speed was lower than 70 m/s, and the axial static pressure gradient was above 0.005.

The agent model was connected to a box-type balance below the test section floor through a support, as depicted in Figure 2a. The balance was fixed to the turntable below the wind tunnel with the balance support table. The variable lateral sliding angle mechanism of the lower turntable altered the angle of attack of the model. The wall boundary effect was mitigated by installing fairing pads. The pad height was 20 mm, the maze groove sealed the pad and the model to prevent the pad and the model from coming into contact and transmitting force, the maze groove clearance was 5 mm, and the pad was secured to the circular floor.

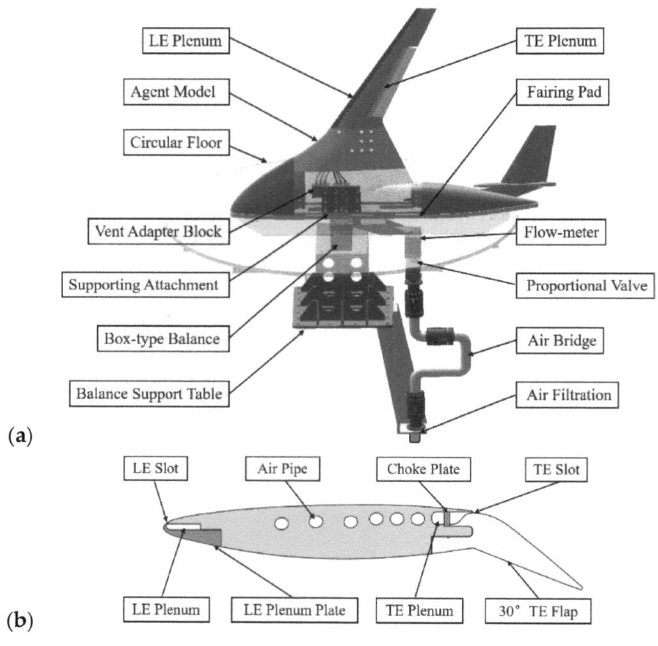

Figure 2. Schematic diagram of the test model: (**a**) general assembly; (**b**) wing internal profile.

2.2. Test Model

The test model was a half-model of an HWB aircraft, including the nose, fuselage, main wing, vertical tail, horizontal tail, and flap, with a half-span of 994 mm and an average aerodynamic chord of 512 mm. The flap was a seamless, simple flap; the flap chord length was approximately 33%c; the flap offset was set to 0°, 30°, 40°, and 60°; and each component was created through aluminum metal processing. Two blowing plenums were arranged in the wing, and blowing control was achieved by controlling the compressed air supply. The total length of the LE slot was 685 mm, being approximately 0.1 mm high; the total length of the TE slot was 377 mm, being approximately 0.4 mm high; and these blowing slots were locked with a span-to-array screw hole (spacing approximately 40 mm).

Based on the relevant research [23–25], it is known that the effect of blowing control near the point of flow separation is suitable. The LE and TE slots were set at 3%c and 63%c, respectively, from the LE, and these slots were typically tangent to the surface. The TE of the main wing was fixed with a hinge, and the TE slot was easily established at the position where the main wing and flap were connected [13]. This will be introduced into the Numerical Calculation section to determine the position of the LE slot.

2.3. Air Supply Device and Method

Before entering the model air supply pipeline, compressed air passes through air filtration, the air bridge, the proportional valve, and the high-precision flow meter, as shown in Figure 1. The air bridge mainly comprised two transverse flexible joints and one flexible vertical joint. The flexible vertical joint could absorb the expansion of the two transverse flexible joints. The two transverse flexible joints could also absorb the expansion of the flexible vertical joint, thus eliminating the influence of the supply pipe on the balance force measurement. Two proportional valves accurately controlled the flow rate of the main wing's leading and trailing edge jet; two high-precision flow meters measured the flow rate of blowing. The measurement accuracy was approximately 0.01 g/s.

Compressed air was connected to the model supply pipe via polyurethane pipes, and two closed four-way vent adapter blocks were fixed inside the model. Three supply pipes connected the LE and TE plenums to the two vent adapter blocks. The flap was connected to the TE plenum through a row of screws arrayed in the span direction. The flow in each plenum passes through a choke plate to achieve a uniform flow along the span of the blown slot. Sealing strips and glue were used to seal the gap between the flaps and the wing, as seen in Figure 2b.

A photograph of the test model is shown in Figure 3, with the screw holes on the model's surface all filled in the test.

Figure 3. A front view of the test model.

The dimensionless momentum coefficient is expressed as the jet strength, as shown in Equation (1).

$$C_\mu = \frac{m_j U_j}{qS} = \frac{m_j}{qS}\left\{\frac{2\gamma R T_0}{\gamma - 1}\left[1 - \left(\frac{P_a}{P_0}\right)^{\frac{\gamma-1}{\gamma}}\right]\right\}^{\frac{1}{2}} \tag{1}$$

The mass flow rate m_j in the gas supply line is measured with the flow control unit, q is the test velocity pressure, S is the model reference area, and C_μ can be found for a given pressure ratio.

3. Numerical Calculation

3.1. Calculation Models and Methods

The position of the LE slot was determined via CFD, and we analyzed the combined blowing control mechanism. The calculation software is based on the NNW-FlowStar open-source fluid engineering software developed by the China Aerodynamic Research and Development Center (CARDC) [26]. The control equation is a three-dimensional viscous compressible unsteady Reynolds-averaged Navier–Stokes equation; thus, the control equation takes the following integral form:

$$\frac{\partial}{\partial t}\int\int\int Q\mathrm{d}V + \int\int f\cdot n\mathrm{d}S = 0 \tag{2}$$

where t is a time quantity, V denotes the volume of the control body, S denotes the surface area of the control body, Q is a conserved variable, f is the sum of the inviscid and viscous fluxes through the surface S, and n is the outer normal unit vector of the control body surface S. The control equations were discretized using the finite volume method, with the fluxes in Roe [27] format numerically and in second-order backward differential format in time, the spatial discretization format being a second-order accuracy windward format, and the k-ω SST turbulence model was used to close the mean Reynolds equation. The dimensionless time step $\Delta t^* = \Delta t \times U/c = 0.002$, and the LU-SGS (Lower-Upper Symmetric Gauss–Seidel) implicit time advance algorithm was used. The calculation of the region entrance, exit, and upper and lower boundary distances of the aircraft was 40 times the chord length, and the first boundary layer grid thickness was approximately 1×10^{-5} m. The model surface mesh and the symmetrical surface part of the mesh are shown in Figure 4. The model surface satisfies the no-slip boundary condition, the jet boundary is the velocity inlet boundary, and the calculated angle of attack range is 0°~10°. Figure 5 shows the distribution of y+ on the wing surface mesh. In the figure, y+ is less than 5 for the mesh, and in most areas, y+ remains around 1, indicating a complete solution of the boundary layer. The size of the CFD model remains the same as that of the test model. The calculations use symmetrical surface boundary conditions and do not consider the boundary layer effects of the floor; thus, the fairing pad is not included in the CFD model. The far field has a velocity inlet set at 50 m/s, and the outlet boundary condition is set to a static pressure outlet, while all other boundary conditions in the far field are set to free-slip wall boundary conditions. The jet slots are all set to velocity inlet conditions, and the LE momentum coefficient and TE momentum coefficient are calculated to be 0.03 and 0.005, respectively. The numerical Reynolds number is consistent with the real wind tunnel test at approximately 1.75×10^6.

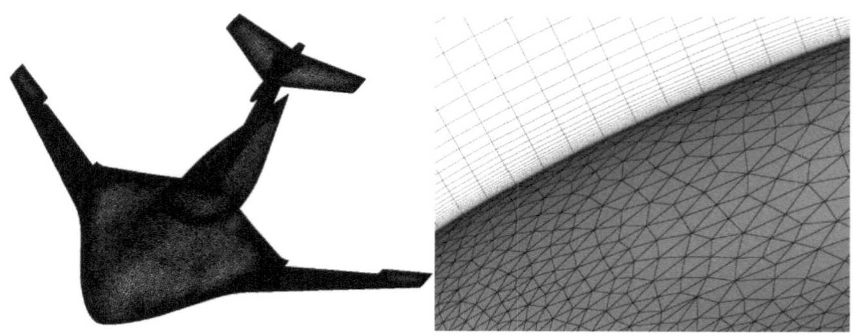

Figure 4. CFD mesh of the HWB model.

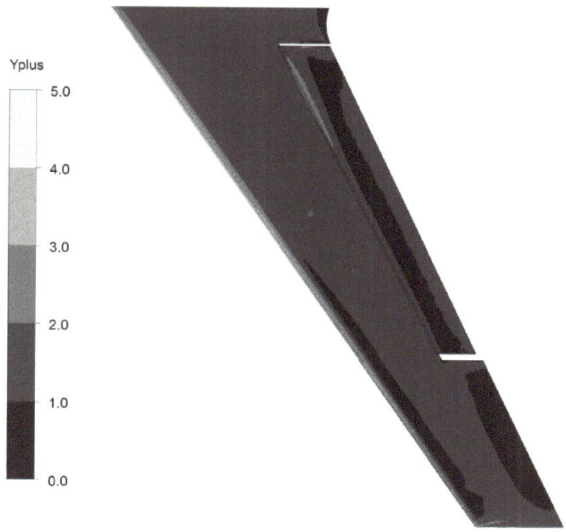

Figure 5. Wing surface grid y+ distribution.

3.2. Calculation Validation

To analyze the impact of the mesh size and ensure the independence of the grid configuration in regard to the calculation results, three grids of different densities were generated for the model cruise state and compared with the experimental data.

The error between the numerical simulations and experimental data gradually decreased as the grid size increased. The lift coefficients calculated with grid sizes of 5 million, 10 million, and 15 million had errors of 2.1%, 0.8%, and 0.6%, respectively, compared to the experimental results. As the medium-density grid closely approximated the denser grid and was in good agreement with the test results, we used a 10-million-number grid to improve the computational efficiency.

Figure 6 compares the wind tunnel test and CFD results at a wind speed of 50 m/s and Reynolds number of approximately 1.75×10^6. For the cruise and 30° flap state, the linear section lift results of the wind tunnel test and CFD are more consistent, indicating that the CFD results are reliable.

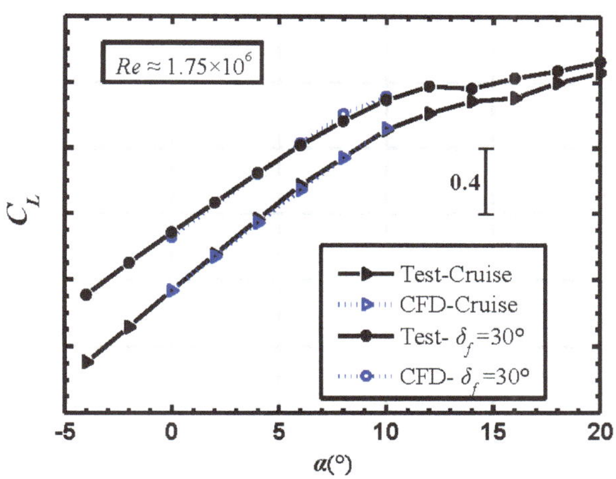

Figure 6. Comparison of wind tunnel test and CFD results.

3.3. Position of the LE Slot

The lift characteristics corresponding to different LE blowing positions obtained via CFD can be seen in Figure 7, where $C_{\mu l}$ and $C_{\mu t}$ are the momentum coefficients of LE and TE blowing, respectively.

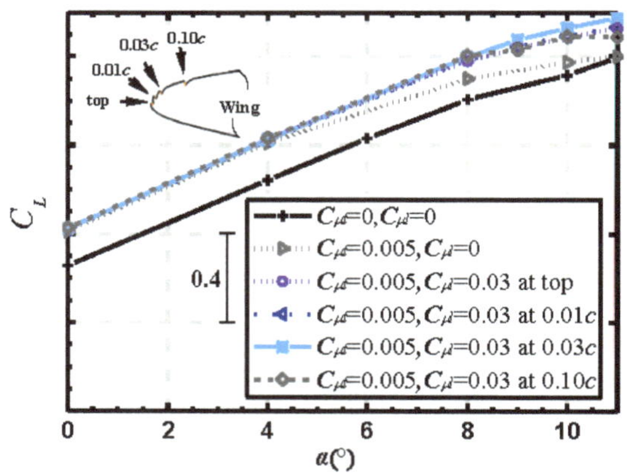

Figure 7. Lift coefficients at different LE jet positions.

It can be seen that when $\alpha \leq 8°$, the influence of different LE blowing positions on the lift coefficient is small. With the increasing angle of attack, the lift effect is better when the LE slot is at 3%c; thus, this location was chosen. Figure 8 presents the pitch moment coefficient results before and after combined blowing control. Compared to the uncontrolled state, combined blowing increases the nose-down moment by 48%, which is similar to the results obtained from the high-lift configuration used in the HWB described in [28]. The reason for this may be that the pitch moment center of gravity of the HWB layout is relatively forward, which inevitably leads to additional nose-down moments after increasing the lift. As the research on horizontal tailplane balancing is complex, with many influencing factors, it is not the focus of this study. Additionally, no elevator deflection was carried out during the research process. Therefore, this pitch moment coefficient result is only briefly explained in the calculation section of this paper.

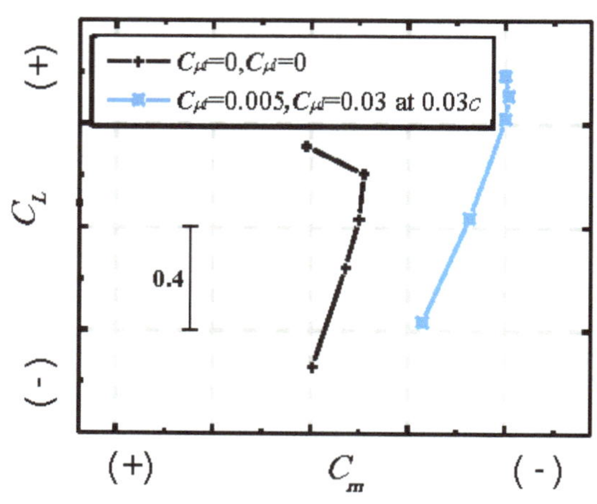

Figure 8. Pitching moment coefficients before and after combined blowing control.

3.4. Combinatorial Blowing Control Process

To analyze the combined blowing control process and understand the blowing control mechanism, the calculated results of the shear stresses and flow lines on the wing surface for different momentum coefficients during the process from uncontrolled to stable blowing control are given in Figure 9 for the state of 30° flaps and a 10° angle of attack. In natural conditions, the cross-flow developed at the flaps moves in the direction of the wing tip, and there is a clear separation between the flaps and the wing tip surface. As combined control is applied, the tip separation is first restrained and the separation gradually moves in the direction of the flow, with the cross-flow being impeded by the blowing and gradually decreasing in distance along the span. As the injection momentum is further increased, no significant cross-flow occurs, and the combined blowing effectively suppresses flow separation.

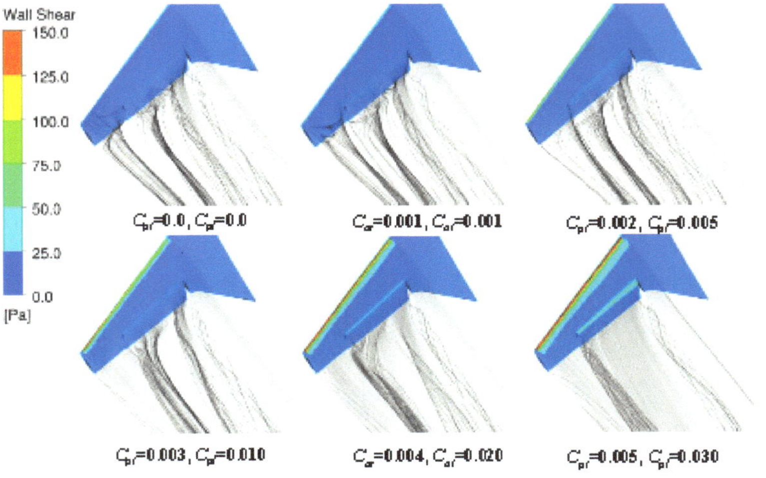

Figure 9. Wall shear and streamlines at different momentum coefficients.

Figure 10 shows the lift coefficient results at different moments for a flap angle of 30° and 10° attack of angle, where the combined blowing starts at the moment t = 100 ms and the momentum coefficient gradually becomes steady around 0.03 and 0.005 after the moment t = 300 ms. Around t = 200 ms, there is a tendency for the lift coefficient to decrease and then increase, consistent with previous research [29]; however, this phenomenon is not analyzed with respect to the flow field. The lift coefficient gradually stabilizes after approximately 200 ms, when the momentum of the combined blowing injection stabilizes. In order to identify the reasons for variations in the lift coefficients, an analysis of the evolution process of the flow field resulting from combined blowing excitation was conducted.

Figure 11 shows the time-dependent induced flow field in a typical chordal cross-section after using combined blowing. Before t = 200 ms, a significant low-speed separation is evident on the upper surface of the flap, and the flow line depicts the separation vortex gradually moving downstream with time. At t = 200 ms, the separation region is squeezed by the injection, and this leads to a slight increase in the separation compared to the period before, which may be the reason for the decrease in the lift coefficient. When t = 300 ms, the effect of combined blowing becomes pronounced, and the low-speed separation region is flattened and concentrated near the TE of the flap. At t = 400 ms, separation is effectively restrained, and the high-speed region on the suction surface of the wing spreads in the flow direction from the leading edge. When t = 500 ms, the induced flow field stabilizes, and the induced injection velocity exhibits minimal variation over time.

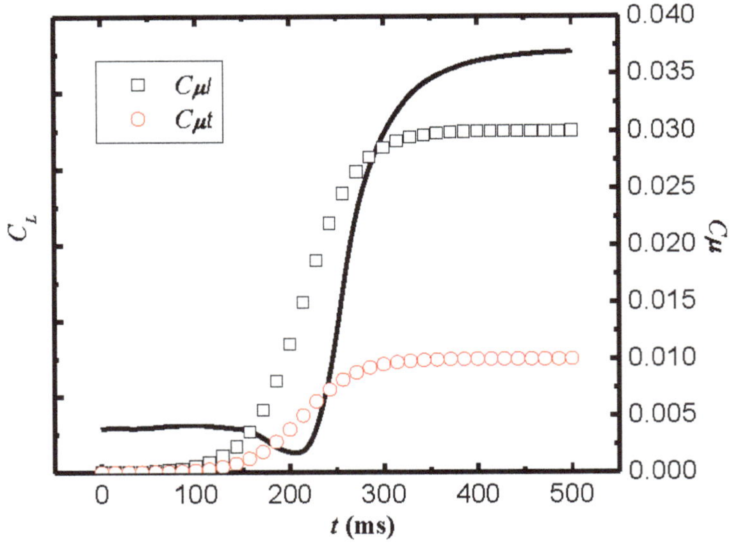

Figure 10. Lift coefficient variation curve during combined blowing control.

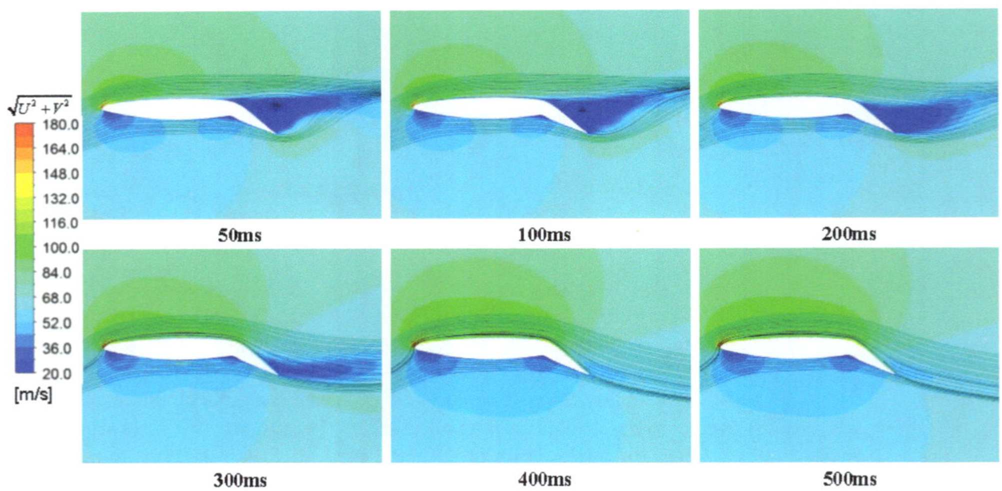

Figure 11. Combined-blowing-induced flow field change process.

Figure 12 shows the variation in the transient vortex field in a typical spanwise cross-section of the aircraft. At t = 200 ms, the vortex break-up causes a number of vortices with different directions and a low strength. As the combined blowing momentum coefficient gradually increases and stabilizes, the low-intensity vortices gradually disappear. After t = 400 ms, the injected momentum significantly enhances the vortex strength and delays the vortex break-up, which, in turn, enhances its stability and induces the stabilization of the flow field.

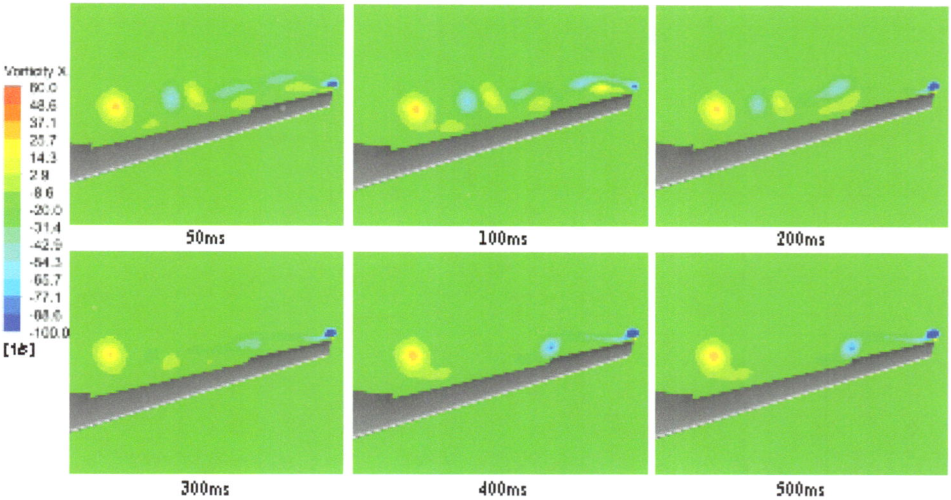

Figure 12. Combined-blowing-induced flow field vortices' change process.

3.5. Surface Pressure and Velocity Distribution

To compare the pressure changes on the aircraft surface with and without control and to provide a reference for the adhesive position of the thread in the flow visualization test, the pressure coefficients and Mach number clouds for different spanwise profiles of the wing without control and combined blowing control are provided in Figures 13 and 14. Combining the results of these figures at an angle of attack of 10°, there is separation at the wing tip in natural conditions, the suction peak at the wing tip is weakened, and the flow velocity in the flap and wing tip region is lower. When combined blowing control is applied, the injection weakens the effect of the cross-flow and, thus, the buildup of the boundary layer at the wing tip, and the pressure gradient distribution in the flow direction is restored. Moreover, the range of the low-speed region is significantly reduced after the injection of momentum, and the slope of the lift line remains basically the same before and after combined blowing control is applied (see Figure 7). Based on these results, the density of the visual slit slightly increased to an appropriate degree at the TE, LE, and wingtip locations.

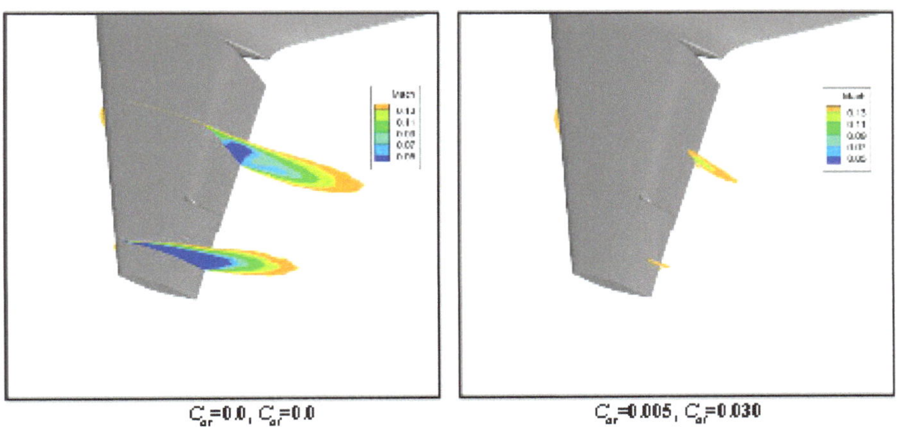

Figure 13. Mach number contours for different spanwise profiles of the wing (Y/Span = 60%, 90%).

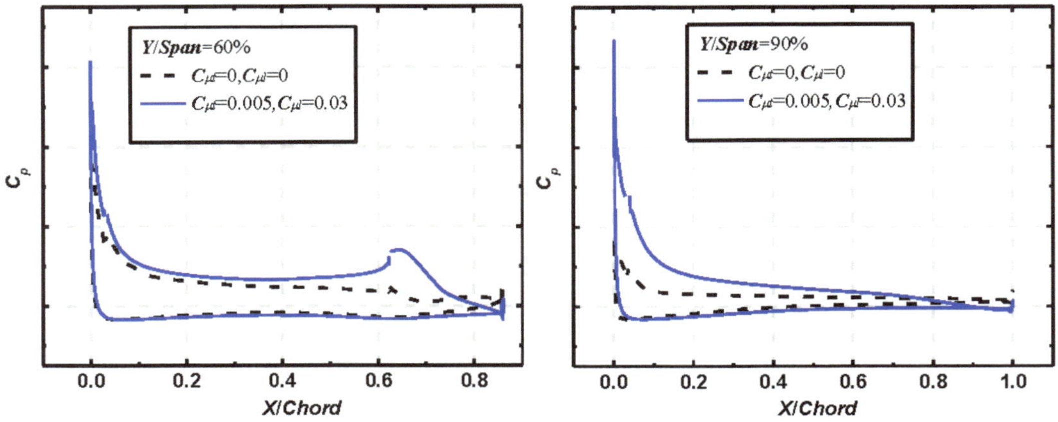

Figure 14. Pressure coefficients for different spanwise profiles of the wing.

4. Test Results and Discussion

4.1. Data Repeatability

Under the condition of a wind speed of 50 m/s (Reynolds number 1.75×10^6), seven repeatability verification tests were conducted. Table 1 shows the maximum mean square deviation of the seven repeatability test results.

Table 1. Repeatability test data.

$\alpha/°$	σ_{CL}	σ_{CD}	σ_{Cm}
−4	0.001260	0.000286	0.000426
−2	0.001170	0.000320	0.000538
0	0.000899	0.000354	0.000442
2	0.000839	0.000348	0.000320
4	0.000940	0.000313	0.000492
6	0.000746	0.000324	0.000527
8	0.000936	0.000346	0.000522
10	0.000761	0.000430	0.000353

The mean square deviation is the square root of the ratio of the sum of the squares for the deviations of the test values x from the average value μ to the number of observations N. It is defined here in Equation (3) as:

$$\sigma_{SD} = \sqrt{\frac{1}{N}\sum_{i=1}^{N}\left(x^{(i)} - \mu\right)^2} \quad (3)$$

The maximum mean square deviations of the lift and drag coefficients do not exceed 0.002 and 0.0005, respectively, and this accuracy meets the wind tunnel test specification.

4.2. Uncontrolled Performance

Figure 15 shows the test model's lift and drag coefficient curves for different flap deflections at a wind speed of 50 m/s. The lift coefficient of the HWB with flap angles of 30° and 60° increased by approximately 0.223 and 0.382, respectively, compared to the cruise configuration (α = 8°). With the increase in the flap angle, the separation vortex on the suction surface of the flap gradually increased, resulting in a lack of increase in the lift with the enhancement of the curvature, while the drag coefficient increased significantly. Compared to the cruise configuration, the stall angle was increased by approximately 2° at the flap angle of 30° and by approximately 4° at the flap angles of 40° and 60°. Similar to the conclusion in [18], the lift enhancement effect of the seamless, simple flap is limited and causes a forward stall angle, accompanied by higher drag, which requires a combination of flow control methods to further improve the lift.

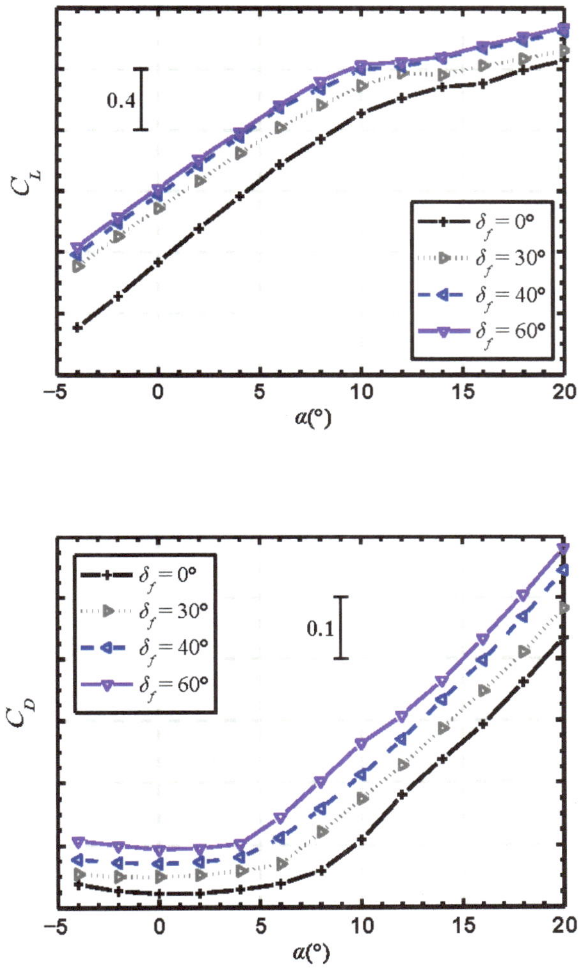

(a)

(b)

Figure 15. Basic aerodynamic performance with variable flap skewness: (**a**) C_L~α; (**b**) C_D~α.

In Figure 16, when the angle of attack is 4° and the flap angle is 60°, the suction surface of the flap is completely separated due to the lack of effective slit flow. The separated region on the main wing is mainly concentrated in the area near the LE covering 1/3 of the wing root and the middle section when the angle of attack is 8°. With the increase in the angle of attack, the reverse pressure gradient is further enhanced, resulting in the expansion of the separation on the main wing from the LE to the entire outer wing section, with a strong separation forming in the middle section. As shown in Figure 15, separation reduces the slope of the lift linear segment and increases the drag.

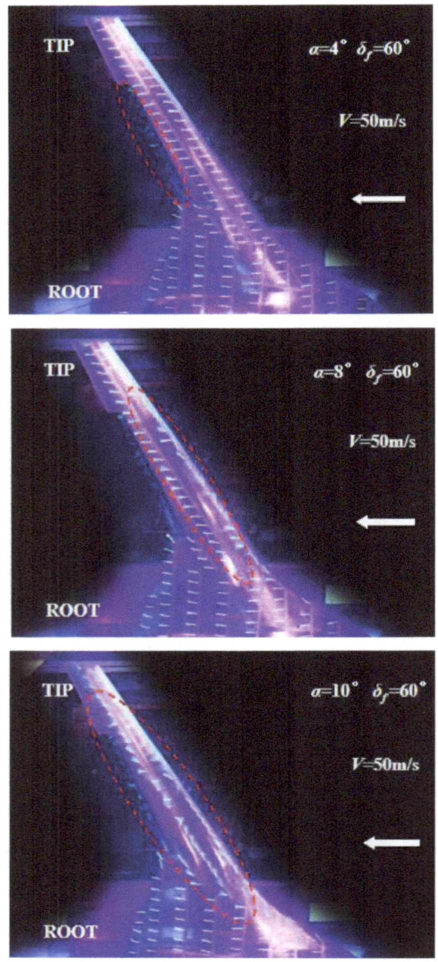

Figure 16. Flow visualization on suction side (uncontrolled, arrows indicate the direction of incoming flow): (**a**) $\alpha = 4°$; (**b**) $\alpha = 8°$; (**c**) $\alpha = 10°$.

In contrast, the 30° flap angle has a significant lift enhancement effect, a larger stall angle of attack, and a smaller drag coefficient; thus, the following section focuses on the study of blowing lift enhancement technology based on a 30° flap angle.

4.3. *Effect of TE Blowing Alone*

With no incoming wind and a 30° flap angle, the additional force and moments caused by different TE blowing momentum coefficients are shown in Figure 17. When the momentum coefficient is a relatively large, at 0.04, the additional lift, drag, and pitch moment loads are 1.7%, 3.8%, and 3.7%, respectively, compared to aerodynamic loads when there is a wind speed of 50 m/s. Since the changes in the aerodynamic coefficients caused by the additional loads are relatively small and also affect the actual flight state, the following test results are not deduced from the additional force and moments.

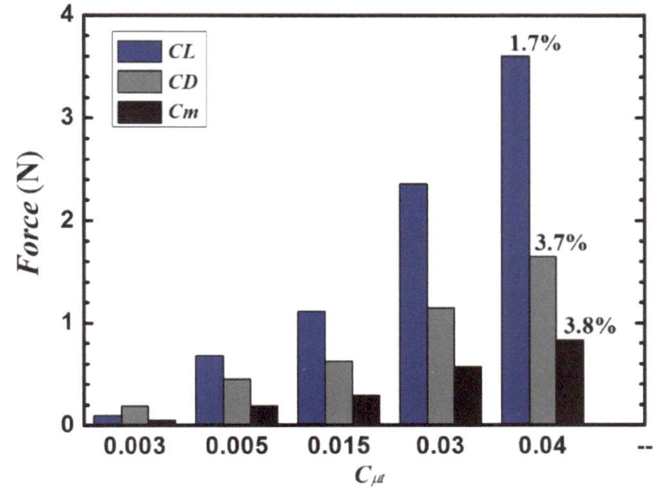

Figure 17. Influence of blowing control on force or moment measurement results.

When the wind speed is 50 m/s and the flap angle is 30°, the lift and drag coefficients of different TE blowing momentum coefficients change with the angle of attack, as presented in Figure 18. With the increase in the momentum coefficient, the linear section lift gradually increases, and the linear lift enhancement effect at the uncontrolled 40° flap angle is reached when the momentum coefficient is 0.005. The momentum coefficient of 0.015 increases the lift by approximately 0.215 ($\alpha = 4°$). At the same time, over the entire range of the test angle of attack, the drag coefficient with TE blowing is approximately 0.02, smaller than that under an uncontrolled 40° flap. The lift–drag ratio of the small flap angle with TE blowing is significantly higher than that of the uncontrolled large flap angle, but with the increase in the angle of attack, the lift–drag ratio of the two states tends to be consistent (see Table 2).

Table 2. Comparison of lift–drag ratio between TE blowing and uncontrolled state.

$\alpha/°$	*L/D*		$\Delta L/D$	$\frac{\Delta L/D}{(L/D)_{\delta_f=40°}}$
	$\delta_f = 30°$ ($C_{\mu t} = 0.015$)	$\delta_f = 40°$ (Uncontrolled)		
4	12.908	9.342	3.566	38.2%
6	11.124	8.472	2.651	31.3%
8	8.368	6.791	1.577	23.2%
10	6.512	5.659	0.853	15.1%
12	5.169	4.526	0.643	14.2%
14	4.146	3.837	0.310	8.1%

In summary, TE blowing alone can significantly enhance the linear section of the lift and provide a significant lift-to-drag ratio advantage over the uncontrolled large flap deflection state prior to stall. However, as the blowing momentum coefficient increases, the stall gradually advances.

A significant flap suction surface separation occurs at a wind speed of 50 m/s and an angle of 4° with no control applied. A TE blowing momentum coefficient of 0.005 at a 4° angle of attack alone can restrain flap separation (see Figure 19).

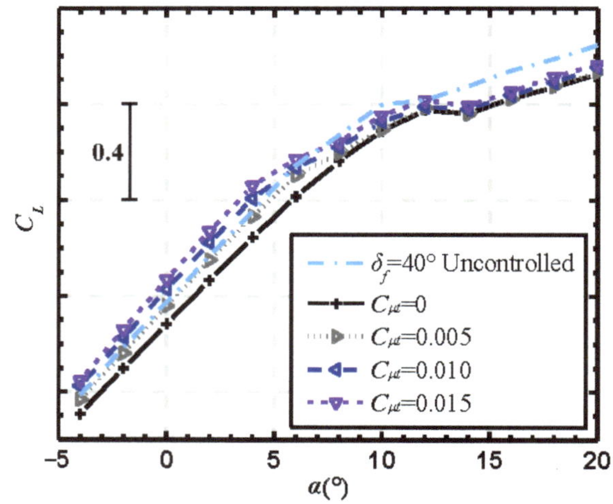

(**a**)

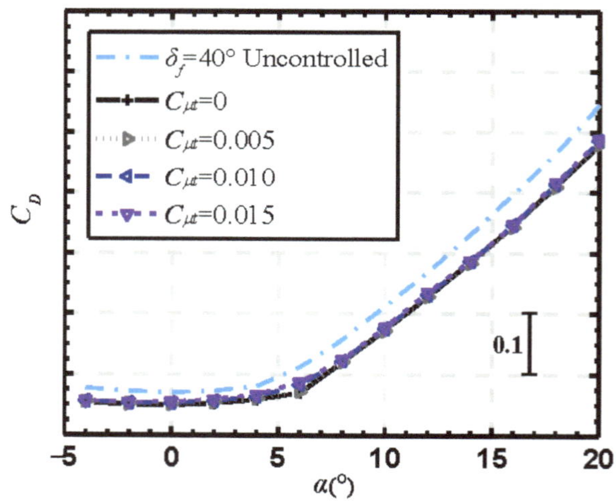

(**b**)

Figure 18. Lift and drag coefficient curves for different momentum coefficients (δ_f = 30°): (**a**) C_L~α; (**b**) C_D~α.

In addition, with a flap deflection of 30° and a 6° angle of approach, without blowing control, there is a significant separation on both the flap and the main wing suction surfaces, as shown in Figure 19c, with the spanwise flow starting at the transition between the wing and the fuselage (see the solid yellow line) and significant backflow at the LE (see the dashed red line). This indicates that although TE blowing can effectively suppress flap airflow separation, there is still a significant separation at the LE. Thus, further research is needed to investigate the effect of blowing control near the LE on lift performance.

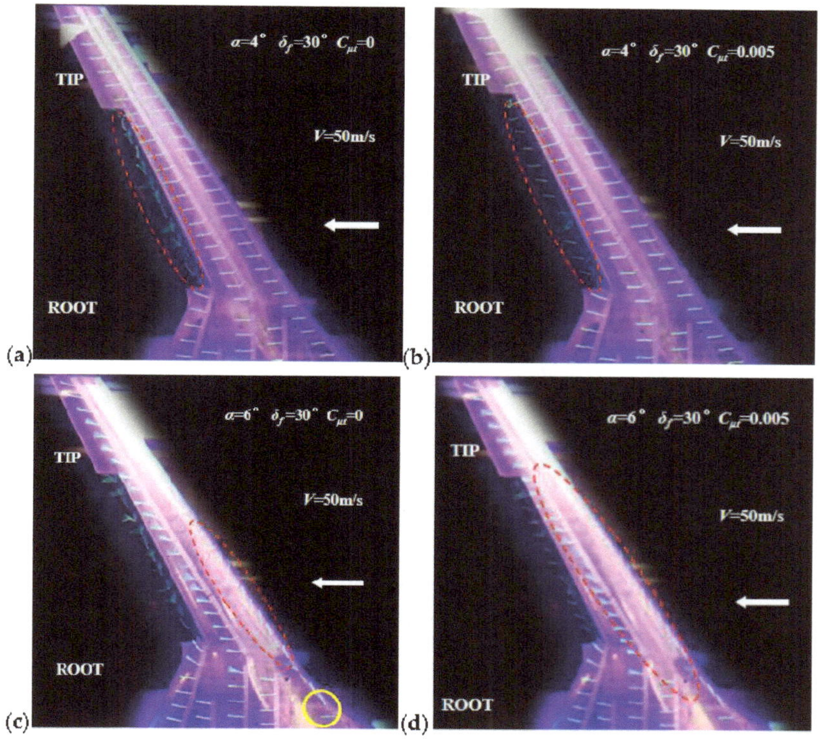

Figure 19. Flow visualization using silk on the suction side (TE blowing alone, arrows indicate the direction of incoming flow): (**a**) $\alpha = 4°$, $C_{\mu t} = 0$; (**b**) $\alpha = 4°$, $C_{\mu t} = 0.005$; (**c**) $\alpha = 6°$, $C_{\mu t} = 0$; (**d**) $\alpha = 6°$, $C_{\mu t} = 0.005$.

4.4. Effect of LE Blowing Alone

Figure 20 shows the characteristic lift curves of the model after applying LE blowing control alone at a flap deflection angle of 30°. The lift linear segment did not change significantly compared with the uncontrolled state. However, the stall delay effect was significant, with a maximum increase in the lift coefficient of approximately 0.16 ($\alpha = 16°$) and an increase in the stall angle from 12° to 18°.

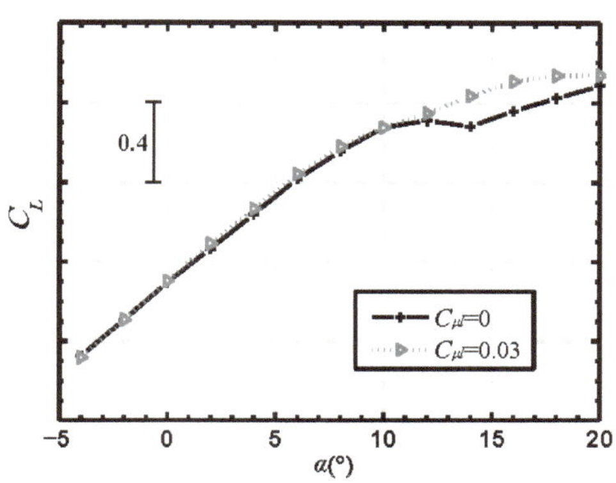

Figure 20. Lift coefficient curves for LE blowing control and uncontrolled state ($\delta_f = 30°$).

In summary, the effect of TE blowing alone on the aerodynamic characteristics of the model is mainly observed before the stall, while the effect of LE blowing alone on the lift is mainly observed after the stall. Therefore, blowing at the leading and trailing edges of the main wing can be used for flow control to achieve an increased lift linear segment while delaying stall.

4.5. Combined Blowing Control

Whether the injection is supplied using engine-induced air or other equipment, the achievement of maximum lift enhancement at a minimal energy cost is expected. When the wind speed is 50 m/s and the flap deflection angle is 30°, blowing control is applied simultaneously to both the leading and trailing edges of the main wing. Combining the results in Figure 18a, TE blowing uses a relatively small momentum coefficient of 0.005 to ensure that the leading edge has enough momentum to suppress the larger separation region of the main wing suction surface.

The resulting lift and drag coefficient curves are shown in Figure 21. The lift linear segment during combined blowing and TE blowing alone is basically the same, and after the angle of attack of 6°, the combined blowing significantly delays the stall, with a maximum lift enhancement of approximately 0.19 ($\alpha = 14°$). The lift linear segment under combined blowing reaches the effect of gain in the 40° flap uncontrolled state, and after the angle of attack of 10°, the combined blowing continues to increase the lift coefficient. As shown in Figure 21b, the drag reduction at an angle of attack of 8° is approximately 0.024, which is approximately 0.06 lower than the drag coefficient at 40° flaps without control.

Table 3 shows the lift-to-drag ratios for 30° flaps with combined blowing control and 40° flaps without control. Compared to the larger uncontrolled flap deflection state, the smaller flap deflection state with control applied maintains a specific advantage in the 4° to 12° angle of attack range due to the delayed stall effect of combined blowing, with increments of 2.2 or more.

Table 3. Comparison of lift–drag ratio between combined blowing and uncontrolled state.

$\alpha/°$	**L/D** $\delta_f = 30°$ ($C_{\mu t} = 0.005$ $C_{lt} = 0.030$)	**L/D** $\delta_f = 40°$ (Uncontrolled)	$\Delta L/D$	$\frac{\Delta L/D}{(L/D)_{\delta_f=40°}}$
4	13.110	9.342	3.768	40.3%
6	13.107	8.472	4.635	54.7%
8	11.148	6.790	4.357	64.2%
10	9.016	5.659	3.357	59.3%
12	6.777	4.526	2.251	49.7%
14	5.240	3.837	1.403	36.6%

The separation of the flap is effectively suppressed when TE blowing control is used with $C_{\mu t} = 0.005$, but the separation area of the main wing suction surface is extensive, and stall occurs, as can be seen in Figure 22a. For $C_{\mu t} = 0.015$ and $C_{\mu l} = 0.03$, except for a slight separation at the unarranged blown slot at the wing root, no significant separation is observed between the wing tip and the middle section of the main wing, effectively delaying stall.

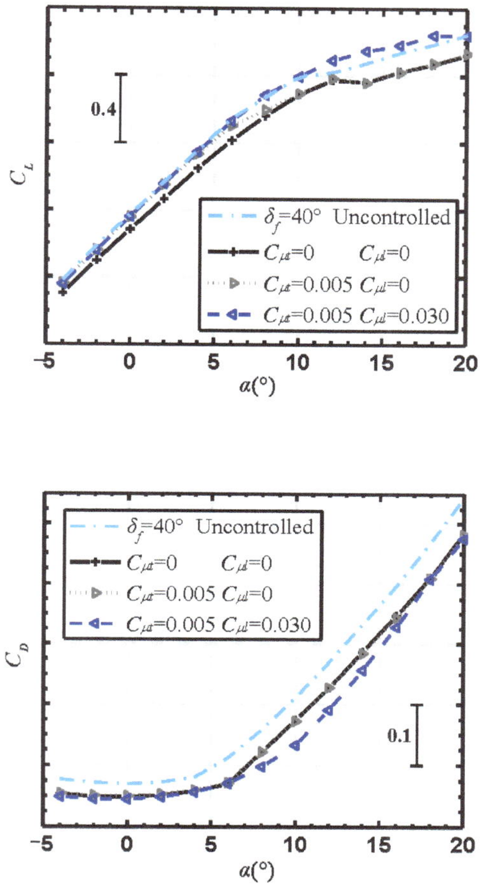

(a)

(b)

Figure 21. Lift coefficient curve based on combined blowing (δ_f = 30°): (**a**) C_L~α; (**b**) C_D~α.

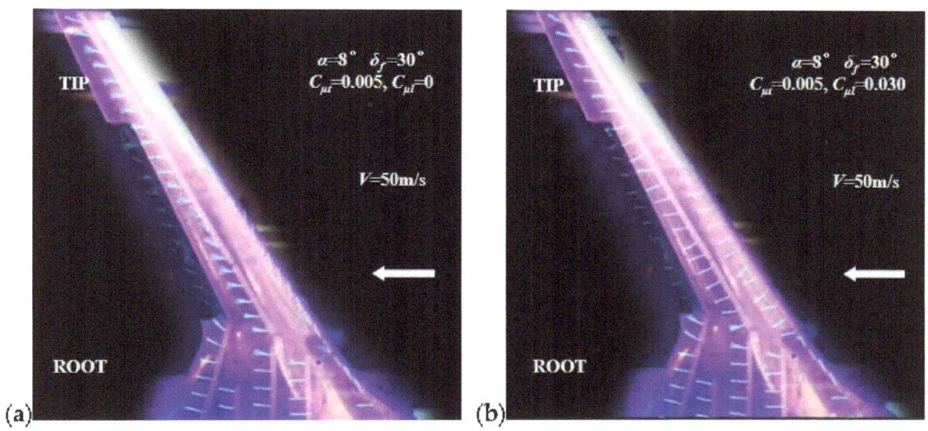

Figure 22. Flow visualization using silk on the suction side (combined blowing, arrows indicate the direction of incoming flow): (**a**) $C_{\mu t}$ = 0.005, $C_{\mu l}$ = 0; (**b**) $C_{\mu t}$ = 0.005, $C_{\mu l}$ = 0.030.

5. Conclusions

In this paper, the CFD method was used to analyze the combined jet control mechanism for the HWB aircraft. Through wind tunnel tests, the investigation first analyzed a seamless flap without an exposed structure for HWB applications. Then, the control effects of TE and LE blowing were investigated separately. Finally, a more efficient combined blowing flow control method was proposed. The results show that:

The combination of blowing prevents the development of cross-flow and weakens the accretion of the boundary layer at the wing tips. The injection of momentum delays vortex break-up, and the combination of blowing effectively restrains flow separation.

Seamless, simple flaps have a limited effect on lift enhancement. The premature separation of the flap suction surfaces leads to earlier stalls and must be combined with blowing control to exploit the increasing lift capability.

The effect of the blown deflected TE method on the model's aerodynamic characteristics is mainly observed before the stall. In contrast, the blown LE method mainly increases the lift after the stall.

The combined blowing technique increases the lift in the linear section while delaying the stall angle. The maximum lift increment is approximately 0.19, and the stall angle of attack is increased from 12° to 18°, with a drag reduction effect. The controlled flap deflection of 30° yields an increase in the lift-to-drag ratio of more than 2.2 compared to the uncontrolled flap deflection of 40° in an angle of attack range of 4° to 12°.

Author Contributions: Conceptualization, J.P. and W.W.; methodology, J.P.; software, W.W.; data curation, J.P.; formal analysis, C.Q.; writing—original draft preparation, J.P.; writing—review and editing, Q.S.; supervision, X.W. and X.Z.; project administration, W.W.; funding acquisition, W.W. All authors have read and agreed to the published version of the manuscript.

Funding: This research received no external funding.

Data Availability Statement: Not applicable.

Conflicts of Interest: The authors declare no conflict of interest.

References

1. Liebeck, R.H. Design of the Blended Wing Body Subsonic Transport. *J. Aircr.* **2004**, *41*, 10–25. [CrossRef]
2. Panagiotou, P.; Antoniou, S.; Yakinthos, K. Cant angle morphing winglets investigation for the enhancement of the aerodynamic, stability and performance characteristics of a tactical Blended-Wing-Body UAV. *Aerosp. Sci. Technol.* **2022**, *123*, 107467. [CrossRef]
3. Larkin, G.; Coates, W. A design analysis of vertical stabilisers for Blended Wing Body aircraft. *Aerosp. Sci. Technol.* **2017**, *61*, 51–61. [CrossRef]
4. Graham, W.R.; Hall, C.A.; Morales, M.V. The potential of future aircraft technology for noise and pollutant emissions reduction. *Transp. Policy* **2014**, *34*, 36–51. [CrossRef]
5. Ammar, S.; Legros, C.; Trépanier, J.Y. Conceptual design, performance and stability analysis of a 200 passengers Blended Wing Body aircraft. *Aerosp. Sci. Technol.* **2017**, *65*, S1270963816312640. [CrossRef]
6. Xin, Z.; Chen, Z.; Gu, W.; Zhang, M.; Zhang, B. Externally blown elevon applied for the longitudinal control of blended wing body transport with podded engines. *Aerosp. Sci. Technol.* **2019**, *93*, 105324. [CrossRef]
7. Hooker, J.R. Design of a Hybrid Wing Body for Fuel Efficient Air Mobility Operations at Transonic Flight Conditions. In Proceedings of the 52nd Aerospace Sciences Meeting, National Harbor, MD, USA, 13–17 January 2014; American Institute of Aeronautics and Astronautics: Reston, VA, USA, 2014.
8. Wick, A.T.; Hooker, J.R.; Walker, J.; Chan, D.T.; Plumley, R.; Zeune, C. *Hybrid Wing Body Performance Validation at the National Transonic Facility*; NASA Technical Report AIAA-2017-0099; American Institute of Aeronautics and Astronautics: Reston, VA, USA, 2017.
9. Peter, M.; Shmilovich, A.; Douglas, S. *Refined AFC-Enabled High-Lift System Integration Study*; NASA Technical Report NASA/CR–2016-219170; National Aeronautics and Space Administration: Hampton, VA, USA, 2016.
10. McLean, J.; Crouch, J.; Stoner, R.; Sakurai, S.; Seidel, G.E.; Feifel, W.M.; Rush, H.M. *Study of the Application of Separation Control by Unsteady Excitation to Civil Transport Aircraft*; NASA Technical Report NASA-CR-1999-209338; American Institute of Aeronautics and Astronautics: Reston, VA, USA, 1999.
11. Smith, D.; Dickey, E.; Vonklein, T. The ADVINT Program. In Proceedings of the 3rd AIAA Flow Control Conference, San Francisco, CA, USA, 5–8 June 2006.
12. Bumazzi, M.; Radespiel, R. Design and Analysis of a Droop Nose for Coanda Flap Applications. *J. Aircr.* **2014**, *51*, 1567–1579.

13. Radespiel, R.; Burnazzi, M. *Active Flow and Combustion Control*; Springer: Berlin, Germany, 2015; pp. 101–114.
14. Rao, R. Acquisition of Japanese Aircraft likely in 2015. *Int. Aerosp. Rev. Anal.* **2015**, *17*, 41.
15. Lichtwardt, J.; Paciano, E.; Marshall, D.; Jameson, K.K. STOL Performance of Cal Poly's AMELIA. In Proceedings of the 2013 AIAA Atmospheric Flight Mechanics Conference, Grapevine, TX, USA, 7–10 January 2013; American Institute of Aeronautics and Astronautics: Reston, VA, USA, 2013.
16. Collins, S.W.; Westra, B.W. Wind Tunnel Testing of Powered Lift, All-Wing STOL Model. *Aeronaut. J.* **2009**, *113*, 323–331. [CrossRef]
17. Milholen, W.; Jones, G.; Cagle, C. NASA high Reynolds number circulation control research-overview of CFD and planned experiments. In Proceedings of the AIAA Aerospace Sciences Meeting, Orlando, FL, USA, 4–7 January 2010; American Institute of Aeronautics and Astronautics: Reston, VA, USA, 2010.
18. Lin, J.C.; Melton, L.; Hannon, J.; Andino, M.Y.; Koklu, M.; Paschal, K.C.; Vatsa, V.N. Wind Tunnel Testing of High Efficiency Low Power (HELP) Actuation for Active Flow Control. In Proceedings of the AIAA Scitech 2020 Forum, Orlando, FL, USA, 6–10 January 2010; American Institute of Aeronautics and Astronautics: Reston, VA, USA, 2020.
19. Imamura, T.; Ura, H.; Yokokawa, Y.; Yamamoto, K. *A Far-Field Noise and Near-Field Unsteadiness of a Simplified High-Lift-Configuration Model(Slat)*; AIAA Paper, no. 1239; American Institute of Aeronautics and Astronautics: Reston, VA, USA, 2009.
20. Jameson, K. Part 1: The Wind Tunnel Model Design and Fabrication of Cal Poly's AMELIA 10 Foot Span Hybrid Wing-Body Low Noise CESTOL Aircraft. In Proceedings of the AIAA Atmospheric Flight Mechanics Conference, Portland, OR, USA, 8–11 August 2011; American Institute of Aeronautics and Astronautics: Reston, VA, USA, 2011.
21. Hooker, J.R.; Wick, A.T.; Hardin, C.J. *Commercial Cargo Derivative Study of the Advanced Hybrid Wing Body Configuration with Over-Wing Engine Nacelles*; NASA Technical Report NASA/CR-2017-219653; National Aeronautics and Space Administration: Hampton, VA, USA, 2017.
22. Chan, D.T.; Hooker, J.R.; Wick, A.; Plumley, R.W.; Zeune, C.H.; Ol, M.V.; DeMoss, J.A. Transonic Semispan Aerodynamic Testing of the Hybrid Wing Body with Over Wing Nacelles in the National Transonic Facility. In Proceedings of the 55th AIAA Aerospace Sciences Meeting, Grapevine, TX, USA, 9–13 January 2017; American Institute of Aeronautics and Astronautics: Reston, VA, USA, 2017.
23. Galbtaith, M.C. Numerical simulations of a high-lift airfoil employing active flow control. In Proceedings of the 44th AIAA Aerospace Sciences Meeting and Exhibit, Reno, NV, USA, 9–12 January 2006.
24. Shmilovich, A.; Yadlin, Y. Flow control for the systematic buildup of high-lift systems. *J. Aircr.* **2008**, *45*, 1680–1688. [CrossRef]
25. Bauer, M.; Peltzer, I.; Nitsche, W.; Gölling, B. *Active Flow Control on an Industry-Relevant Civil Aircraft Half Model*; Springer: Berlin, Germany, 2010.
26. Chen, J.; Wu, X.; Zhang, J.; Li, B.; Jia, H.; Zhou, N. FlowStar: General unstructured-grid CFD software for National Numerical Wind tunnel (NNW) Project. *Acta Aeronaut. Astronaut. Sin.* **2021**, *42*, 9–30.
27. Roe, P.L. Approximate Riemann solvers, parameter vectors, and difference schemes. *J. Comput. Phys.* **1981**, *43*, 357–372. [CrossRef]
28. Wick, A.T.; Hooker, J.R.; Clark, C.M. Powered Low Speed Testing of the Hybrid Wing Body. In Proceedings of the 55th AIAA Aerospace Sciences Meeting, Grapevine, TX, USA, 9–13 January 2017.
29. Chen, C.; Zakharin, B.; Wygnanski, I.J. On the parameters governing fluidic control of separation and circulation. In Proceedings of the 46th AIAA Aerospace Sciences Meeting and Exhibit, Reno, NV, USA, 7–10 January 2008.

actuators

Article

Numerical Investigation on the Evolution Process of Different Vortex Structures and Distributed Blowing Control for Dynamic Stall Suppression of Rotor Airfoils

Guoqiang Li [1,2], Shihe Yi [1], Binbin Li [3,*] and Xin Zhang [2]

[1] College of Aerospace Science and Engineering, National University of Defense Technology, Changsha 410000, China; cardcl@126.com (G.L.); yishihe@nudt.edu.cn (S.Y.)
[2] Low Speed Aerodynamics Institute, China Aerodynamics Research and Development Center, Mianyang 621000, China; lookzx@mail.ustc.edu.cn
[3] School of Civil Engineering and Architecture, Southwest University of Science and Technology, Mianyang 621000, China
* Correspondence: libinbin-8@163.com

Abstract: The influencing characteristic for the evolution mechanism of a dynamic stall vortex structure and distributed blowing control on rotor airfoils was investigated. Based on the moving-embedded grid method, the finite volume scheme, and Roe's FDS scheme, a simulation method for the unsteady flow field of a pitch-oscillating airfoil was established. The flow field of the NACA63-218 airfoil was calculated using Reynolds-averaged Navier–Stokes equations. The evolution processes of different vortex structures during dynamic stall and the principal controlled vortex mechanism affecting aerodynamic nonlinearity were analyzed based on the pressure contours C_p and Q of the flow field structure and the spatiotemporal evolution characteristics of the wall pressure distribution. The research indicated that dynamic stall vortices (DSVs) and shear layer vortices (SLVs) were the major sources of the increase in aerodynamic coefficients and the onset of nonlinear hysteresis. Building upon these findings, the concept of distributed blowing control for DSVs and shear layer vortices (SLVs) was introduced. A comparative analysis was conducted to assess the control effectiveness of dynamic stall with different blowing locations and blowing coefficients. The results indicated that distributed blowing control effectively inhibited the formation of DSVs and reduced the intensity of SLVs. This led to a significant decrease in the peak values of the drag and pitch moment coefficients and the disappearance of secondary peaks in the aerodynamic coefficients. Furthermore, an optimal blowing coefficient existed. When the suction coefficient C_μ exceeded 0.03, the effectiveness of the blowing control no longer showed a significant improvement. Finally, with a specific focus on the crucial motion parameters in dynamic stall, the characteristics of dynamic stall controlled by air blowing were investigated. The results showed that distributed air blowing control significantly reduced the peak pitching moment coefficient and drag coefficient. The peak pitching moment coefficient was reduced by 72%, the peak drag coefficient was reduced by 70%, and the lift coefficient hysteresis loop area decreased by 46%. Distributed blowing jet control effectively suppressed the dynamic stall characteristics of the airfoil, making the unsteady load changes gentler.

Keywords: rotor airfoil; dynamic stall vortex; evolutionary process; distributed blowing; flow control

Citation: Li, G.; Yi, S.; Li, B.; Zhang, X. Numerical Investigation on the Evolution Process of Different Vortex Structures and Distributed Blowing Control for Dynamic Stall Suppression of Rotor Airfoils. *Actuators* **2024**, *13*, 30. https://doi.org/10.3390/act13010030

Academic Editor: Luigi de Luca

Received: 26 November 2023
Revised: 7 January 2024
Accepted: 9 January 2024
Published: 11 January 2024

1. Introduction

The dynamic stall of an airfoil refers to the strong unsteady and nonlinear flow phenomenon caused by lifting components such as the wing or rotor blade when the angle of attack undergoes periodic or abrupt dynamic changes; this leads to widespread airflow separation above the airfoil surface [1,2]. The forward flight speed, noise, and vibration levels of a helicopter are greatly affected by the dynamic stall characteristics of the airfoil of the helicopter rotor [3]. Dynamic stall leads to aerodynamic issues in the helicopter rotor, including an increase in the required power, a sudden drop in lift, reverse torque, and a sudden increase in blade vibrations, which limits the helicopter's flight speed and maneuverability. These serve as the primary factors restricting the improvement of the aerodynamic performance in helicopter rotors [4]. Therefore, research on the dynamic stall characteristics of a rotor airfoil to reveal the evolution process of dynamic stall has had a guiding role in improving the aerodynamic performance of rotors and improving the overall performance of helicopters. It had always been a research hotspot and cutting-edge issue in relation to the unsteady aerodynamics of helicopter rotors [5,6].

The mechanism of dynamic stall on a helicopter rotor airfoil is intricate, and its suppression poses significant challenges, making it a focal point of attention in the field of helicopter aerodynamics. In the 1940s, Himmelskamp [7] first observed the phenomenon of dynamic stall in experiments. However, due to the relatively low demands for aircraft maneuverability at the time and limited analytical methods, this phenomenon was not extensively investigated. It was not until the 1960s, following a helicopter rotor experiment, that dynamic stall garnered widespread attention, initiating subsequent research on this phenomenon. Ham [8] was among the first to provide a theoretical description of the development process of dynamic stall. From the 1970s onward, McCroskey et al. [9] conducted a substantial number of experimental studies on the dynamic stall phenomenon, contributing to a deeper understanding of the formation mechanism of dynamic stall. Carr, L.W. [10,11] delineated that the fundamental characteristics of dynamic stall encompass the presence of intricate unsteady separation and large-scale vortex structures in the flow field, resulting in the manifestation of pronounced nonlinear hysteresis properties in aerodynamics. Geissler et al. [12] conducted experimental and numerical studies on the dynamic stall characteristics on the OA312 airfoil and found that the development, shedding, and accumulation of leading-edge vorticity had an important impact on the dynamic stall characteristics of the airfoil. Wang et al. [13] employed particle image velocimetry (PIV) technology to measure the transport velocity of leading-edge vortices on OA209 and SC1095 airfoils. The study indicated that the transport velocity of leading-edge vortices was primarily influenced by the oscillation frequency of the airfoil. As the oscillation frequency increases, the transport velocity of the leading-edge vortices is also promoted accordingly. Ekaterinaris et al. [14] conducted a thorough summary of the numerical methods and research outcomes employed in the study of airfoils' dynamic stall. The results indicated that factors such as the reduced frequency, amplitude angle, average angle of attack, pivot location, and free-stream Mach number during airfoil pitching motion had a direct impact on the strength, development, and shedding of the dynamic stall vortices. Additionally, the geometric shape of the airfoil also influenced the dynamic stall characteristics of the airfoil. Naughton et al. [15] used PIV phase-locked averaging technology and surface pressure testing technology to measure the unsteady aerodynamic characteristics of wind turbine airfoils under dynamic stall. Analysis of the measurement results showed that there were four stall types: (1) trailing-edge stall; (2) trailing-edge stall with a secondary vortex; (3) trailing-edge stall with a separated vortex; and (4) the stall induced by a leading-edge separated vortex. Airfoil dynamic stall is a very complex process, including the formation and evolution of different vortex structures. Ullah et al. [16] conducted a study of dynamic stall on pitching swept finite-aspect-ratio wings using lifetime PSP. Gardner, A.D. et al. [17] undertook an investigation of the three-dimensional dynamic stall on an airfoil using fast-response pressure-sensitive paint. Disotell, K.J. et al. [18] conducted a study on global surface pressure measurements of the static and dynamic stall on a wind turbine airfoil at a low Reynolds number. Due to the complexity and high cost of wind tunnel experiments on airfoil dynamic stall and the limitations imposed by the measurement equipment and technology, research has been typically confined to limited operating conditions. With the advancement in computational fluid dynamics, numerical simulation gradually became a crucial approach to studying the dynamic stall characteristics of airfoils [19–25]. Visbal and Garmann [26–29] studied the effect of sweep and unsweep on the dynamic stall of a pitching finite-aspect-ratio wing and undertook an analysis of the dynamic stall on a pitching airfoil using high-fidelity large-eddy simulations.

Many scholars, both domestically and internationally, have conducted extensive numerical simulations and experimental research on the flow control of rotor airfoil dynamic stall, achieving significant breakthroughs. Flow control can be categorized into passive control and active control based on whether external energy injection is required. Passive flow control methods include leading-edge deformation [30–33], a waveform leading edge [34], trailing-edge deflection flaps [35–37], Gurney flaps [38–40], and vortex generators [41,42], among others. These methods primarily control dynamic stall through two pathways: (1) shape deformation, causing changes in the wing surface pressure distribution; and (2) altering the flow state over the wing surface to inhibit the formation and development of dynamic stall vortices. While these control methods performed well on fixed wings, they were less practical on helicopter rotors. On the one hand, this was because they faced a rapidly changing working environment, where passive control struggled to consistently maintain high control efficiency. The active flow control methods included air blowing control [43–47], a synthetic jet [48], and plasma control technology [49–54], among others. The blowing control technology involved injecting high-momentum gas into the boundary layer to reduce the flow instability [55], suppress flow separation, and consequently delay dynamic stall. Plasma and synthetic jet control have simple structures and small volumes, but inhibiting the dynamic stall vortices of airfoils under high Mach numbers and extreme conditions is challenging. Implementing continuous and efficient control is difficult. Air blowing control, on the other hand, is a widely researched solution with advantages such as high intensity and high maturity. Using a high-speed airstream to inject energy into the flow field near the airfoil controls the generation of dynamic stall vortices without introducing additional "waste drag". Adjusting the opening and closing of the blowing not only effectively controls the retreating blades of the rotor [56,57] but also does not interfere with the advancing blades. It is expected to be a promising technical means to solve the dynamic stall problem of airfoils and rotor blades.

Although there has been some research on blowing control of rotor airfoil dynamic stall, both domestically and internationally, the focus has mainly been on the study of the blowing airflow control methods. There has been a lack of research on the evolution processes of different vortex structures and the mechanisms of the main controlling vortices in rotor airfoil dynamic stall. Additionally, there is limited research on the blowing control of rotor airfoil dynamic stall under high wind speeds and high-frequency oscillations. This study establishes a numerical simulation method for the unsteady flow field of a pitching oscillation airfoil based on the overset grid technology, the finite volume method, and Roe's FDS scheme. By solving the Reynolds-averaged Navier–Stokes equations, the flow field around the NACA63-218 airfoil under a high wind speed is simulated. Using an analysis of the spatiotemporal evolution characteristics of the flow structure and wall pressure distribution based on the pressure coefficient (C_p) and Q contours, this study investigates the evolution processes of different vortex structures during dynamic stall and the main controlling vortex mechanism influencing the aerodynamic nonlinear hysteresis. This study aims to provide the necessary theoretical foundation for exploring more efficient and practical flow control methods and developing predictive methods for the evolution of dynamic stall. Building upon this, the concept of distributed blowing control is introduced, and the control effects of dynamic stall are compared for different blowing locations and blowing coefficients. Finally, focusing on the important motion parameters in airfoil dynamic stall, the control characteristics of distributed blowing control are studied under multiple operating conditions.

2. Geometric Model and Numerical Methods

2.1. Dynamic Chimera Grid System

The numerical model employed the NACA63-218 airfoil, as previously utilized in the XH-59A helicopter [58]. To simulate the unsteady aerodynamic characteristics during airfoil pitching oscillation, dynamic chimera grid technology was used. This facilitated the movement of the airfoil grid and the exchange of information between the airfoil grid and the background grid. Figure 1 depicts a local view of the grid for the NACA63-218 airfoil. The grid exhibits the favorable characteristics of orthogonality and conformity to the airfoil geometry. The airfoil was subjected to a no-slip wall condition and the surrounding region of the background grid utilized a pressure far-field condition. The grid had 38,000 nodes for the airfoil and 68,000 nodes for the component. The y+ value was about 1 for the first layer of the grid, and the normal grid spacing near the wall increased outward at a rate of 1.2. The chord length (*c*) of the airfoil was 0.35 m (consistent with the chord length dimension of the wind tunnel test model), the inflow velocity was M = 0.3, and the Reynolds number, based on the chord length, was $R_e = 2.41 \times 10^6$. The atmospheric pressure was 101,325 Pa and the ambient temperature was $T = 288.15$ K. In this case, the computational domain is depicted in Figure 2.

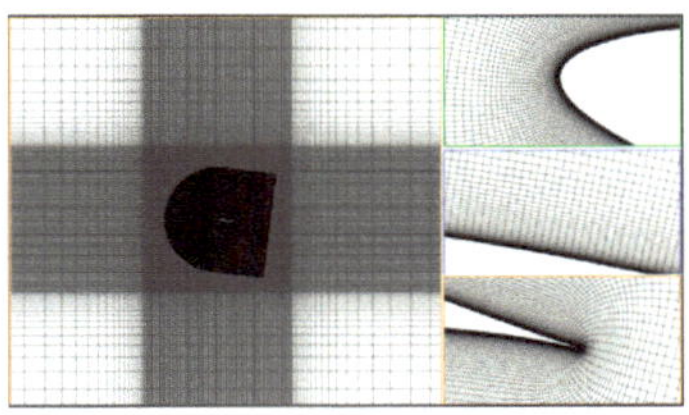

Figure 1. C-type grid around NACA63-218 airfoil.

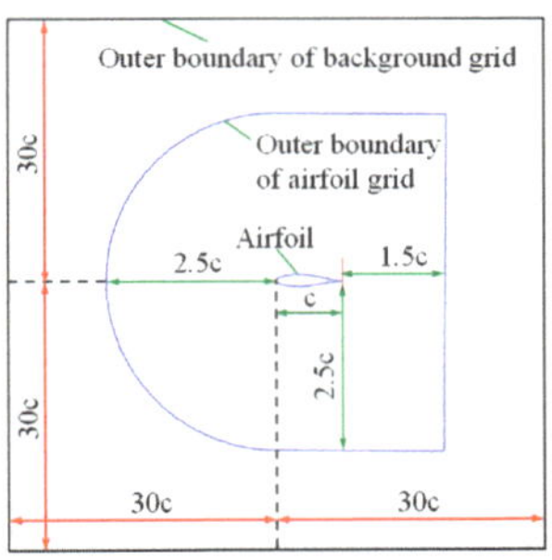

Figure 2. Schematic diagram of computational domain.

2.2. Governing Equations and Numerical Simulation Methods

To simulate the flow field around the rotor airfoil, unsteady Reynolds-averaged Navier–Stokes (RANS) equations are applied in this research, and the integral form of the governing equations is described as:

$$\frac{\partial}{\partial t}\int_{\Omega} W d\Omega + \oint_{\Omega}(F_c - F_v)dS = 0 \tag{1}$$

The conserved variable is $W = [\rho\ \rho u\ \rho v\ \rho E]^{T}$, where u and v are the Cartesian velocity components, ρ is the density, and E is the total energy per unit mass. Ω is the control volume and S is the boundary of the control volume. F_c and F_v are the vectors of the convective fluxes and viscous fluxes, which are expressed as:

$$F_c = \begin{bmatrix} \rho V_r \\ \rho u V_r + n_x p \\ \rho v V_r + n_y p \\ \rho H V_r + V_t p \end{bmatrix},\ F_v = \begin{bmatrix} 0 \\ n_x \tau_{xx} + n_y \tau_{xy} \\ n_x \tau_{yx} + n_y \tau_{yy} \\ n_x \Theta_x + n_y \Theta_y \end{bmatrix} \tag{2}$$

where p is the pressure of the airflow and H is the total enthalpy per unit mass. $V_r = V - V_t$, where V is the velocity normal to the surface element dS and V_t is the contravariant velocity. n_x and n_y are the normal vector components of the control surface. τ_{ij} is the viscous stress term and Θ_i is the term used to describe the heat conduction in the fluid.

In this study, the highly accurate SST k-ω turbulence model for capturing the flow separation phenomena was used. The discretization method utilized was the finite volume method, while pressure–density coupling was achieved using an implicit algorithm. The convective fluxes were discretized using the low-dissipation Roe's FDS format, and the convective terms were discretized using a second-order upwind scheme. The time advancement scheme employed was a second-order implicit method. This study primarily investigated the dynamic stall control of an airfoil during pitch motion. The airfoil rotates about a pivot point located at 0.25 c along the chord, where c denotes the chord length. The airfoil undergoes forced pitching oscillation in a sinusoidal manner. The pitch motion was described by the following equation:

$$\alpha(t) = \alpha_0 + \alpha_m \sin(2kt) \quad (3)$$

In the given equation, α_0 represented the average angle of attack, α_m was the amplitude of the angle of attack, and k was the reduction frequency, describing the extent of the unsteady effects of the oscillatory motion on the incoming flow. A higher value of k was a more significant unsteady effect. The expression for k could be formulated as follows:

$$\mathrm{k} = \pi \mathrm{fc}/\mathrm{V}_\infty \quad (4)$$

Here, f was the oscillation frequency of the airfoil, c was the chord length of the airfoil, and V_∞ was the free-stream velocity.

2.3. Method Validation

Based on the dynamic chimera grid technology, the dynamic stall aerodynamic characteristics of the NACA63-218 airfoil were numerically validated using the computational fluid dynamics (CFD) method. Figure 3 presents a comparison between the calculated and experimental values of the dynamic aerodynamic coefficients for the airfoil. The conditions for both the calculations and experiments were as follows: the airfoil's chord length was $c = 0.35$ m, the free-stream Mach number was 0.1, the angle of attack followed the variation pattern $\alpha = 14^\circ + 10^\circ \sin(2kt)$ with a reduction frequency of $k = 0.097$, and the Reynolds number was $R_e = 7.0 \times 10^5$ (Due to limitations in the experimental setup and wind tunnel, the wind speed did not cover a higher range). The figure demonstrates the basic consistency between the numerical calculations and experimental results, particularly regarding the lift coefficient during the airfoil's pitching-up phase and the simulation of an unsteady load during dynamic stall. These findings validated the effectiveness of the proposed numerical simulation method in accurately capturing the unsteady aerodynamic characteristics on the NACA63-218 airfoil under dynamic stall conditions. Although a wind tunnel test is a reliable means of obtaining airfoil aerodynamics data, the test model is surrounded by the wall of the tunnel, and the constraints of the wall will interfere with the flow and aerodynamic forces of the airfoil. Even though the study tries its best to minimize this interference, there will still be some impact, as shown in Figure 3. Considering that the main research object is the improvement of the dynamic stall characteristics of airfoils using a blowing jet, and the comparative analysis is on the difference in the performance changes in airfoils under aerodynamic actuation opening and closing conditions, it is reasonable to believe that the data analysis and conclusions obtained in this study are reliable. The dynamic stall flow studied is notoriously three-dimensional [59]; however, two-dimensional simulations can also accurately reflect the changes in aerodynamic curves during dynamic stall processes [60].

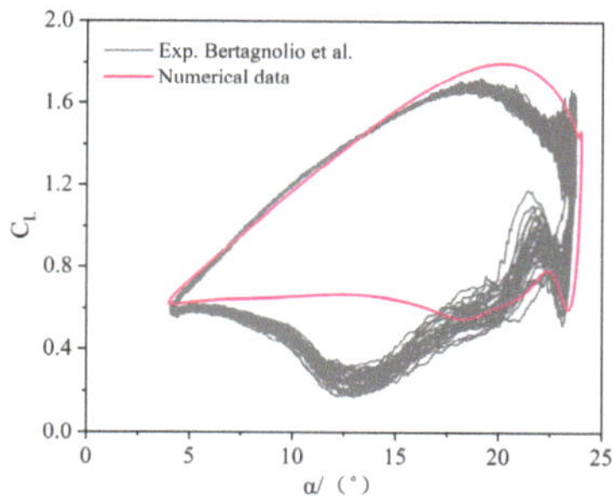

(**a**) Lift coefficient

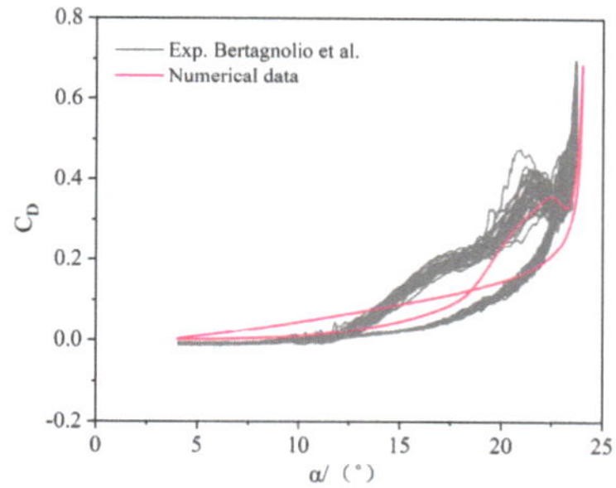

(**b**) Drag coefficient

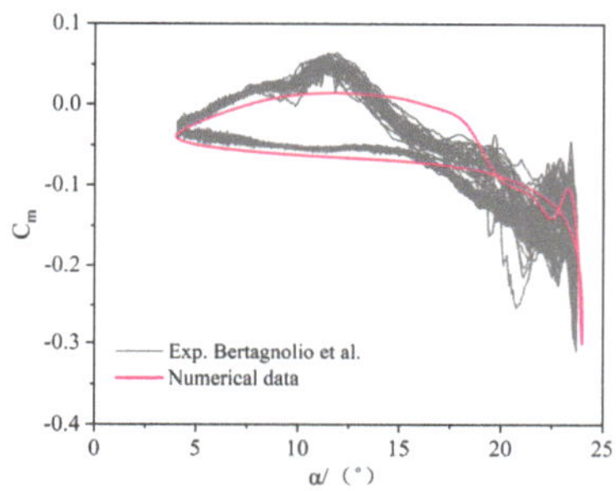

(**c**) Pitching moment coefficient

Figure 3. Comparison of calculated aerodynamic coefficients of airfoils with the literature [61].

3. Evolution Process of the Different Vortex Structures of Dynamic Stall and the Active Control Vortex Mechanism of Nonlinear Aerodynamics

3.1. Analysis of the Aerodynamic Characteristics of Dynamic Stall

Numerical simulations were conducted to investigate the dynamic stall aerodynamic characteristics on the NACA63-218 airfoil. The computational conditions represented the typical operating state of a trailing-edge rotor blade: M = 0.3, the angle of attack varied according to $\alpha = 12^\circ + 10^\circ\sin(2kt)$ with a reduction frequency of $k = 0.129$, and the Reynolds number was $R_e = 2.41 \times 10^6$. To accurately simulate the unsteady aerodynamic forces during the dynamic stall of the airfoil, the oscillation period was divided into 600 time steps. Each sub-iteration consisted of 200 steps and the calculations were performed for two cycles. In Figure 4, the lift, drag, and moment coefficient curves of the airfoil are displayed, which were calculated using the SST k-ω turbulence model. Figure 4 includes six different moments labeled A to F.

From the figure, the following can be observed: (1) The upstroke process: along point A to point B, the lift coefficient changed linearly with the angle of attack, the lift line slope $C_L{}^\alpha$ was 0.0891, and the changes in the drag coefficient and pitching moment coefficient were less than 0.2. Along point B to point C, the angle of attack increased from 20.57° to 21.62°, the lift coefficient increased nonlinearly, but the slope decreased slightly, and the drag coefficient and pitching moment coefficient increased rapidly. Along point C to point D, the angle of attack changed from 21.67° to 22°, the lift coefficient continued to increase nonlinearly, the angle change was only 0.33°, and the drag coefficient and pitching moment coefficient suddenly increased to the maximum value. (2) The downstroke process: Along point D to point E, the angle of attack changed from 22° to 20.41°, the angle change was only 1.6°, the lift coefficient dropped sharply by approximately 75%, and the drag coefficient and pitching moment coefficient decreased by 80%. Along point E to point F, the lift coefficient was restored, the drag coefficient and pitching moment coefficient fluctuated, and the aerodynamic force appeared to have a secondary peak. Along point F to point A, the lift force, drag coefficient, and pitching moment coefficient gradually returned to the initial state.

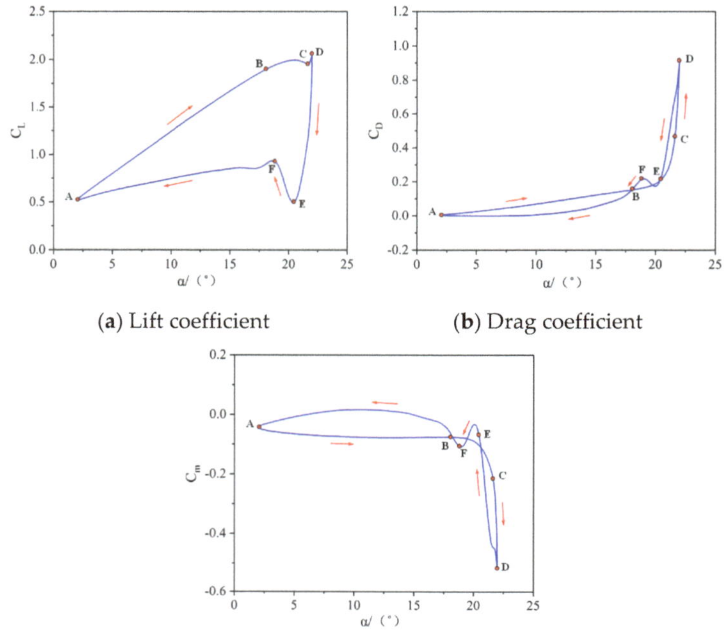

(**a**) Lift coefficient (**b**) Drag coefficient

(**c**) Pitching moment coefficient

Figure 4. Aerodynamic force coefficients of NACA63-218 airfoil.

The complex variations in the aerodynamic characteristics during the dynamic stall of airfoils reflects the intricate evolution of different vortex structures. To further study the main controlling mechanisms of the nonlinear hysteresis of aerodynamic forces caused by different vortex structures in the dynamic stall process, it was necessary to analyze the flow characteristics related to the generation, convection, separation, and evolution of different vortex structures during dynamic stall.

3.2. *Evolution Process of Different Vortex Structures during Dynamic Stall*

Based on the analysis of the spatiotemporal evolution of the flow field structure and surface pressure distribution, the evolution process of different vortex structures (leading-edge vortex, trailing-edge vortex, shear vortex, dynamic stall vortex, secondary vortex) in dynamic stall could be identified, in addition to the active control vortex mechanism that affected the aerodynamic nonlinear hysteresis. Figure 5 shows the C_p contour, Q contour, and surface pressure distribution within a cycle of the airfoil's dynamic stall process.

From the figure, it was observed that based on the flow structures depicted by the pressure contour C_p and the variation in surface pressure distribution, the flow characteristics within a cycle can be divided into the following stages:

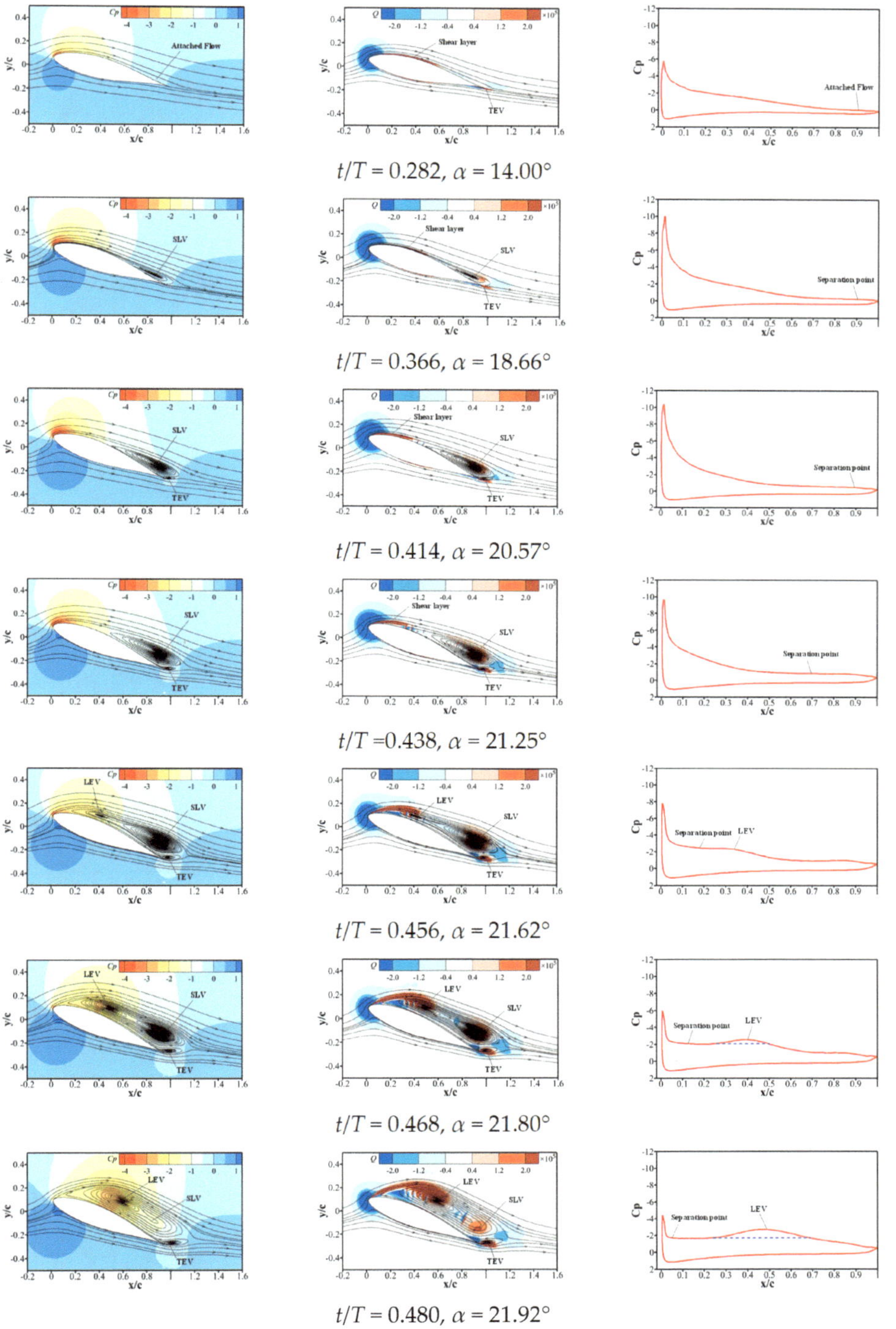

Figure 5. *Cont.*

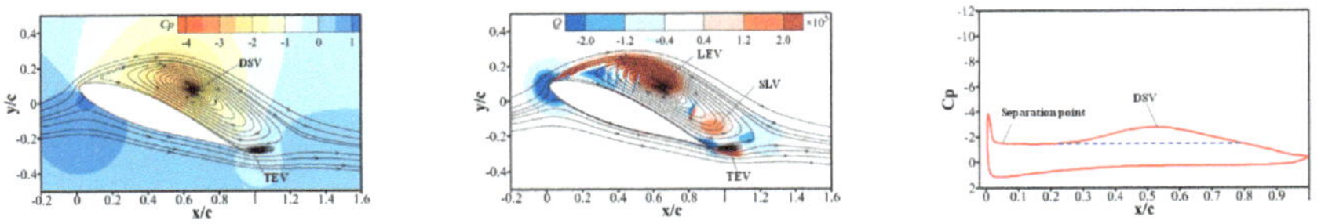

$t/T = 0.486$, $\alpha = 21.96°$

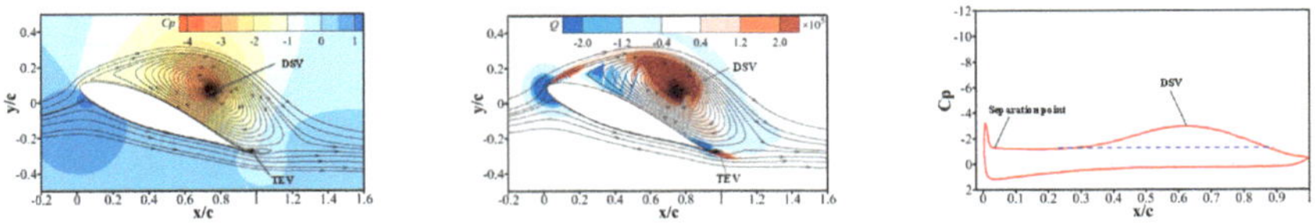

$t/T = 0.498$, $\alpha = 22.00°$

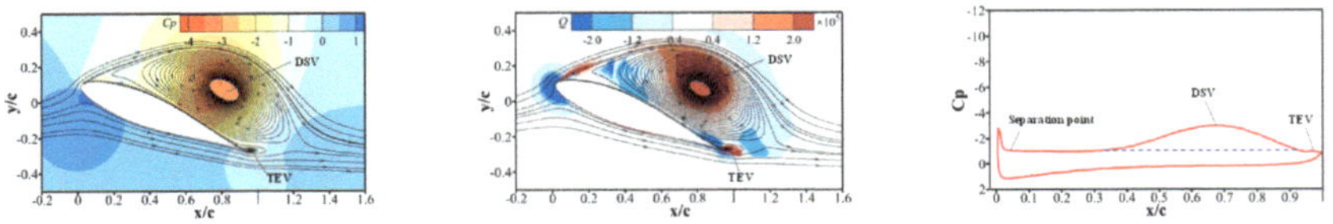

$t/T = 0.510$, $\alpha = 21.98°$

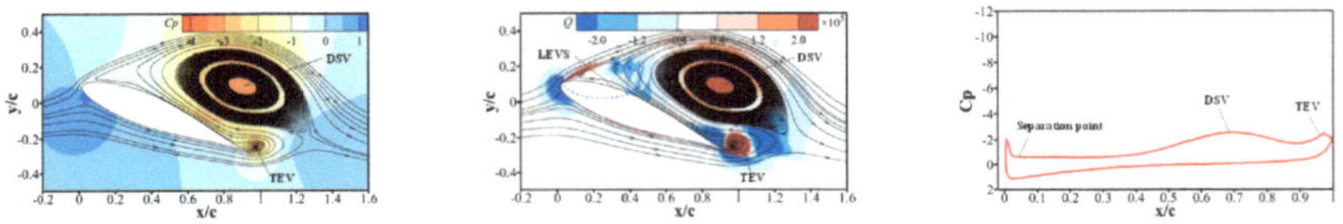

$t/T = 0.528$, $\alpha = 21.85°$

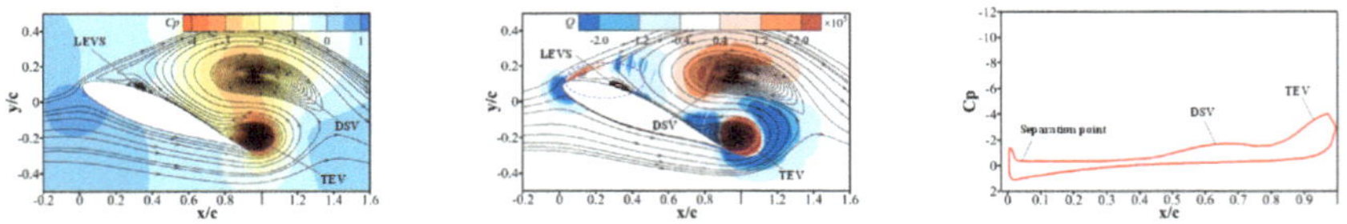

$t/T = 0.546$, $\alpha = 21.59°$

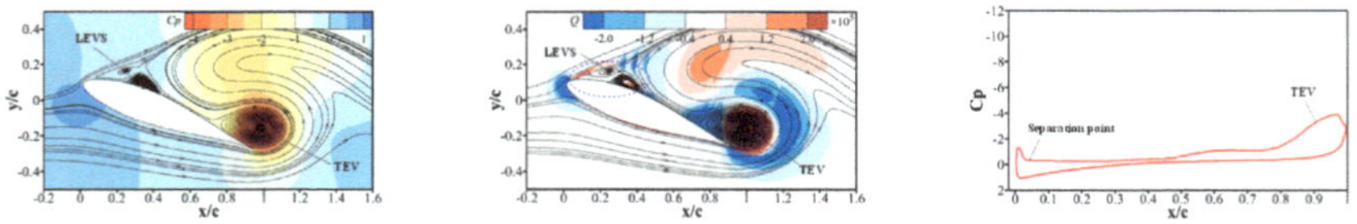

$t/T = 0.558$, $\alpha = 21.34°$

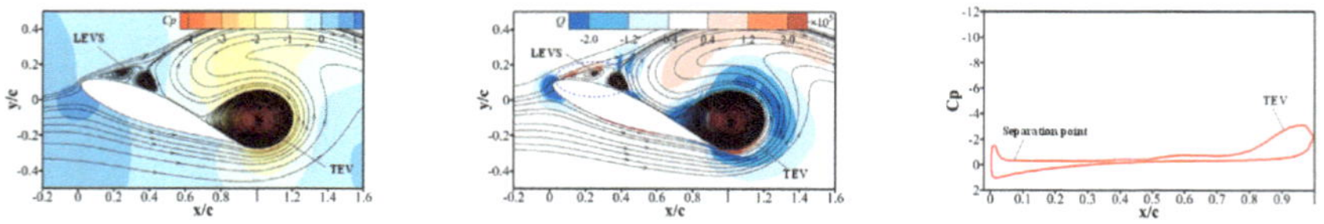

$t/T = 0.570$, $\alpha = 21.05°$

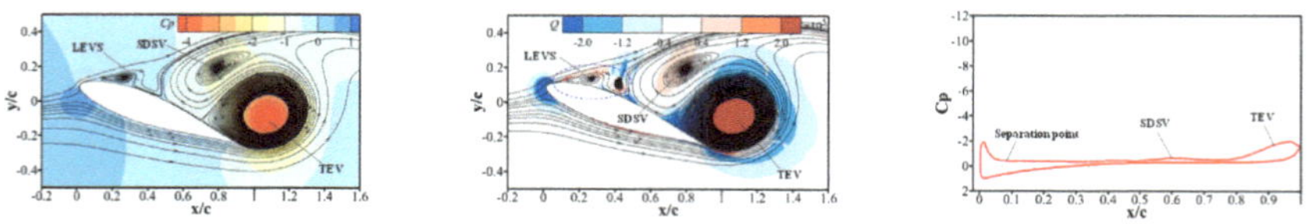

$t/T = 0.582$, $\alpha = 20.70°$

Figure 5. *Cont.*

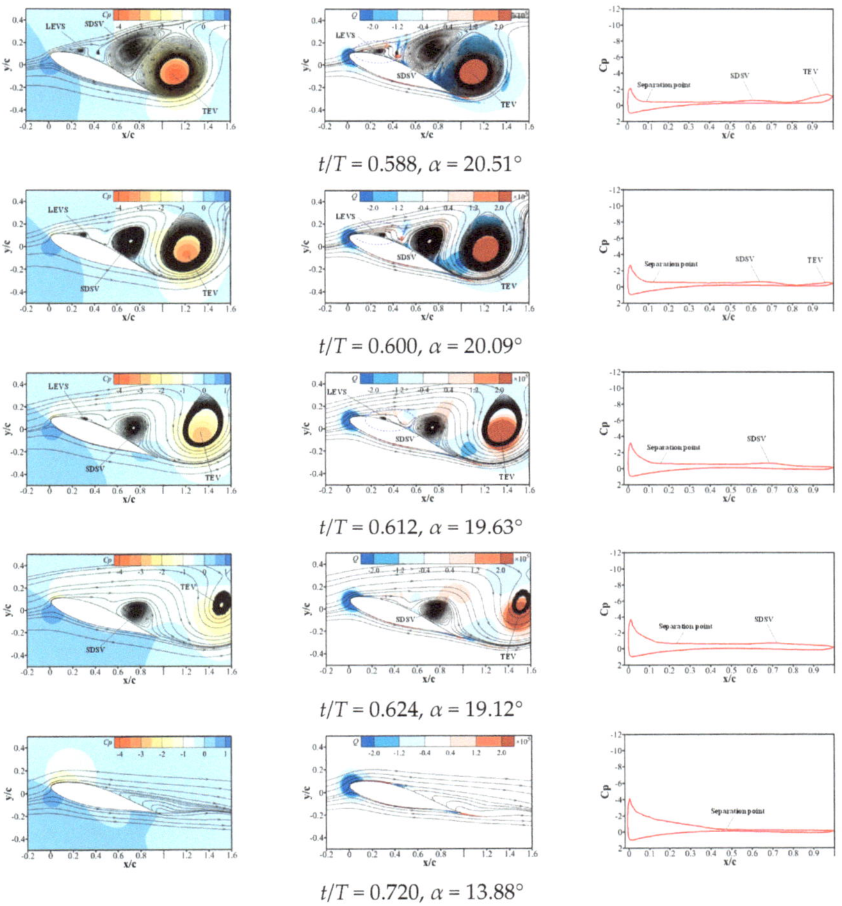

Figure 5. Evolution process of C_p flow structure and pressure distribution within a cycle.

Upstroke process: (1) As the angle of attack increased to 14°, it was shown that the flow remained attached without any separation of the flow structure depicted from the C_p contour, the Q contour, or the surface pressure distribution. The lift coefficient increased linearly with the angle of attack, while the changes in the drag coefficient and pitching moment coefficient were relatively small. (2) As the angle of attack continued to increase, when $\alpha = 18.66°$, the C_p contour and the Q contour showed a small range of flow separation at the trailing edge, forming an SLV (shear layer vortex). The pressure distribution was characterized by a "pressure plateau" with local small-scale vortex separation at the trailing edge, but the peak value of the pressure coefficient at the upper airfoil edge continued to increase. (3) As the angle of attack continued to increase, such as from 20.57° to 21.25°, the strength of the SLV gradually increased (seen from the Q contour), and a large-scale flow separation occurred at the trailing edge. The pressure distribution was characterized by a "pressure plateau", indicating large-scale vortex separation at the trailing edge. At the sharp trailing edge of the lower airfoil, a trailing-edge vortex (TEV) was formed due to the shearing effect of the airflow, but its strength was weak. The flow structure showed that there were two clear structures at the trailing edge. The pressure distribution was characterized by the vortex separation point moving upstream from 0.90 c to 0.61 c at a velocity of $0.069V_{\infty}$. The peak value of the pressure coefficient on the upper surface of the leading edge continued to increase. As a result, the lift coefficient exhibited nonlinear growth, although the slope slightly decreased. Due to the dominant influence of the SLV, the drag coefficient and pitching moment coefficient increase. (4) As the angle of attack continued to increase, when $\alpha = 21.62°$, under the influence of an adverse pressure gradient, vorticity accumulated at the position $x/c = 0.35$ on the leading edge, forming a leading-edge vortex (LEV). As the angle of attack increased from $\alpha = 21.62°$ to $\alpha = 21.92°$, the LEV rapidly merged with the SLV (seen from the C_p contour and the Q contour). The strength of the LEV increased and there was a sudden burst of vorticity. The trailing-edge vortex (TEV) remained relatively unchanged. The flow structure showed the presence of three vortex structures. The pressure distribution indicated that, at the position $x/c = 0.35$, there was a convex increase in the pressure coefficient. The "pressure plateau" underwent a sudden jump, transitioning from $C_p \approx 1$ to $C_p \approx 2$. The separation point moved from 0.68 c to 0.18 c and the peak pressure coefficient on the leading edge was reduced by 50%. The nonlinear increase in the lift coefficient was attributed to the vortex lift generated by the LEV. (5) As the angle of attack increased from $\alpha = 21.92°$ to $\alpha = 22°$, the vortex at the leading edge started to roll up, forming a larger-scale vortex structure that covered the entire upper surface and moved toward the trailing edge, which became the dynamic stall vortex (DSV). The trailing-edge vortex (TEV) remained relatively unchanged and the flow structure exhibited the presence of two vortex structures. The pressure distribution was characterized by a continuous increase in the area of the bulging region in the pressure coefficient. At the same time, the peak value of the pressure coefficient on the leading edge of the upper airfoil edge exhibited a continuous decrease. The DSV moved downstream at a speed of $0.13V_\infty$, which was approximately one-third of the free-stream velocity. This downstream migration of the DSV caused the pressure center to shift toward the rear. The DSV's dominant control and induction effect contribute to a nonlinear increase in the lift coefficient. Additionally, the drag coefficient and pitching moment coefficient increased to their maximum values, resulting in a significant nose-down moment.

Downstroke process: (1) From the maximum angle of attack of 21.98° to 21.59°, the DSV left the airfoil. The pressure distribution was characterized by a gradual decrease in the pressure coefficient at the $x/c = 0.68$ position. At the trailing edge of the airfoil, the flow structure showed that the strength of the TEV was enhanced, and the airflow was adsorbed from the lower surface to the upper surface. The pressure distribution was characterized by an increase in the pressure coefficient at the trailing edge, and the pressure coefficient of the lower airfoil ($x/c = 0.9$~1.0) The pressure coefficient decreased and the peak value of the pressure coefficient and the position of the separation point on the upper airfoil remained basically unchanged. Therefore, the lift coefficient, drag coefficient, and pitching moment coefficient gradually decreased. (2) From 21.34° pitching down to 21.05°, the flow structure showed that the DSV dissipated rapidly, the vortex structure basically disappeared completely, and the intensity of the TEV increased. In addition, the vortex scale increased to two to three times that of the upstroke process and formed a negative impact on the upper wing surface, a strong induction effect, and a reduced pressure coefficient. Simultaneously, a leading-edge vortex system (LEVS) formed at the leading edge of the upper surface, and the vortex structure gradually became more complex. The lift coefficient, drag coefficient, and pitching moment coefficient all decreased sharply, reaching 75% of their maximum values. (3) Descending from 20.70° to 20.09°, the flow structure showed the gradual disappearance of the LEVS, and the scale of the TEV continued to increase. It strongly induced low-energy fluid in the dissipation region of the dynamic stall vortex, forming a secondary dynamic stall vortex (SDSV). During this stage, the intensity of the SDSV continued to increase and gradually approached the airfoil surface. The pressure distribution was characterized by a secondary convexity in the pressure coefficient at $x/c = 0.65$, but with a lower peak. At the trailing edge, although the scale of the TEV increased, the peak of the pressure coefficient decreased. (4) From 19.63° to 19.12°, the flow structure showed that the SDSV migrated downstream along the airfoil, exhibiting secondary stall, and the aerodynamic force and pitch moment coefficients showed fluctuations with a secondary peak in aerodynamic force. The TEV gradually separated from the trailing-edge position and the dissipation disappeared. The pressure distribution was characterized by an increase in the pressure coefficient peak at the leading edge of the upper surface and the separation point gradually moved from 0.10c along the leading edge to 0.28 c at a speed of 0.123 $V\infty$. (5) Decreasing from 13.88° to the minimum angle of attack, at this stage, the pressure distribution was characterized by the separation point rapidly moving to 0.68 c at a speed of 0.069 $V\infty$. In addition, the pressure coefficient at the leading edge of the upper surface gradually recovered. Therefore, the lift coefficient began to recover but was lower than the corresponding upstroke value at the same angle of attack. This was mainly due to the separation on the upper surface of the airfoil (complete attachment during the upstroke). The drag coefficient was lower than the corresponding upstroke value, while the pitching moment coefficient was larger than the corresponding upstroke value. The delayed movement of the separation point resulted in an asymmetric pressure distribution, which was a major factor influencing the nonlinear hysteresis of the aerodynamics.

In conclusion, the complex nonlinear hysteresis characteristics of aerodynamics in dynamic stall arose from the evolution process of different vortex structures and motion characteristics during dynamic stall. During the pitching-up phase, under the influence of an adverse pressure gradient, a separated leading vortex (SLV) formed at the trailing edge. With the continuous increase in the angle of attack, due to the sustained supply of vorticity from the bottom layer to the boundary layer, vorticity accumulated at the airfoil's leading edge, forming a leading-edge vortex (LEV). As the angle of attack continued to increase, the LEV and SLV quickly merged, and the vorticity underwent a rapid burst, resulting in convection toward the trailing edge to form a DSV. The aerodynamic coefficients and pitching moment coefficient increased to their maximum values. During the pitching-down phase, the DSV gradually dissipated, and the strength and scale of the TEV increased. This induced the low-energy fluid in the dynamic stall vortex dissipation region to form an SDSV. This led to a secondary stall and the aerodynamic coefficients exhibited a secondary peak. The DSV and SLV were the main control vortex mechanisms influencing the increase in the aerodynamic coefficients, pitching moment coefficient, and nonlinear hysteresis.

Note: The TEV was a vortex structure near the trailing edge. The airflow was adsorbed from the lower surface to the upper surface and the shear layer rolled up to form a vortex. The vortex was fixed near the trailing edge, so it was called a trailing-edge vortex. The SLV was the vortex formed by the shear layer separation due to the flow detachment at a high Reynolds number, also known as a shear separation vortex. The LEV could be considered the initial stage of the DSV. As the LEV gradually strengthened, when it could not maintain its current scale, it detached from the leading edge of the airfoil and migrated downstream, forming the dynamic stall vortex. The SDSV was a secondary vortex structure, reformed due to the inductive effect of the TEV on the low-energy fluid in the dissipation region of the DSV.

Figure 6 shows the variation in the pitching moment coefficient within a cycle of the dynamic process. From the figure, it can be observed that during the pitch-up phase, the LEV quickly merged with the SLV, forming a DSV. This was an important source of the increase in aerodynamic force and pitching moment coefficient, leading to a peak in the aerodynamic force. During the pitch-down phase, the formation of an SDSV resulted in a secondary peak in both the aerodynamic force and pitching moment coefficients.

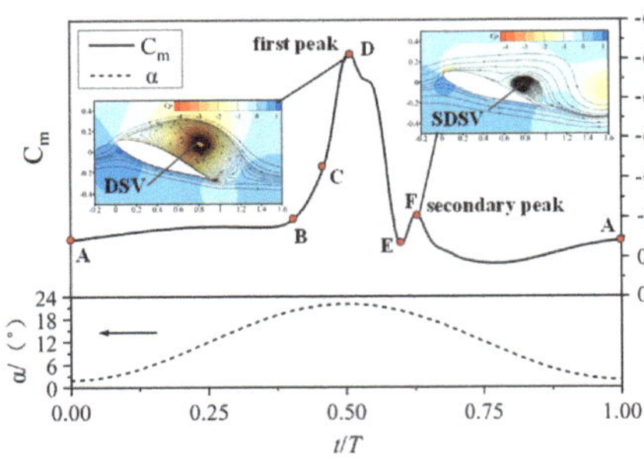

Figure 6. Characteristics of pitch torque coefficient variation within a cycle (The "A–F" here is consistent with the "A–F" in Figure 4).

4. Air-Blow Control of Dynamic Stall

4.1. Proposal of Air-Blow Control

Based on the analysis of the spatiotemporal evolution characteristics of the flow structure and surface pressure distribution of the pressure contour C_p, the SLV and the DSV were the main control vortices that increased the aerodynamic force and pitching moment coefficient during the dynamic stall process. Therefore, without changing the aerodynamic shape and structural form of the airfoil, if it was necessary to control the negative damping of dynamic stall, control measures needed to be designed based on the formation and convergence of the DSV and SLV. In response to control of the DSV and SLV, the concept of distributed air blow control was proposed, as shown in Figure 7. The leading edge of the airfoil was equipped with three distributed jet outlets, located at 5% c, 10% c, and 30% c from the leading edge, where c was the chord length. The gap at the outlet h was 1 mm and the ratio of h to c was 0.002857. The outlet angle was inclined at an angle θ of 30° to the local airfoil surface. The blowing positions at 5% c and 10% c mainly controlled the generation of an LEV and suppressed the formation of the DSV. The blowing position at 30% c mainly controlled the strength of the SLV. To adapt to engineering applications, the jet outlets were set as pressure outlet boundary conditions, and by altering the pressure values at the jet outlets, control over the jet blowing velocity could be achieved.

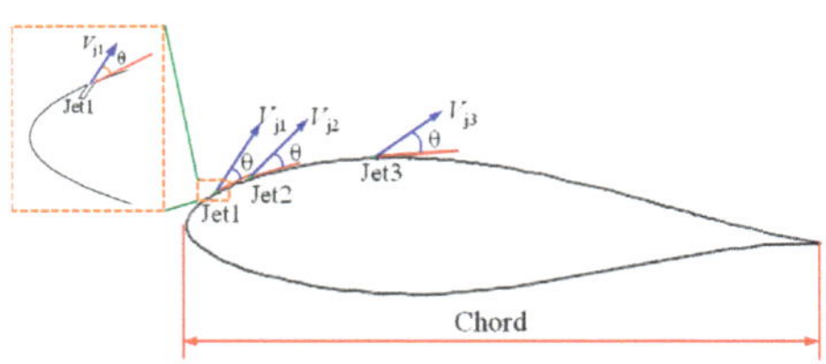

Figure 7. Leading-edge distributed blowing control.

The blowing momentum coefficient C_μ was used to measure the strength of the blowing and it was defined as follows:

$$C_\mu = \frac{\dot{m}_j V_j}{\frac{1}{2}\rho_\infty V_\infty^2 c} \tag{5}$$

$$V_j = \sqrt{\frac{2\gamma RT_j}{\gamma - 1}[1 - (\frac{P_j}{P_0})^{\frac{\gamma}{\gamma-1}}]} \tag{6}$$

Here, $\dot{m}_j$ was the mass flow rate of the jth jet, and $\dot{m}_j = \rho_j V_j h$, where h was the width of the jet exit, ρ_j was the density of the jth jet, V_j was the jet velocity at the exit (expansion to the jet velocity when isentropically expanded to the incoming static pressure), ρ_∞ was the density of the incoming flow, V_∞ was the incoming flow velocity, and c was the chord length of the airfoil. P_0 was the total pressure of the incoming flow, T_j and P_j were the total temperature and total pressure, respectively, at the jet exit, R was the gas constant, and γ was the specific heat ratio.

It is shown in Figure 8 that the aerodynamic coefficient comparison under the control of different combinations of blowing positions resulted in M = 0.3, $R_e = 2.41 \times 10^6$, and an angle of attack variation pattern $\alpha = 12° + 10° \sin(2kt)$, with $k = 0.129$, the total pressure at the jet outlet P_j = 110,545 P_a, the total temperature T_j = 288.15 K, and the blowing positions at x/c = 5%, 10%, and 30%, respectively.

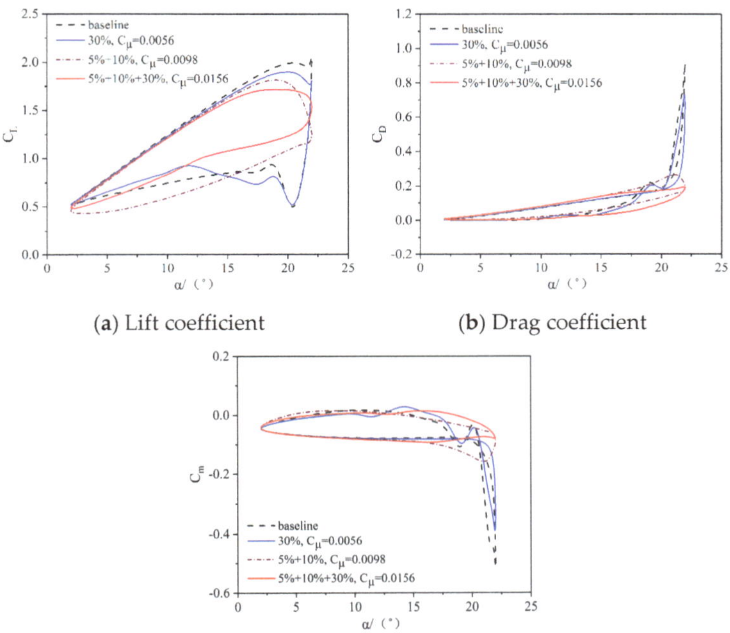

(**a**) Lift coefficient (**b**) Drag coefficient

(**c**) Pitching moment coefficient

Figure 8. Aerodynamic coefficients of combined control with different blowing positions.

It was shown that the combination control of the blowing positions had a significant impact on the aerodynamics of dynamic stall. During the blowing control, both the drag and pitching moment coefficients of the airfoil decreased. Within the linear range of lift, the combined control over the air blowing at different positions had a minor impact on the lift coefficient at the same angle of attack, and the effect on the slope of the lift curve in the linear range was also minimal. Comparison of different combinations of air blowing positions: (1) When the air blowing position was at x/c = 30%, compared to the baseline state, the change in the lift coefficient at the stall angle of attack was the most significant. At the maximum angle of attack, the lift coefficient still showed a nonlinear increase. The reduction in the peak values of drag and pitching moment coefficients was 20%. (2) When the air blowing positions were at x/c = 5% and 10%, the reduction in the peak values of the drag and pitching moment coefficients could reach 70%. However, around an angle of attack of 19°, the lift coefficient experienced a sharp decrease. In the downward pitching phase, compared to the baseline state, there was an increase in the hysteresis area of the lift coefficient curve. (3) When the air blowing positions were at x/c = 5%, 10%, and 30%, compared to the baseline state, the hysteresis curve of the lift coefficient changed smoothly with the variation in the angle of attack, and the hysteresis loop area decreased by 41%. Around an angle of attack of 19°, there was no fluctuation in the drag and pitching moment coefficients, and the reduction in the peak values of the drag and pitching moment coefficients could reach 80%. The secondary peaks in the aerodynamic coefficients disappeared and the hysteresis loop area of the lift coefficient decreased by 41%.

Figure 9 shows the comparison results of the C_p contour, the Q contour, and the surface pressure distribution controlled by different blowing position combinations when the maximum angle of attack $\alpha = 22°$: (1) When the blowing position was at $x/c = 30\%$, from the pressure distribution, it could be seen that the blowing position was downstream of the separation point, and the inhibitory effect on the aerodynamics was relatively weak. However, due to the influence of the jet vortex, the strength of the DSV decreased, the area of the pressure distribution bulge decreased, and the peak values of the drag and pitch moment coefficients decreased. As was seen from the pressure distribution C_p, the strength and scale of the DSV decreased, but it still remained the main controlling vortex affecting the increase in the aerodynamic forces and pitch moment coefficients. Due to the inductive effect of the DSV, vortex lift was generated, and the lift coefficient curve still exhibited a small-range nonlinear increase. (2) When the blowing positions were at $x/c = 5\%$ and 10%, it could be observed from the pressure distribution C_p that the DSV was suppressed, and the flow structure exhibited two vortex structures, the SLV and TEV. Due to the strong induction effect of the SLV on the TEV, the intensity of the TEV was enhanced. From the pressure distribution, it could be observed that the region of pressure rise disappeared. The pressure distribution indicated a "pressure plateau" caused by the large-scale vortex separation on the upper surface and the separation point was located at $x/c = 0.33$. This indicated that the SLV was the main controlling vortex affecting the increase in the aerodynamic force and pitching moment coefficients at this time. Due to the dominant controlling effect of the SLV, the lift coefficient began to experience stall around 19°, dropping sharply. The reduction in the drag and pitching moment coefficients reached 70%. (3) When the blowing positions were at $x/c = 5\%$, 10%, and 30%, from the pressure distribution C_p, it could be observed that there were two vortex structures in the flow, the SLV and TEV. Compared with $x/c = 5\%$ and 10%, the blowing jet at $x/c = 30\%$ suppressed the intensity of the EV due to the inhibition of the SLV. Therefore, the large-scale flow separation was inhibited. From the pressure distribution, it could be observed that the separation point had moved to $x/c = 0.56$. This indicated that the distributed blowing control ($x/c = 5\%$, 10%, and 30%) not only controlled the formation of the DSV but also had a controlling effect on the SLV, demonstrating a dual control effect.

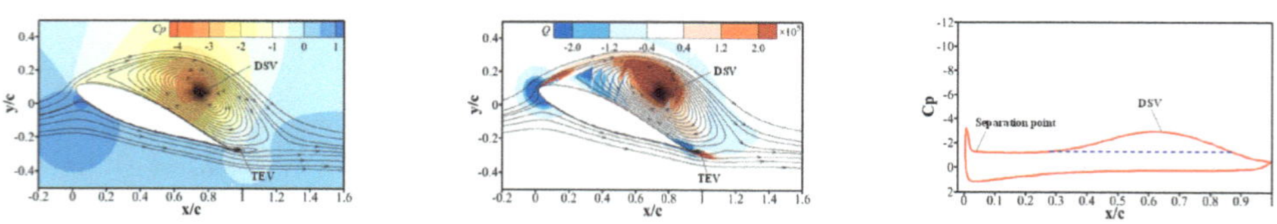

(**a**) Baseline

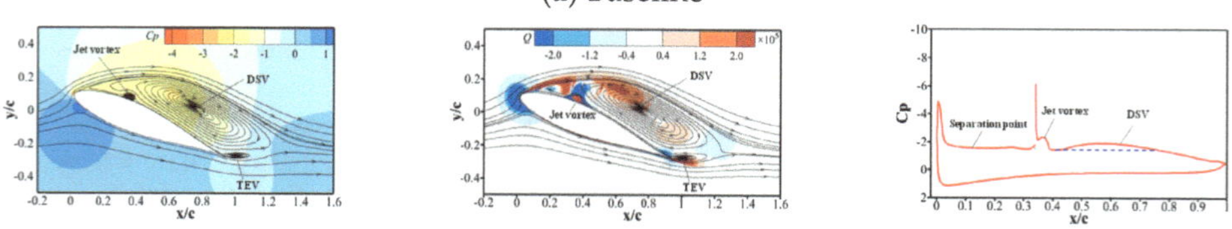

(**b**) $x/c = 30\%$, $C_\mu = 0.0056$

Figure 9. *Cont.*

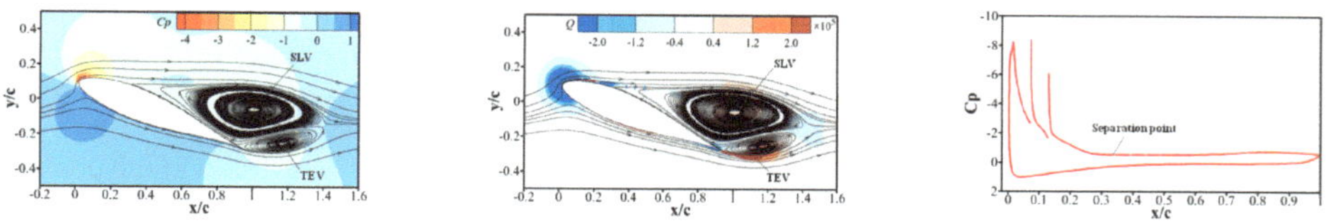

(**c**) $x/c = 5\% + 10\%$, $C_\mu = 0.0098$

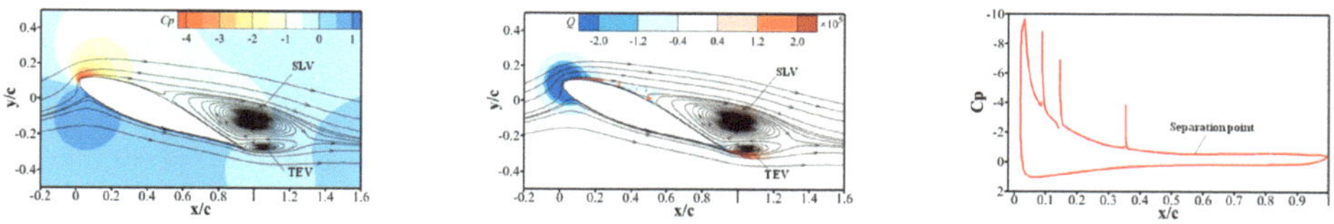

(**d**) $x/c = 5\% + 10\% + 30\%$, $C_\mu = 0.0156$

Figure 9. C_p contour, Q contour, and pressure distribution of the airfoil under combined control state at different blowing positions.

4.2. Effect of the Blowing Coefficient

The comparison of aerodynamic coefficients of different blowing coefficients can be seen in Figure 10, with M = 0.3, $R_e = 2.41 \times 10^6$, the change pattern of the angle of attack $\alpha = 12° + 10°\sin(2kt)$, $k = 0.129$, the temperature at the jet outlet $T_j = 288.15$ K, and the blowing positions at $x/c = 5\%$, 10%, and 30%. When the total pressure at the jet nozzle exit P_j changed, the variation in the stagnation ring area $C_{L.S}$, the peak lift coefficient $C_{L.peak}$, the peak drag coefficient $C_{D.max}$, and the peak pitch moment coefficient $C_{m.min}$ were compared with different blowing coefficients C_μ for control.

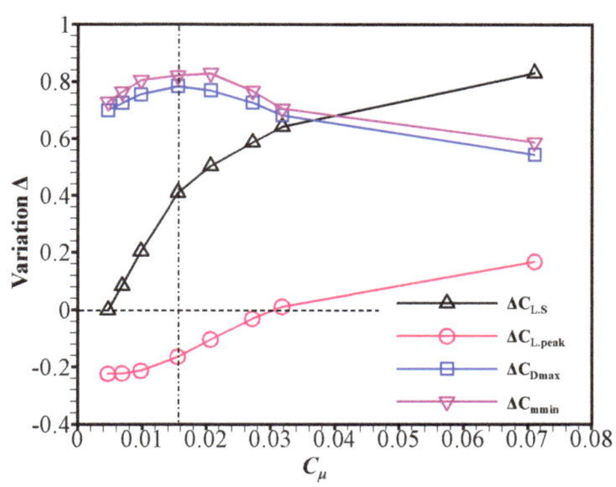

Figure 10. Comparison of aerodynamic coefficients of different blowing coefficients.

From the graph, it was observed that with an increase in the blowing coefficient C_μ, corresponding to an increase in the jet exit pressure P_j, the peak variation in the drag coefficient $\Delta C_{D.max}$ and the peak variation in the pitch moment coefficient $\Delta C_{m.min}$ initially showed a gradual increase around $C_\mu = 0.016$, reaching the maximum value. The variation in $\Delta C_{D\cdot max}$ and $\Delta C_{m.min}$ reached 80%. With the continued increase in the blowing coefficient (C_μ), the variations in the peak drag and pitch moment coefficients gradually decreased. Around $C_\mu = 0.07$, the variations in $\Delta C_{D\cdot max}$ and $\Delta C_{m.min}$ dropped to less than 60%. This indicated the existence of an optimal blowing coefficient. Beyond this upper limit, the effect of the blowing control no longer showed a significant improvement. With the increase in the blowing coefficient (C_μ), significant changes were observed in the variation in the stagnation ring area ($\Delta C_{L.S}$) and the peak lift coefficient ($\Delta C_{L.peak}$) when C_μ was less than 0.03. However, when C_μ exceeded 0.03, the changes in the stagnation ring area ($\Delta C_{L.S}$) and the peak lift coefficient ($\Delta C_{L.peak}$) became less pronounced.

In order to compare the aerodynamic coefficient variations under different jet exit pressures, the results of the aerodynamic coefficient variations for blowing coefficients (C_μ) of 0.0069, 0.0156, 0.0272, and 0.0711 can be seen in Figure 11. From the graph, it can be observed that when the blowing coefficient (C_μ) was 0.0069, despite a significant reduction in the drag and pitch moment coefficients, there was a decrease in the peak lift coefficient during the upstroke and the lift coefficient during the downstroke. Additionally, the area of the stagnation ring increased. When the blowing coefficient (C_μ) was 0.016, there was a significant decrease in the peak values of the drag and pitch moment coefficients, an increase in the lift coefficient, and a smooth variation in the aerodynamic coefficients with an angle of attack. The area of the stagnation ring also decreased. When the blowing coefficient was 0.0711, there was a 16% increase in the peak lift coefficient, but the lift curve also underwent a significant shift. Compared to C_μ = 0.016, at the maximum angle of attack, the angle of attack changed by only 1°. Furthermore, the lift coefficient showed a sharp decrease, followed by a rapid recovery, indicating a prominent peak discontinuity. The peak values of the drag and pitch moment coefficients increased, exhibiting the characteristics of sharp peaks in both the upward and downward directions, with rapid variations. Rapid changes in aerodynamic forces could have a significant impact on the control system and are an important source of control loads for helicopter swashplate manipulation. The comparison of the aerodynamic coefficient variations under different jet exit pressures illustrates the existence of an optimal blowing coefficient. Beyond this upper limit, not only did the effectiveness of the blowing control fail to show a significant improvement but it also led to excessive control loads.

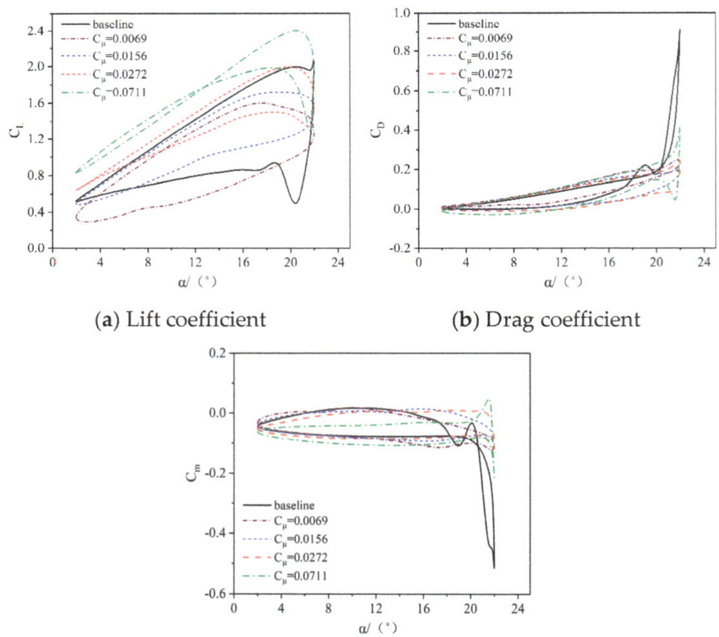

(**a**) Lift coefficient (**b**) Drag coefficient

(**c**) Pitching moment coefficient

Figure 11. Comparison of aerodynamic coefficients under different blowing coefficients.

5. Effect of Motion Parameters

To implement distributed blowing control in engineering applications, it was crucial to delve into its operational scope. This study focused on the dynamic stall control effectiveness of distributed blowing control under multiple operating conditions, particularly examining the motion parameters during dynamic stall. The blowing locations were situated at x/c = 5%, 10%, and 30%, with a specified jet exit total pressure P_j = 110545Pa and total temperature T_j = 288.15 K.

The reference conditions were as follows: M = 0.3, $R_e = 2.41 \times 10^6$, and the variation pattern of the angle of attack (α) followed $\alpha = 12^\circ + 10^\circ \sin(2kt)$, with $k = 0.129$. The sensitivity parameter analysis involved changing only the parameter under consideration, while keeping the other parameters consistent with the reference conditions.

5.1. Mean Angle of Attack

Simulations were performed for three scenarios with the average angles of attack (α_0) set at 10°, 12°, and 14°. Figure 12 illustrates the impact of the distributed air blowing control on the dynamic stall characteristics of the airfoil under different average angles of attack, while maintaining a constant total pressure at the jet outlet. From the graph, it could be observed that the variation in the average angle of attack had a minimal impact on the blowing coefficient. With the other conditions held constant, at an average angle of attack of 10°, the lift coefficient increased with the angle of attack. Near the maximum angle of attack, the lift coefficient curve exhibited a sharp decrease, indicating that the dynamic stall was primarily controlled by the SLV. Therefore, the impact on the pressure distribution was limited, resulting in lower peak values for the drag and pitch moment coefficients. The stall curve exhibited a clockwise "∞"-shaped pattern. At an average angle of attack of 12°, the dynamic stall characteristics of the airfoil intensified. With the variation in the angle of attack, the lift coefficient gradually increased. Near the maximum angle of attack, the lift coefficient decreased due to the onset of separation on the upper surface of the wing. With a further increase in the angle of attack, the lift coefficient exhibited nonlinear growth. The drag and pitch moment coefficients sharply increased and the area under the stall curve expanded. The DSV was the primary controlling vortex, leading to an increase in the aerodynamic coefficient peaks. In the pitching-down phase, unlike at lower average angles of attack, the influence of the TEV contributed to a secondary dynamic stall process, resulting in significant fluctuations in the aerodynamic coefficients. With the continued increase in the average angle of attack, the dynamic stall characteristics intensified. This indicated that the strength of the DSV also increased with higher average angles of attack. After implementing distributed blowing control, at an average angle of attack of 10°, the dynamic stall vortex primarily controlled by the SLV was suppressed. The vortex-induced lift associated with the dynamic stall vortex diminished, resulting in a reduction in the peak lift coefficient, a decrease in the stagnation ring area, and a 50% reduction in the peak values of the drag and pitch moment coefficients. The area of the lift stagnation ring decreased by 47%. At an average angle of attack of 12°, it could be observed from the lift coefficient curve that the nonlinear increase near the maximum angle of attack disappeared. This indicated that distributed blowing control had mitigated the formation of the DSV. At this time, the SLV became the primary controlling vortex influencing the peak aerodynamic coefficients. The peak values of the drag and pitch moment coefficients decreased by 80% and the area of the lift stagnation ring decreased by 41%.

The C_p contour and Q contour of the airfoil at different mean angles of attack can be seen in Figure 13. It can be observed that (1) at an average angle of attack of 10 degrees, the SLV was the primary controlling vortex structure influencing the dynamic stall, playing a major role in the aerodynamic forces during dynamic stall, albeit with a lower peak. With the implementation of blow control, the intensity of the SLV weakened, resulting in a reduction in the peak drag and pitch moment coefficients. (2) At average angles of attack of 12° and 14°, the formation of the DSV became the primary controlling vortex structure influencing the increase in aerodynamic forces, leading to a sharp increase in the peak drag and pitch moment coefficients. After the implementation of blow control, the formation of the LEV was suppressed, leading to the disappearance of the DSV. The secondary SLV became the primary controlling vortex influencing the aerodynamic forces. Due to the relatively minor impact of the SLV on the surface pressure distribution, the peak aerodynamic coefficient was lower.

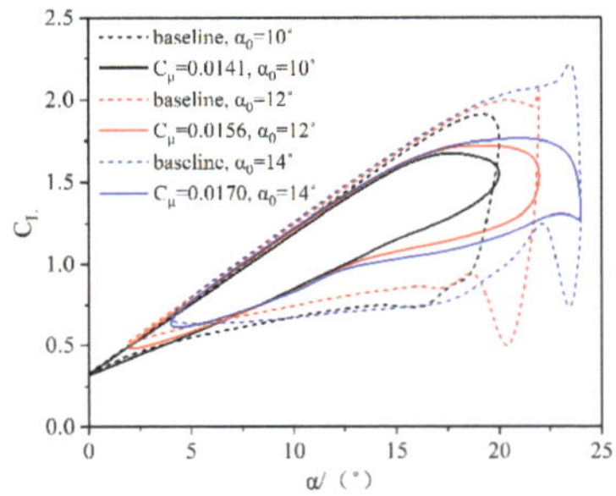

(**a**) Lift coefficient

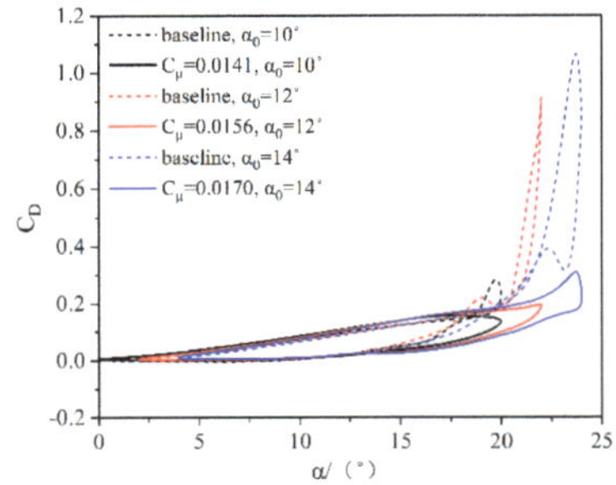

(**b**) Drag coefficient

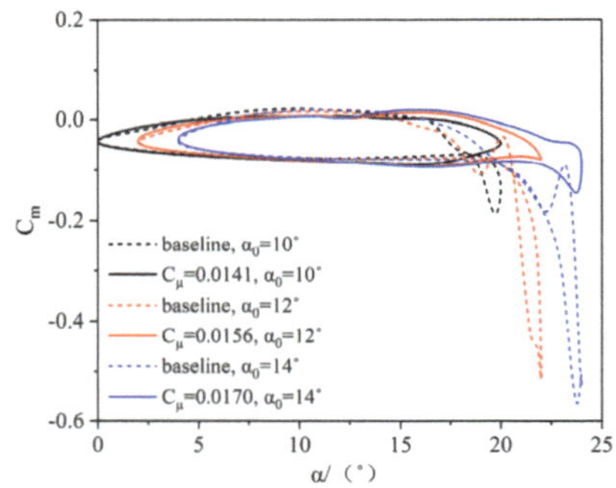

(**c**) Pitching moment coefficient

Figure 12. Aerodynamic coefficients at different mean angles of attack.

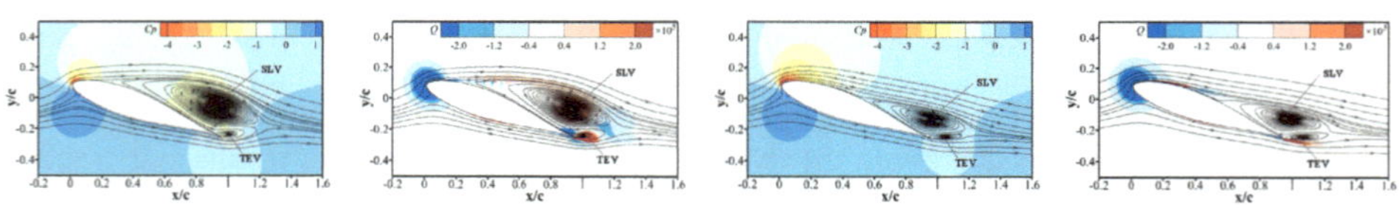

(**a**) $\alpha_0 = 10°$

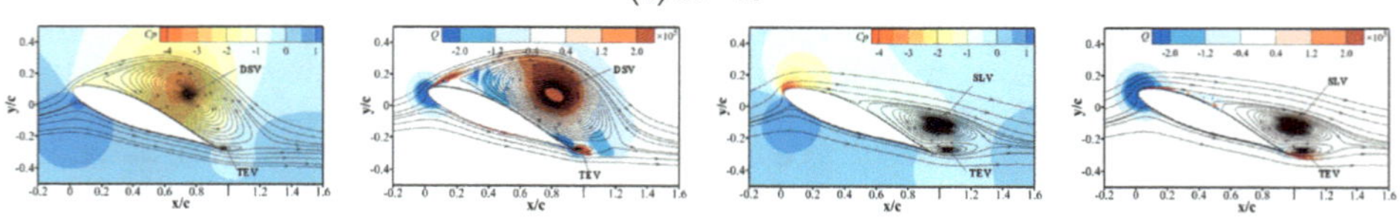

(**b**) $\alpha_0 = 12°$

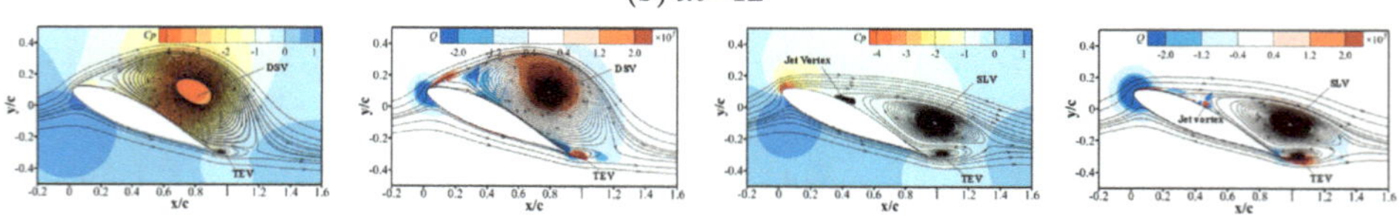

(**c**) $\alpha_0 = 14°$

Figure 13. C_p contour and Q contour of the airfoil at different mean angles of attack (baseline is in left column and controlled in right).

5.2. Amplitude of the Angle of Attack

A comparison of the aerodynamic forces during dynamic stall for different amplitudes of the angle of attack (α_m) of 8°, 10°, and 12° can be seen in Figure 14, with a constant jet exit total pressure. The comparison was based on distributed blowing control under different amplitudes of the angle of attack. From the curves in Figure 14, it can be observed that, with an increase in the amplitude of the angle of attack, the dynamic stall characteristics of the airfoil continually intensified. The peak values of the drag and pitch moment coefficients increased and the stall effect became more pronounced. The area under the stall curve increased with the growth of the angle of attack amplitude. At a low-amplitude angle of attack (α_m = 8°), the nonlinear increase in the lift coefficient did not occur, and the peak aerodynamic coefficients during dynamic stall were primarily controlled by the SLV. At large amplitudes of the angle of attack (α_m = 10° and 12°), near the maximum angle of attack, the lift coefficient exhibited a nonlinear increase with an increase in the angle of attack. The peak values of the drag and pitch moment coefficients sharply increased and the peak aerodynamic coefficients were primarily controlled by the DSV. Distributed blowing control could suppress the occurrence of the DSV and the nonlinear peak in the aerodynamic forces, thereby mitigating load fluctuations. At different amplitudes of the angle of attack, distributed blowing control could suppress the peak values of the drag and pitch moment coefficients. Under a small-amplitude angle of attack (α_m = 8°), the peak drag coefficient decreased by 46%, the peak pitch moment coefficient decreased by 74%, and the hysteresis loop area of the lift coefficient was reduced by 50%. At a large-amplitude angle of attack (α_m = 12°), the peak drag coefficient decreased by 70%, the peak pitch moment coefficient decreased by 72%, and the hysteresis loop area of the lift coefficient was reduced by 46%.

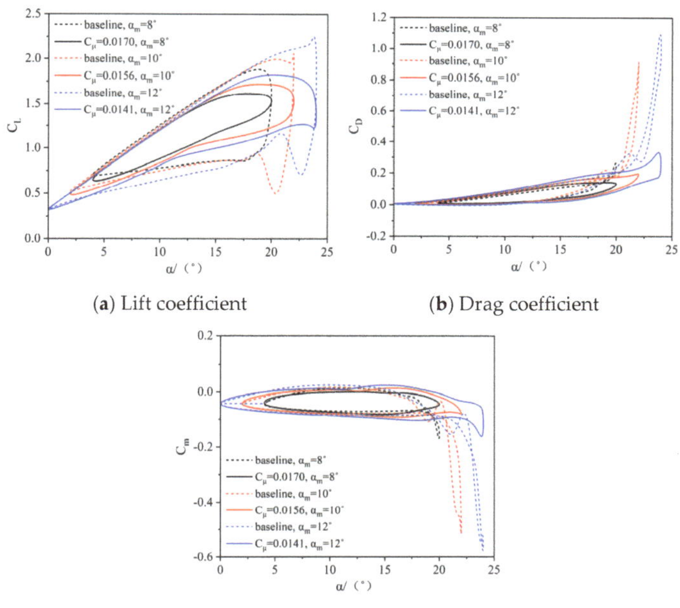

(**a**) Lift coefficient (**b**) Drag coefficient

(**c**) Pitching moment coefficient

Figure 14. Aerodynamic coefficients at different amplitudes of angle of attack.

Figure 15 presents the C_p contour and the Q contour of the airfoil at different amplitudes of angle of attack. The following insights could be drawn from Figure 15: When the amplitude of angle of attack was 8°, the SLV served as the primary controlling vortex structure affecting the dynamic stall; it played a key role in the aerodynamic forces during dynamic stall. With the implementation of combined blow control, the weakening of the SLV intensity led to a reduction in the peak values of the drag and pitch moment coefficients. When the angle of attack amplitude was 10° and 12°, the DSV gradually formed, serving as the primary controlling vortex structure influencing the increase in the aerodynamic forces.

This led to a sharp increase in the peak values of the drag and pitch moment coefficients. With the application of blow control, the formation of the LEV was suppressed, causing the disappearance of the DSV structure. The SLV became the primary controlling vortex structure affecting the aerodynamic forces. As the impact of the SLV on the surface pressure distribution was relatively small, the peak aerodynamic coefficients were lower.

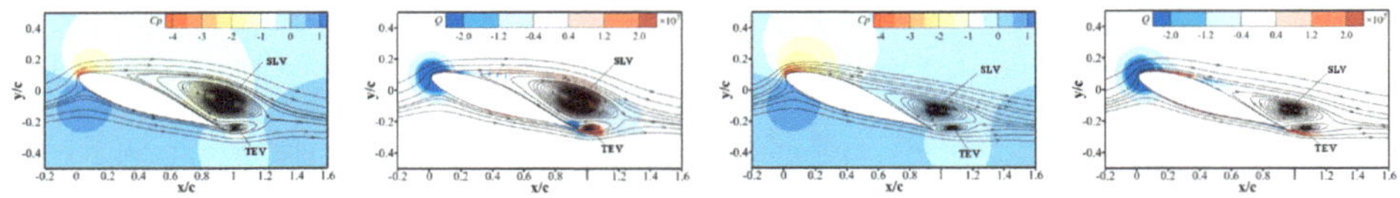

(**a**) $\alpha_{\rm m} = 8°$

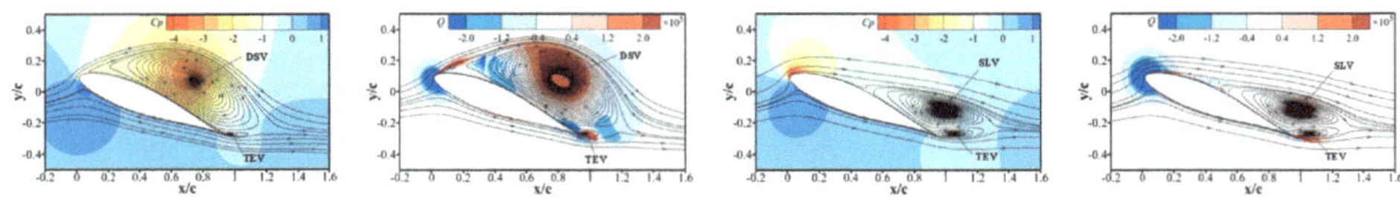

(**b**) $\alpha_{\rm m} = 10°$

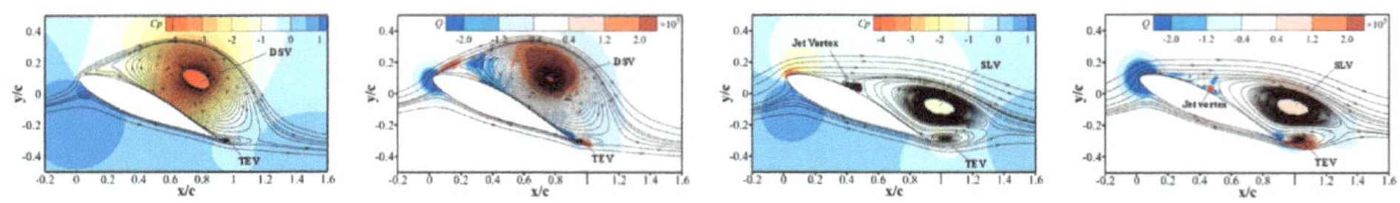

(**c**) $\alpha_{\rm m} = 12°$

Figure 15. C_p contour and Q contour of the airfoil at different amplitudes of angle of attack (baseline is in left column and controlled in right).

5.3. Reduced Frequency

Reduced frequency is a dimensionless quantity commonly used in unstable aerodynamic conditions, and it is one of the parameters that defines the instability level of the problem [62]. Simulation of the dynamic stall was conducted for three different reduced frequencies (k = 0.097, 0.129, 0.162). Figure 16 provides the coefficients at different reduced frequencies, the drag coefficient, and the pitch moment coefficient for the dynamic stall of the airfoil under a constant jet exit total pressure with distributed blowing control. From the graph, it can be observed that with a greater reduced frequency, the severity of the dynamic stall increased. The hysteresis loop area significantly enlarged, and the peak aerodynamic load, along with the corresponding angles of attack, also increased. Near the maximum angle of attack, the lift coefficient curve exhibited distinct variations. When the reduced frequency was k = 0.097, there was no nonlinear increase in the lift coefficient. Compared to high-frequency oscillations, the peak values of the drag and pitch moment coefficients were lower, and the increase in the peak aerodynamic coefficients was primarily controlled by the SLV. With a greater reduced frequency, around the maximum angle of attack, at a reduced frequency k = 0.129, the lift coefficient exhibited nonlinear growth. The peak values of the drag and pitch moment coefficients sharply increased and the DSV was the primary cause of the peak aerodynamic coefficients. When the reduced frequency was k = 0.162, the increase in oscillation frequency delayed the merging of the LEV and SLV. The lift coefficient exhibited a sudden nonlinear increase near the maximum angle of attack. Subsequently, during the downstroke process, the lift coefficient decreased, and the hysteresis loop in the lift coefficient curve exhibited closure near the maximum angle of attack. With an increase in reduced frequency, at the same angle of attack, the stall phenomenon became more pronounced in the airflow. During the downstroke process, the strength and scale of the TEV increased, and the position of the SDSV shifted forward. The amplitude of nonlinear fluctuations in the aerodynamic coefficients significantly increased and the angle of attack for nonlinear fluctuations advanced. This led to a slower reattachment of the airflow during the downstroke process, resulting in an increased hysteresis area in the aerodynamic characteristic curve. At different reduced frequencies, distributed blowing control effectively suppressed the peak values of the drag and pitch moment coefficients, resulting in a reduction of 78% in the peak values of the drag and pitch moment coefficients. Additionally, the hysteresis loop area of the lift coefficient was decreased by 40%.

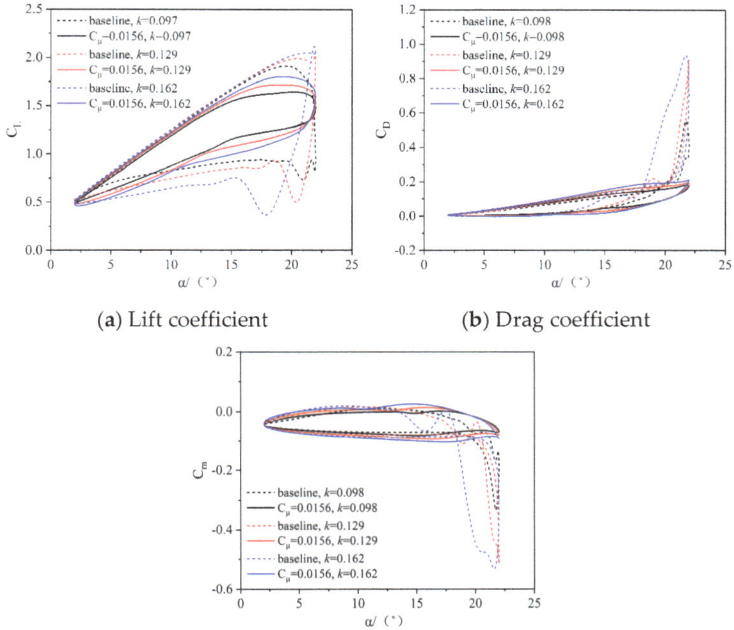

(**a**) Lift coefficient (**b**) Drag coefficient

(**c**) Pitching moment coefficient

Figure 16. Aerodynamic coefficients at different reduced frequencies.

The C_p contour and Q contour under the peak pitch moment coefficient state with combined blow control are shown in Figure 17, with variations in reduced frequency. The following insights can be drawn from Figure 17: When the reduced frequency was k = 0.097, the DSV served as the primary controlling vortex structure influencing the aerodynamic forces during dynamic stall. The peak values of the drag and pitch moment coefficients significantly increased. Distributed blowing control could suppress the occurrence of the DSV and the nonlinear peak in aerodynamic forces. With an increase in reduced frequency, the intensity of the DSV gradually strengthened, leading to a sharp increase in the peak values of the drag and pitch moment coefficients. With the application of blow control, the formation of the LEV was suppressed, leading to the disappearance of the DSV. The SLV became the primary controlling vortex structure influencing the increase in the aerodynamic forces. Due to the relatively minor impact of the SLV on the surface pressure distribution, the peak aerodynamic coefficients were lower.

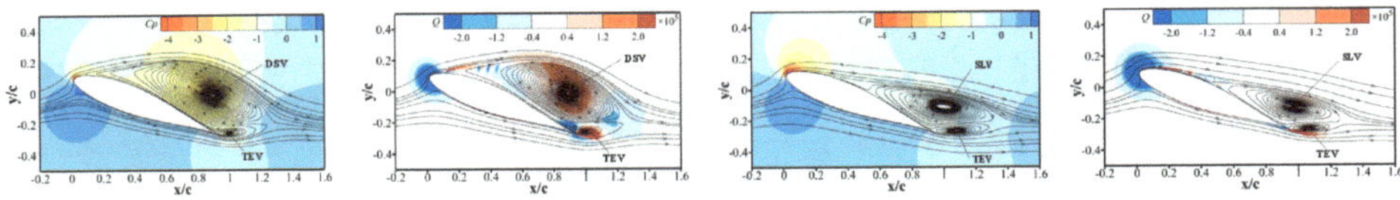

(**a**) $k = 0.098$.

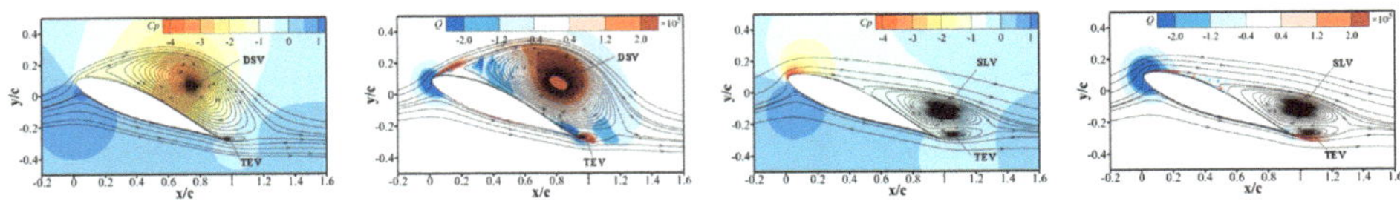

(**b**) $k = 0.129$

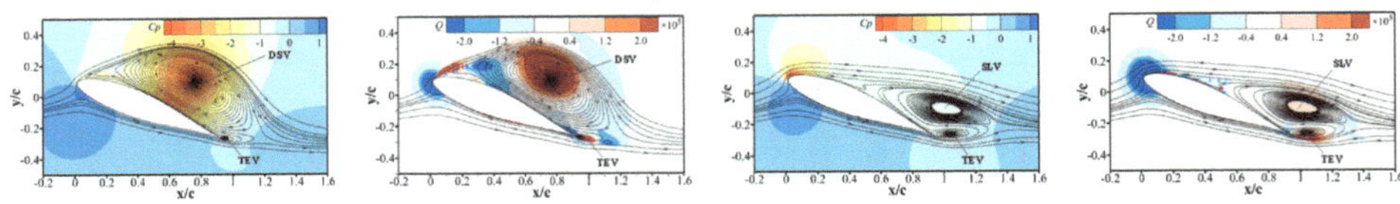

(**c**) $k = 0.162$

Figure 17. C_p contour and Q contour of the airfoil under different reduced frequencies (baseline is in left column and controlled in right).

6. Conclusions

Based on motion-nested grid technology, the SST k-ω turbulence model was used to numerically simulate the flow field characteristics of the NACA63-218 airfoil dynamic stall by solving the unsteady Reynolds-averaged Navier–Stokes equation. Based on the spatiotemporal evolution characteristics of the flow field structure and wall pressure distribution, the evolution process of the dynamic stall vortex structure and the primary controlling vortex mechanism influencing the nonlinear aerodynamic effects were revealed. In addressing the suppression of dynamic stall in airfoil behavior, the concept of distributed blowing flow control was introduced. The study explored the control effectiveness of distributed blowing positions and blowing coefficients in dynamic stall, as well as the dynamic stall suppression effect of distributed blowing control under various operating conditions. The conclusions were as follows:

(1) The evolution and motion characteristics of the DSV and SLV during dynamic stall led to nonlinear hysteresis in the aerodynamics. In the upstroke process, separation first occurred at the trailing edge, forming the SLV. As the angle of attack gradually increased, vorticity accumulated at the leading edge of the airfoil, forming the LEV. With a further increase in the angle of attack, the LEV and the SLV rapidly merged, causing a vortex eruption. This vorticity convection was downstream, forming the DSV, resulting in a sudden spike in aerodynamic forces and pitch moment coefficients to their maximum values. In the downstroke process, the strength of the TEV increased, inducing low-energy fluid in the dissipation region of the DSV, forming the SDSV. This resulted in a secondary dynamic stall and the aerodynamic coefficients exhibited a secondary peak phenomenon, although with relatively smaller magnitudes.

(2) To control the formation and merging of the SLV and DSV, the concept of distributed blowing control was introduced. The study investigated the control effectiveness of changes in the distributed blowing positions and blowing coefficients on the aerodynamics. The results indicated that distributed blowing control could suppress the formation of the DSV and reduce the intensity of the SLV, thereby lowering the aerodynamic peak values during dynamic stall. There existed an optimal blowing coefficient, and when the blowing coefficient C_μ exceeded 0.03, the effectiveness of blowing control no longer significantly improved.

(3) The control effectiveness of distributed blowing control on dynamic stall was investigated under various operating conditions. The results indicated that under different motion parameters, distributed blowing control could suppress the formation of the DSV, transforming the primary controlling vortex structure in the flow field into the SLV. When distributed blowing control was applied at the positions x/c = 5%, 10%, and 30% with a blowing momentum coefficient C_μ of 0.0156, it reduced the peak values of the drag coefficient by 70% and the pitch moment coefficient by 72%, and decreased the hysteresis loop area of the lift coefficient by 46%. Moreover, near the maximum stall angle, distributed blowing control could mitigate stall characteristics, resulting in a smoother variation in unsteady loads.

Author Contributions: Conceptualization, G.L. and S.Y.; methodology, G.L.; software, B.L.; validation, G.L., B.L. and X.Z.; resources, G.L.; data curation, B.L.; writing—original draft preparation, G.L. and B.L.; writing—review and editing, G.L. and B.L.; project administration, G.L.; funding acquisition, G.L. All authors have read and agreed to the published version of the manuscript.

Funding: This work was supported by the pre research project, China Aerodynamics Research and Development Center (50906030601), and exploratory research project, China Aerodynamics Research and Development Center (JK20211A020092).

Data Availability Statement: Data are contained within the article.

Conflicts of Interest: The authors declare no conflicts of interest.

References

1. Leishman, G.J. *Principles of Helicopter Aerodynamics with CD Extra*; Cambridge University Press: Cambridge, UK, 2006; pp. 761–763.
2. Mulleners, K.; Raffel, M. Dynamic stall development. *Exp. Fluids* **2013**, *54*, 1469. [CrossRef]
3. Gardner, A.D.; Jones, A.R.; Mulleners, K.; Naughton, J.W.; Smith, M.J. Review of rotating wing dynamic stall: Experiments and flow control. *Prog. Aerosp. Sci.* **2023**, *137*, 100887. [CrossRef]
4. Conlisk, A.T. Modern helicopter rotor aerodynamics. *Prog. Aerosp. Sci.* **2001**, *37*, 419–476. [CrossRef]
5. Zi, K.A.N.; Daochun, L.I.; Tong, S.H.E.N.; Xiang, J.; Zhang, L. Aerodynamic characteristics of morphing wing with flexible leading-edge. *Chin. J. Aeronaut.* **2020**, *33*, 2610–2619.
6. Corke, T.C.; Thomas, F.O. Dynamic Stall in Pitching Airfoil: Aerodynamic Damping and Compressibility Effects. *Annu. Rev. Fluid. Mech.* **2015**, *47*, 479–505. [CrossRef]
7. Himmel, S.H. *Profile Investigations on a Rotating Production Airscrew*; Ministry of Aircraft: Gottingen, Germany, 1947; p. 832.
8. Harri, F.D.; Pruyn, R.R. Blade Stall-Half Fact, Half Fiction. *J. Am. Helicopter Soc.* **1968**, *13*, 27–48. [CrossRef]
9. McCroskey, W.J.; Carr, L.W.; McAlister, K.W. Dynamic Stall Experiments on Oscillating Airfoils. *AIAA J.* **1976**, *14*, 57–63. [CrossRef]
10. Carr, L.W. Progress in analysis and prediction of dynamic stall. *J. Aircr.* **1988**, *25*, 6–17. [CrossRef]
11. McAlister, K.W.; Carr, L.W.; McCroskey, W.J. *Dynamic Stall Experiments on the NACA 0012 Airfoil*; Technical Report No. NASA-TP-1100; NASA: Washington, DC, USA, 1978.
12. Geissler, W.; Raffel, M.; Dietz, G.; Mai, H. Helicopter aerodynamics with emphasis placed on dynamic stall. In *Wind Energy: Proceedings of the Euromech Colloquium*; Springer: Berlin/Heidelberg, Germany, 2007; pp. 199–204.
13. Wang, Q.; Zhao, Q.J. Experiments on unsteady vortex flow field of typical rotor airfoils under dynamic stall conditions. *Chin. J. Aeronaut.* **2016**, *29*, 358–374. [CrossRef]
14. Ekaterinaris, J.A.; Platzer, M.F. Computational prediction of airfoil dynamic stall. *Prog. Aerosp. Sci.* **1997**, *33*, 759. [CrossRef]
15. Naughton, J.; Strike, J.; Hind, M.; Magstadt, A.; Babbitt, A. Measurements of Dynamic Stall on the DU Wind Turbine Airfoil Series. In Proceedings of the 69th Annual Forum of the American Helicopter Society, Phoenix, AZ, USA, 21–23 May 2013.
16. Ullah, A.H.; Fabijanic, C.; Refling, W.; Estevadeordal, J. A Lifetime-PSP Study of Dynamic Stall on Pitching Swept Finite-Aspect-Ratio Wings. In Proceedings of the AIAA SCITECH 2022 Forum, San Diego, CA, USA & Virtual, 3–7 January 2022. AIAA Paper No. 2022-1959.
17. Gardner, A.D.; Klein, C.; Sachs, W.; Henne, U.; Mai, H.; Richter, K. Investigation of three-dimensional dynamic stall on an airfoil using fast-response pressure-sensitive paint. *Exp. Fluids* **2014**, *55*, 1807. [CrossRef]
18. Disotell, K.J.; Nikoueeyan, P.; Naughton, J.W.; Gregory, J.W. Global surface pressure measurements of static and dynamic stall on a wind turbine airfoil at low Reynolds number. *Exp. Fluids* **2016**, *57*, 82. [CrossRef]
19. Larsson, J.; Kawai, S.; Bodart, J.; Bermejo-Moreno, I. Large eddy simulation with modeled wall-stress: Recent progress and future directions. *Mech. Eng. Rev.* **2016**, *3*, 15–00418. [CrossRef]
20. Khalifa, N.M.; Rezaei, A.S.; Taha, H.E. Comparing the performance of different turbulence models in predicting dynamic stall. In Proceedings of the AIAA Scitech 2021 Forum, Virtual Event, 11–15 & 19–21 January 2021. AIAA Paper No. 2021-1651.

21. Xu, Z.; Wu, B.; Gao, C.; Wang, N. Numerical simulation of dynamic stall flow control using a multi-dielectric barrier discharge plasma actuation strategy. *Phys. Plasmas* **2022**, *29*, 103503. [CrossRef]
22. Wang, S.; Ingham, D.B.; Ma, L.; Pourkashanian, M.; Tao, Z. Turbulence modeling of deep dynamic stall at relatively low Reynolds number. *J. Fluids Struct.* **2012**, *33*, 191–209. [CrossRef]
23. Kaufmann, K.; Costes, M.; Richez, F.; Gardner, A.D.; Pape, A.L. Numerical investigation of three-dimensional static and dynamic stall on a finite wing. *J. Am. Helicopter Soc.* **2015**, *60*, 1–12. [CrossRef]
24. Jain, R.; Le Pape, A.; Grubb, A.; Costes, M.; Richez, F.; Smith, M. High-Resolution Computational Fluid Dynamics Predictions for the Static and Dynamic Stall of a Finite-Span OA209 Wing. *J. Fluids Struct.* **2018**, *78*, 126–145. [CrossRef]
25. Costes, M.; Richez, F.; Le Pape, A.; Gaveriaux, R. Numerical Investigation of Three-Dimensional Effects During Dynamic Stall. In Proceedings of the 37th European Rotorcraft Forum, Gallarate, Italy, 13–15 September 2011.
26. Visbal, M.R.; Garmann, D.J. Dynamic stall of a finite aspect-ratio wing. *AIAA J.* **2019**, *57*, 962–977. [CrossRef]
27. Visbal, M.R.; Garmann, D.J. Effect of sweep on dynamic stall of a pitching finite-aspect-ratio wing. *AIAA J.* **2019**, *57*, 3274–3289. [CrossRef]
28. Visbal, M.R.; Garmann, D.J. Analysis of dynamic stall on a pitching airfoil using high-fidelity large-eddy simulations. *AIAA J.* **2018**, *56*, 46–63. [CrossRef]
29. Visbal, M.R.; Benton, S.I. Exploration of high-frequency control of dynamic stall using large-eddy simulations. *AIAA J.* **2018**, *56*, 2974–2991. [CrossRef]
30. Zhong, J.; Li, J.; Guo, P. Effects of leading-edge rod on dynamic stall performance of a wind turbine airfoil. *Proc. Inst. Mech. Eng. Part A* **2017**, *231*, 753–769. [CrossRef]
31. Niu, J.; Lei, J.; Lu, T. Numerical research on the effect of variable droop leading-edge on oscillating NACA 0012 airfoil dynamic stall. *Aerosp. Sci. Technol.* **2018**, *72*, 476–485. [CrossRef]
32. Zhao, G.Q.; Zhao, Q.J. Dynamic stall control optimization of rotor airfoil via variable droop leading-edge. *Aerosp. Sci. Technol.* **2015**, *43*, 406–414. [CrossRef]
33. Martin, P.; Mcalister, K.; Chandrasekhara, M.; Geissler, W. Dynamic Stall Measurements and Computations for a VR-12 Airfoil with a Variable Droop Leading Edge. In Proceedings of the American Helicopter 59th Annual Forum, Phoenix, AZ, USA, 6–8 May 2003.
34. Lorber, P.F.; Carta, F.O.; Covino, A.F., Jr. *An Oscillating Three-Dimensional Wing Experiment: Compressibility, Sweep, Rate, Waveform and Geometry Effects on Unsteady Separation and Dynamic Stall. Report R92-958325-6*; United Technologies Research Center: Farmington, NM, USA, 1992.
35. Raiola, M.; Discetti, S.; Ianiro, A.; Samara, F.; Avallone, F.; Ragni, D. Smart rotors: Dynamic-stall load control by means of an actuated flap. *AIAA J.* **2018**, *56*, 1388–1401. [CrossRef]
36. Zanotti, A.; Grassi, D.; Gibertini, G. Experimental investigation of a trailing edge L-shaped tab on a pitching airfoil in deep dynamic stall conditions. *Proc. Inst. Mech. Eng. Part. G J. Aerosp. Eng.* **2014**, *228*, 2371–2382. [CrossRef]
37. Samara, F.; Johnson, D.A. Dynamic stall on pitching cambered airfoil with phase offset trailing edge flap. *AIAA J.* **2020**, *58*, 2844–2856. [CrossRef]
38. Joo, W.; Lee, B.; Yee, K.; Lee, D.-H. Combining passive control method for dynamic stall control. *J. Aircr.* **2006**, *43*, 1120–1128. [CrossRef]
39. Lee, B.; Yee, K.; Joo, W.; Lee, D.-H. Dynamic Stall Control with Droop Leading Edge and Gurney Flap. *J. Korean Soc. Aeronaut. Space Sci.* **2004**, *32*, 10–17.
40. Chandrasekhara, M.S. Optimum gurney flap height determination for lost-lift recovery in compressible dynamic stall control. *Aerosp. Sci. Technol.* **2010**, *14*, 551–556. [CrossRef]
41. De Tavernier, D.; Ferreira, C.; Vire, A.; LeBlanc, B.; Bernardy, S. Controlling dynamic stall using vortex generators on a wind turbine airfoil. *Renew. Energy* **2021**, *172*, 1194–1211. [CrossRef]
42. Zhu, C.; Chen, J.; Wu, J.; Wang, T. Dynamic stall control of the wind turbine airfoil via single-row and double-row passive vortex generators. *Energy* **2019**, *189*, 116272. [CrossRef]
43. Green, R.B.; Prince, S.A.; Wang, Y.; Khodagolian, V.; Coton, F.N. Delay of dynamic stall using pulsed air-jet vortex generators. *AIAA J.* **2018**, *56*, 2070–2074. [CrossRef]
44. Liu, J.; Chen, R.; You, Y.; Shi, Z. Numerical investigation of dynamic stall suppression of rotor airfoil via improved co-flow jet. *Chin. J. Aeronaut.* **2022**, *35*, 169–184. [CrossRef]
45. Gardner, A.D.; Richter, K.; Mai, H.; Neuhaus, D. Experimental investigation of air jets to control shock-induced dynamic stall. *J. Am. Helicopter Soc.* **2014**, *59*, 1–11. [CrossRef]
46. Dano, B.; Zha, G.C.; Castillo, M. *Experimental Study of Co-Flow Jet Airfoil Performance Enhancement Using Discreet Jets*; Report No.: AIAA-2011-0941; AIAA: Reston, VA, USA, 2011.
47. Lefebvre, A.; Dano, B.; Bartow, W.B.; Difronzo, M.; Zha, G.C. Performance and energy expenditure of coflow jet airfoil with variation of Mach number. *J. Aircr.* **2016**, *53*, 1757–1767. [CrossRef]
48. Zhao, G.; Zhao, Q. Parametric analyses for synthetic jet control on separation and stall over rotor airfoil. *Chin. J. Aeronaut.* **2014**, *27*, 1051–1061. [CrossRef]
49. Kissing, J.; Stumpf, B.; Kriegseis, J.; Hussong, J.; Tropea, J. Delaying leading edge vortex detachment by plasma flow control at topologically critical locations. *Phys. Rev. Fluids* **2021**, *6*, 023101. [CrossRef]

50. Mukherjee, S.; Roy, S. Enhancement of lift and drag characteristics of an oscillating airfoil in deep dynamic stall using plasma actuation. In Proceedings of the 50th AIAA Aerospace Sciences Meeting Including the New Horizons Forum and Aerospace Exposition, Nashville, TN, USA, 9–12 January 2012. AIAA Paper No. 2012-702.
51. Li, G.; Yi, S. Large eddy simulation of dynamic stall flow control for wind turbine airfoil using plasma actuator. *Energy* **2020**, *212*, 118753.
52. Yu, H.; Zheng, J. Numerical investigation of control of dynamic stall over a NACA0015 airfoil using dielectric barrier discharge plasma actuators. *Phys. Fluids* **2020**, *32*, 035103. [CrossRef]
53. Yang, H.S.; Zhao, G.Y.; Liang, H.; Wei, B. Dynamic stall control over an airfoil by NS-DBD actuation. *Chin. Phys. B* **2020**, *29*, 105203. [CrossRef]
54. Ramesh, K.; Gopalarathnam, A.; Granlund, K.; Ol, M.V.; Edwards, J.R. Discrete-vortex method with novel shedding criterion for unsteady aerofoil flows with intermittent leading-edge vortex shedding. *J. Fluid. Mech.* **2014**, *751*, 500. [CrossRef]
55. Ma, Y.Y.; Zhao, Q.J.; Zhao, G.Q. New combinational active control strategy for improving aerodynamic characteristics of airfoil and rotor. *Proc. Inst. Mech. Eng. Part. G J. Aerosp. Eng.* **2020**, *234*, 977–996. [CrossRef]
56. Visbal, M.R.; Garmann, D.J. Investigation of spanwise end effects on dynamic stall of a pitching wing section. *J. Aircr.* **2019**, *56*, 2118–2130. [CrossRef]
57. Gardner, A.D.; Richter, K.; Rosemann, H. Numerical investigation of air jets for dynamic stall control on the OA209 airfoil. *CEAS Aeronaut. J.* **2011**, *1*, 69–82. [CrossRef]
58. Lind, A.H. An Experimental Study of Static and Oscillating Rotor Blade Sections in Reverse Flow. Ph.D. Thesis, University of Maryland, College Park, MD, USA, 2015; pp. 5–10.
59. Khalifa, N.M.; Rezaei, A.; Taha, H.E. On computational simulations of dynamic stall and its three-dimensional nature. *Phys. Fluids* **2023**, *35*, 105143. [CrossRef]
60. Carta, M.; Putzu, R.; Ghisu, T. A comparison of plunging- and pitching-induced deep dynamic stall on an SD7003 airfoil using URANS and LES simulations. *Aerosp. Sci. Technol.* **2022**, *121*, 107307. [CrossRef]
61. Bertagnolio, F.; Johansen, J.; Uglsang, P. *Wind Turbine Airfoil Catalogue*; Risø National Laboratory: Roskilde, Danmark, 2001; pp. 13–16.
62. Jing, S.; Zhao, G.; Zhao, Q. Numerical research on secondary peak of aerodynamic forces of airfoil under dynamic stall. *J. Nanjing Univ. Aeronaut. Astronaut.* **2022**, *54*, 191–202.

actuators

MDPI

Article

Reduced-Order Model Approaches for Predicting Airfoil Performance †

Antonio Colanera *, Eduardo Di Costanzo, Matteo Chiatto and Luigi de Luca

Department of Industrial Engineering, University of Naples Federico II, 80138 Naples, Italy
* Correspondence: antonio.colanera@unina.it

† This paper constitutes an expanded version of the conference paper: Di Costanzo, E.; Colanera, A.; Chiatto, M.; de Luca, L. Cluster-based network reduced order modeling for flow fields around airfoil profiles. In Proceedings of the AIAA AVIATION 2023 Forum, San Diego, CA, USA, 12–16 June 2023.

Abstract: This study delves into the construction of reduced-order models (ROMs) of a flow field over a NACA 0012 airfoil at a moderate Reynolds number and an angle of attack of 8°. Numerical simulations were computed through the finite-volume solver OpenFOAM. The analysis considers two different reduction techniques: the standard Galerkin projection method, which involves projecting the governing equations onto proper orthogonal decomposition modes (POD−ROMs), and the cluster-based network model (CNM), a fully data-driven nonlinear approach. An analysis of the topology of the dominant POD modes was conducted, uncovering a traveling wave pattern in the wake dynamics. We compared the performances of both ROM techniques regarding their prediction of flow field behavior and integral quantities. The ROM framework facilitates the practical actuation of control strategies with significantly reduced computational demands compared to the full-order approach.

Keywords: reduced-order modeling; clustering; airfoils

Citation: Colanera, A.; Di Costanzo, E.; Chiatto, M.; de Luca, L. Reduced-Order Model Approaches for Predicting Airfoil Performance. *Actuators* **2024**, *13*, 88. https://doi.org/10.3390/act13030088

Academic Editors: Hui Tang, Xin Wen and Feng Ren

Received: 24 January 2024
Revised: 21 February 2024
Accepted: 23 February 2024
Published: 26 February 2024

1. Introduction

The quest to enhance aircraft performance, specifically in terms of drag reduction and improving the lift–drag ratio, is a paramount objective in aircraft design (Couto and Bergada [1] Chiatto et al. [2]). To achieve these goals, the incorporation of flow control mechanisms, such as synthetic jets (Ceglia et al. [3], Girfoglio et al. [4], de Luca et al. [5]), has proven beneficial for delaying separation and managing turbulence transition (Tousi et al. [6,7]). Through careful optimization of airfoil shapes and control parameters, lift, drag, and virtual mass can be tailored to meet specific flight envelope requirements (Lahey et al. [8]). However, this optimization process often necessitates many computational simulations, which can be both time-consuming and computationally intensive (Li et al. [9], Reumschüssel et al. [10], Wang et al. [11]).

Modal decomposition techniques have emerged as valuable tools for gaining a deeper understanding of the key coherent flow structures (Taira et al. [12]) and for designing effective control strategies (Rowley and Dawson [13]). The recent increase in computational capabilities has paved the way for the development of various data-driven analytical tools for different aspects of fluid dynamics, including spectral characterization (Schmid [14]), system frequency response (Herrmann et al. [15]), and even the study of two-phase flows (Schmidt and Oberleithner [16]).

Recent research, as highlighted by Stabile et al. [17], emphasizes the relevance of formulating reduced-order models (ROMs) that retain the key features of full-order models (FOMs) while significantly reducing computational demands. This step is crucial for simplifying the governing equations' complexity and enhancing the efficiency of the optimization process (Noack et al. [18]).

As a matter of fact, when employing reduced-order models one encounters several drawbacks, for instance, the introduction of unphysical artifacts and instabilities (Noack et al. [19]), the loss of fidelity at fine temporal and spatial scales (Taira et al. [20]), and the inability to accurately predict rare or extreme events due to nonlinear interactions (Racca and Magri [21]). Cluster-based network modeling (CNM) stands as a robust approach for investigating complex nonlinear dynamics using data (Fernex et al. [22]). CNM relies on an unsupervised machine learning technique to process a dataset of snapshots, reducing them to a small set of representative flow states. Subsequently, statistical methods are employed to construct a reduced-order model that presents a probabilistic representation of the system dynamics. CNM offers a fully automated framework for understanding system dynamics (Hou et al. [23], Colanera et al. [24]), estimating parameters, and implementing model-based control (Wang et al. [25]).

Furthermore, recent advancements in machine learning techniques (Brunton et al. [26]) have demonstrated their effectiveness, even in the prediction of stall behavior (Saetta et al. [27]).

This study aims to develop reduced-order models (ROMs) for analyzing the flow around a specific airfoil configuration. Two distinct types of low-order models were constructed: proper orthogonal decomposition (POD) Galerkin ROMs (Stabile and Rozza [28]), which involve projecting the governing equations onto selected POD modes and cluster-based network models (CNMs) (Kaiser et al. [29] and Li et al. [30]). To the authors' best knowledge, this paper represents an original application by comparing these two model approaches for airfoil configurations. This paper constitutes an expanded version of the conference paper by Di Costanzo et al. [31].

The paper is structured as follows: Section 2 concisely introduces the methodologies used, including both the full-order model (FOM) and ROMs. In Section 3, the results obtained are presented, along with a comparative analysis of the different low-order models.

2. Methodology

This section provides an overview of full-order modeling for unsteady turbulent flow analysis. Additionally, it describes two different approaches to reduced-order modeling. Full-order modeling offers a detailed representation of the flow field but can be computationally expensive. Reduced-order modeling techniques offer a more efficient alternative by capturing essential flow features while reducing computational costs.

2.1. Full Order Model

The present study investigates the behavior of NACA0012 airfoil under two-dimensional, unsteady, and incompressible flow conditions. To tackle this problem, Unsteady Reynolds-averaged Navier–Stokes (URANS) equations were employed. The numerical simulations were conducted using the pimpleFoam solver within the open-source software OpenFOAM. The pimpleFoam employs the PIMPLE algorithm for pressure–velocity coupling, enabling the modeling of a transient flow within a discretized domain, achieved by solving the Reynolds-averaged Navier–Stokes equations.

For the representation of turbulence in this analysis, the well-established $\kappa - \omega$ turbulence model was implemented. In this context, the equations system is denoted as a full-order model (FOM), as detailed in Equations (1) and (2).

$$\nabla \cdot \mathbf{u} = 0, \tag{1}$$

$$\frac{\partial \mathbf{u}}{\partial t} + (\mathbf{u} \cdot \nabla)\mathbf{u} + \nabla p - \nabla \cdot \left(\nu_{\text{eff}}(\nabla \mathbf{u} + [\nabla \mathbf{u}]^T)\right) = 0, \tag{2}$$

where the time variable, t, kinematic pressure, p, and the velocity vector, $\mathbf{u}$, represented as $\mathbf{u} = [u, v, w]$, are considered. The effective kinematic viscosity is denoted as $\nu_{\text{eff}} = \nu + \nu_t$, where ν denotes the kinematic viscosity and ν_t denotes turbulent eddy viscosity. It is worth noting that in this work any additional effect related to the virtual mass has not been considered. The Reynolds number under free-stream conditions is defined as

$Re_\infty = U_\infty c/\nu = 7000$, where U_∞ represents the free-stream velocity, and c is the airfoil chord. The angle of attack, denoted as α, is fixed at $\alpha = 8°$ for this investigation.

The simulations based on the FOM are executed until the values of time-averaged flow variables (such as velocity, drag, lift, and virtual mass when applicable) reach the periodic behavior. Subsequently, snapshots of flow field variables are stored for later use in constructing low-order models.

Utilizing the URANS approach and the $\kappa - \omega$ turbulence model, the FOM describes the unsteady turbulent flow surrounding the NACA0012 airfoil. This forms the basis for subsequent analyses and the creation of reduced-order models.

2.2. POD–ROM

Adopting the methodology outlined in Hijazi et al. [32], one can employ a systematic process that combines the POD technique with Galerkin projections to derive reduced-order models (ROMs) from simulations or experimental data.

Specifically, one can express the truncation of the POD expansions for the discretized fields of velocity $\mathbf{u}$, pressure p, and eddy viscosity ν_t, as follows:

$$\begin{bmatrix} u(x_i, y_i, t) \\ v(x_i, y_i, t) \end{bmatrix} = \tilde{\mathbf{\Phi}}_{\mathbf{u}} \mathbf{a}_r, \quad [p(x_i, y_i, t)] = \tilde{\mathbf{\Phi}}_p \mathbf{b}_r, \quad [\nu_t(x_i, y_i, t)] = \tilde{\mathbf{\Phi}}_{\nu_t} \mathbf{c}_r, \tag{3}$$

where $\tilde{\mathbf{\Phi}}_{\mathbf{u}}$, $\tilde{\mathbf{\Phi}}_p$, and $\tilde{\mathbf{\Phi}}_{\nu_t}$ are, respectively, the POD mode matrices of velocity, pressure, and eddy viscosity. The computation of these mode matrices follows the procedure described by Stabile and Rozza [28], utilizing numerical data derived from the FOM.

The momentum equation is subsequently projected onto the modes represented by $\tilde{\mathbf{\Phi}}_{\mathbf{u}}$, resulting in the following reduced-order system:

$$\frac{d\mathbf{a}_r}{dt} = f_{\mathbf{u}}(\mathbf{a}_r, \mathbf{c}_r, \mathbf{b}_r). \tag{4}$$

Here, $\mathbf{a}_r$, $\mathbf{c}_r$, and $\mathbf{b}_r$ denote the coefficients associated with the velocity, eddy viscosity, and pressure fields, respectively. Additional equations to determine coefficients $\mathbf{b}_r$ and $\mathbf{c}_r$ are derived from the pressure Poisson equation and the eddy viscosity model projections. Initial conditions for $\mathbf{a}_r$ and $\mathbf{c}_r$ are set by projecting $\mathbf{u}(0)$ and $\nu_t(0)$ onto the POD modes. To stabilize the pressure, a velocity space supremizer enrichment strategy is employed, as detailed by Ballarin et al. [33]. The entire procedure is implemented using ITHACA-FV, a C++ library developed in the OpenFOAM framework [28].

The combination of proper orthogonal decomposition (POD) and Galerkin projections presents an effective method for constructing reduced-order models. This approach enables efficient analysis and prediction of flow patterns, substantially lowering computational costs.

2.3. CNM–ROM

The cluster-based network model (CNM) approach depicts the dynamics of a system through a directed network, wherein the clusters (i.e. the nodes) symbolize coarse-grained states of the system. The original snapshots are clustered into a few centroids representing the whole ensemble.

The clusters $\mathcal{C}_k$ are computed using the standard k-means++ algorithm (MacQueen [34]; Lloyd [35]), which minimizes the inner-cluster variance. Consequently, the cost function J is defined as follows:

$$J(\mathcal{C}_1, \ldots, \mathcal{C}_K) = \frac{1}{M} \sum_{m=1}^{M} \|q^m - \mathcal{C}_{k(m)}\|_\Omega^2, \tag{5}$$

where M represents the number of snapshots used for the analysis, q^m denotes the m-th snapshot of velocity components, and $\mathcal{C}_{k(m)}$ denotes the cluster to which q^m belongs. The cluster-affiliation function $k(m)$ associates a snapshot q^m with the index of its nearest centroid:

$$k(m) = \arg\min_i \|q^m - \mathcal{C}_i\|_\Omega, \tag{6}$$

where $\|\cdot\|_\Omega$ denotes a spatial norm. The optimal centroids, $\mathcal{C}_k^\star$, are those that minimize J:

$$(\mathcal{C}_1^\star, \ldots, \mathcal{C}_K^\star) = \arg\min J(\mathcal{C}_1, \ldots, \mathcal{C}_K). \tag{7}$$

After the snapshots are coarse-grained into K clusters, each cluster can be seen as a representative state of the system dynamics. The network nodes are identified as centroids, $\mathcal{C}_k$, of these clusters, calculated as the state averages within each cluster. The transition characteristics between such clusters rely on high-order direct transition probabilities identified from the data [22].

As sketched in Figure 1, CNM utilizes the direct transition matrix Q, which considers only intercluster transitions and ignores the inner-cluster residence probability.

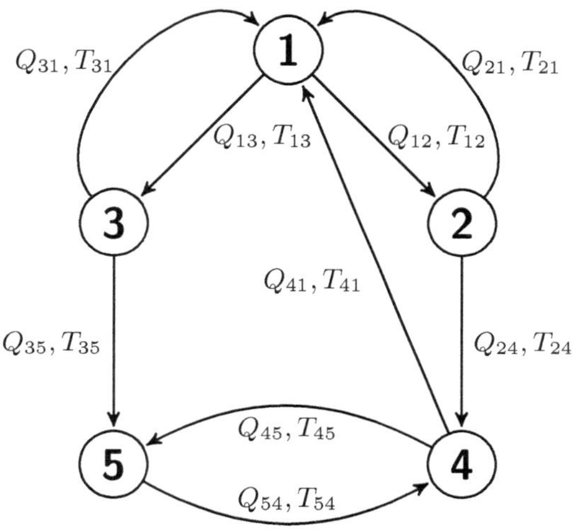

Figure 1. Sketch of a directed network of clusters. The nodes denote cluster centroids, while arrows represent transition directions. Each transition includes a transition probability Q_{ij} and time T_{ij}.

The direct transition probability is inferred from the data as follows:

$$Q_{k,j} = \frac{n_{k,j}}{n_j}, \tag{8}$$

where $Q_{k,j}$ represents the probability of transitioning from cluster $\mathcal{C}_j$ to $\mathcal{C}_k$, and $n_{k,j}$ denotes the number of such transitions, with n_j being the total number of transitions departing from cluster j. Figure 1 shows an example of a directed network of clusters. For a generalized L-order model, the direct transition probability is expressed as $Pr(\mathcal{K}_i = j|\mathcal{K}_{i-1}, ..., \mathcal{K}_{i-L})$ (L being the model order), which is equivalent to using time-delay coordinates. Note that in the CNM framework, $n_{i,i} = Q_{i,i} = 0$.

The residence time $\tau^n = t^{n+1} - t^n$ corresponds to the duration of the transit in a cluster $\mathcal{C}_k$, where t^n and t^{n+1} denote the time of the first and last snapshots to enter and leave $\mathcal{C}_k$. The individual transition time from cluster j to cluster k is defined as half the residence time of both clusters:

$$\tau_{k,j}^n = \frac{\tau^{n-1} + \tau^n}{2} = \frac{t^{n-1} + t^n}{2}. \tag{9}$$

By averaging all $n_{k,j}$ individual transition times from cluster $\mathcal{C}_j$ to $\mathcal{C}_k$, the transition time $T_{k,j} = 1/n_{k,j} \sum_{n=1}^{n_{k,j}} \tau^{n_{k,j}}$ is estimated.

Once the probabilistic model is determined through matrices Q and T, a sequence of centroids, denoted as $\mathcal{K}_0, \mathcal{K}_1, \mathcal{K}_2, \ldots$, can be determined together with with the following time instance:

$$t_0 = 0, \quad t_1 = T_{\mathcal{K}_1\mathcal{K}_0}, \quad t_2 = t_1 + T_{\mathcal{K}_2\mathcal{K}_1} \cdots \tag{10}$$

In this framework, a smooth motion between these discrete jumps is obtained by a linear interpolation.

Within the CNM framework, the N-dimensional state of the system can be represented by various options, including the flow field variables themselves or any other observables. In this study, we consider the POD temporal coefficients as state vector components. Transition probabilities from one cluster centroid to another are estimated based on the training dataset. This estimation enables the computation and reconstruction of the temporal evolution of the reduced state. For further details, the reader is referred to Fernex et al. [22] and Kaiser et al. [29]. In this work, MATLAB (R2020b) was employed for the CNM implementation and test.

3. Results

The methodologies outlined earlier were employed in analyzing 2D numerical data of an unsteady, incompressible flow around an NACA0012 airfoil. These datasets have yielded valuable insights into the flow dynamics and effectiveness of the ROMs. In both the POD-based and CNM−ROM approaches, 400 snapshots were utilized during the training phase, with a sampling time interval of $\Delta t = 0.025$ s.

Figure 2 showcases a time-delay representation of the lift (C_l) and drag (C_d) coefficients with a time-delay parameter of $\tau = 6\Delta t$.

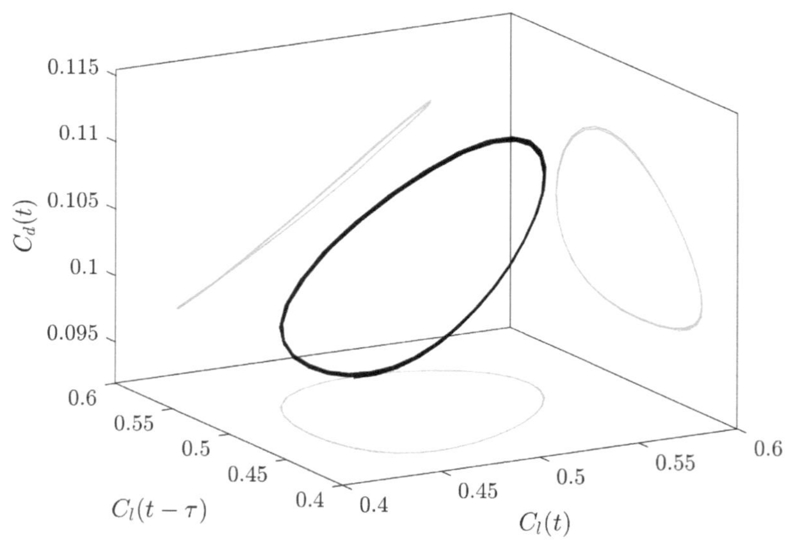

Figure 2. Time-delay representation of the lift/drag coefficients with a $\tau = 6\Delta t$.

It is possible to determine the lift and drag coefficients from the axial (C_x) and the transversal (C_y) force coefficients using the following relationship:

$$\begin{aligned} C_d &= C_x \cos(\alpha) + C_y \sin(\alpha), \\ C_l &= C_x \sin(\alpha) - C_y \cos(\alpha), \end{aligned} \tag{11}$$

where α is the angle of attack. From the inspection of Figure 2, it is possible to recognize the limit cycle behavior of the flow dynamics.

In Figure 3, panels (a) and (b) depict the contours of the time-averaged streamwise velocity component u and the transversal velocity component v.

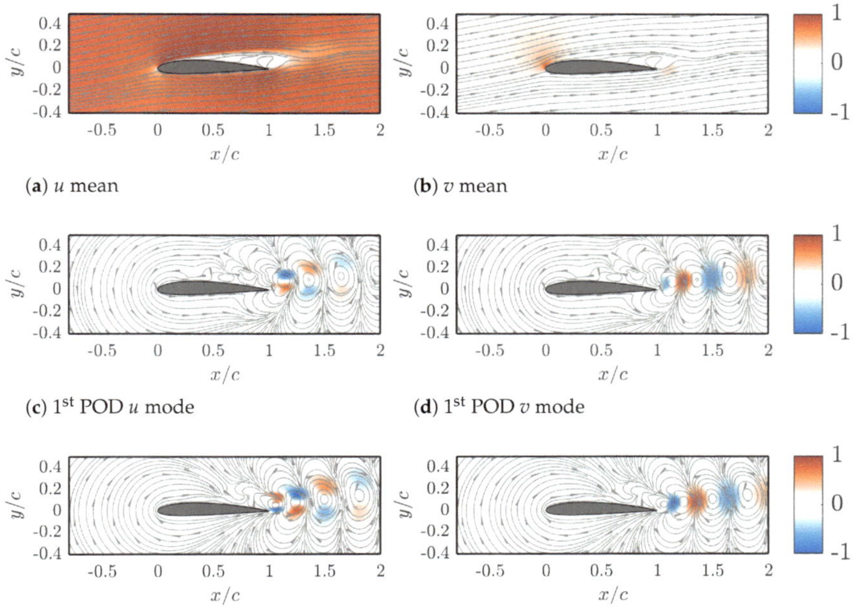

(**a**) u mean (**b**) v mean

(**c**) 1st POD u mode (**d**) 1st POD v mode

(**e**) 2nd POD u mode (**f**) 2nd POD v mode

Figure 3. Mean velocity fields and leading POD modes for a NACA 0012 at $Re_\infty = 7000$ and $\alpha = 8°$. All field variables were normalized concerning their maximum.

These contour plots offer a visual representation of the mean flow characteristics. Panels (c) to (f) showcase the leading velocity modes within the POD framework. The first two modes, in particular, capture the most significant fluctuations in the flow field. It is important to note that the spatial shift in the direction of advection of the POD modes corresponds to the traveling pattern of the emitted eddies.

In Figure 4, velocity contour plots from three methods are compared: the FOM, POD−ROM with 10 velocity modes and 12 pressure and viscosity modes, and CNM−ROM using 10 clusters with a 4th-order model.

This comparison is made at a specific time instant, $t = 1$ second. This analysis allows us to assess the accuracy and fidelity of both the reduced-order models in capturing the flow features. The constructed flow fields obtained from the POD−ROM and CNM−ROM agree with the original FOM results, indicating the effectiveness of the reduced-order models in capturing the essential flow dynamics.

Figure 5 provides a comprehensive comparison of the axial and transversal force coefficients (C_x and C_y) obtained from the FOM, the POD−ROM, and the CNM−ROM at various ROM parameters.

This comparison allows us to evaluate the accuracy and predictive capability of the reduced-order models in capturing the aerodynamic forces acting on the airfoil. The black lines represent the C_x and C_y coefficients obtained from the FOM simulations. In panels (a,b), the blue lines correspond to the C_x and C_y coefficients obtained from the POD−ROM, utilizing 10 velocity modes and 12 pressure and viscosity modes. In panels (c,d), the red lines correspond to the coefficients obtained from a first-order CNM−ROM, while the blue lines correspond to a fourth-order CNM−ROM. The figure demonstrates the capabilities of the CNM−ROM technique in accurately predicting the lift and drag coefficients, surpassing the performance of the POD−ROM. Therefore, the CNM−ROM approach (with a higher order) captures the complex flow phenomena more effectively, resulting in improved predictive accuracy of the aerodynamic forces.

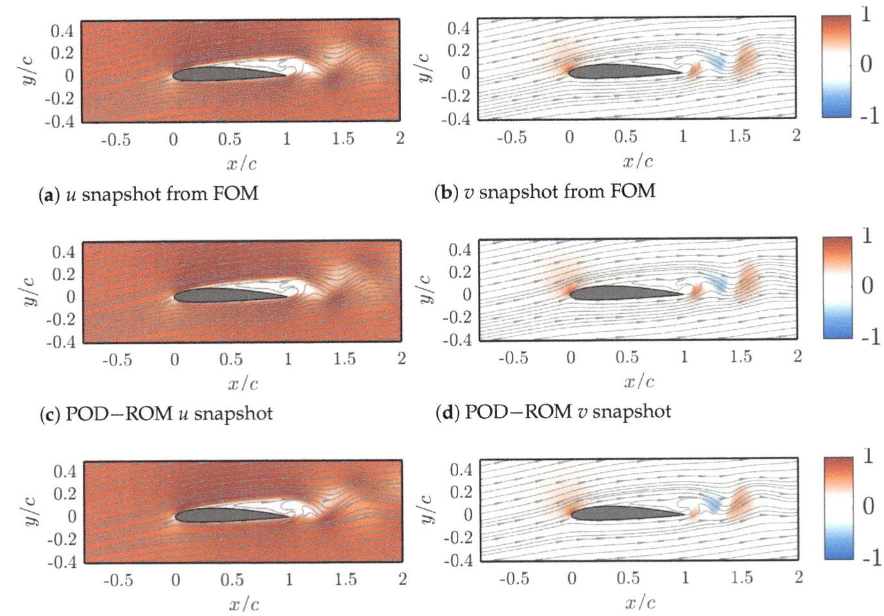

(**a**) u snapshot from FOM (**b**) v snapshot from FOM

(**c**) POD−ROM u snapshot (**d**) POD−ROM v snapshot

(**e**) CNM−ROM u snapshot (**f**) CNM−ROM v snapshot

Figure 4. Snapshots computed by FOM (first line), POD−ROM (second line), and CNM−ROM (third line). All field variables were normalized to their maximum.

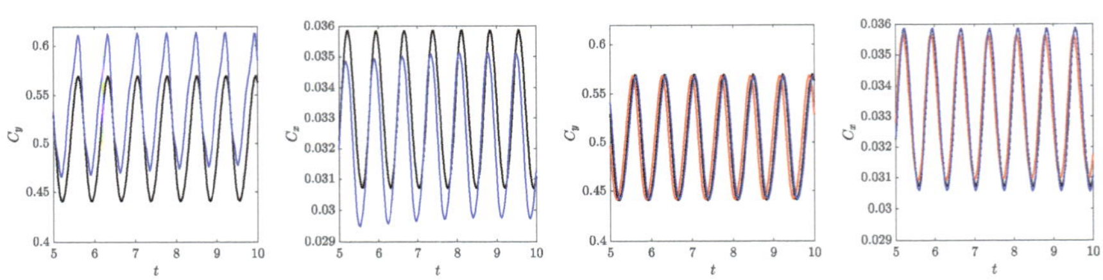

(**a**) C_y POD−ROM (**b**) C_x POD−ROM (**c**) C_y CNM−ROM (**d**) C_x CNM−ROM

Figure 5. Force coefficients computed with FOM, POD−ROM, and CNM−ROM . In all the panels, the black lines refer to the FOM force coefficients. In panels (**a**,**b**), the blue lines correspond to the POD−ROM with 10 velocity modes and 12 pressure and viscosity modes; in panels (**c**,**d**), the red lines correspond to a first-order CNM−ROM, while the blue ones correspond to CNM−ROM with $L = 4$.

To further analyze the statistical properties of the flow, the unbiased autocorrelation function $R(\tau)$ between the FOM, POD−ROM, and CNM−ROM approaches is shown in Figure 6. The unbiased autocorrelation function (Protas et al. [36], Deng et al. [37]) is defined as follows:

$$R(\tau) = \frac{1}{T-\tau}\int_{\tau}^{T} \mathbf{u}(t-\tau)^T \mathbf{u}(t)dt, \qquad \tau \in [0, T), \tag{12}$$

where T is the simulation time and τ denotes the time delay, providing insights into the temporal correlation and persistence of flow features.

Figure 6 demonstrates that the ROM approaches perform quite well in capturing the flow statistical properties and temporal correlations. However, the CNM−ROM exhibits a closer match to the autocorrelation function obtained from the FOM data, indicating its superior capability in reproducing the flow statistics.

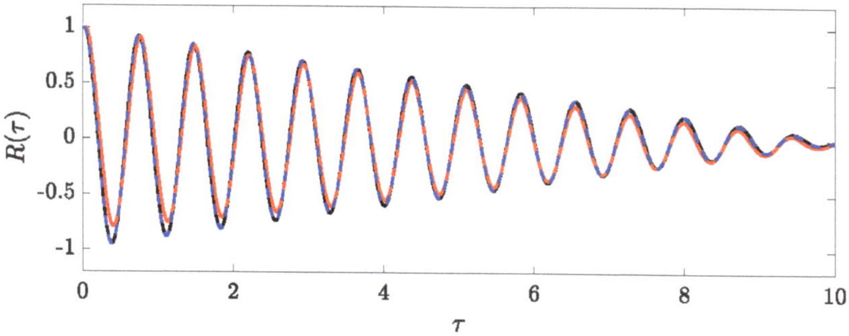

Figure 6. Comparison of the autocorrelation function R between the FOM, POD−ROM, and CNM−ROM. The black line represents the FOM, the red line represents the POD−ROM, and the blue dashed line corresponds to the CNM−ROM.

When considering the first-order CNM, the direct transition probability matrix Q_{ij} and the direct transition times matrix T_{ij} can be examined to describe the dynamics patterns. These matrices provide insights into how frequently and swiftly the system transitions between its various states or clusters. Figure 7 visually represents these matrices in panels (a) and (b), showcasing the direct transition probabilities (Q_{ij}) and times (T_{ij}) of the CNM.

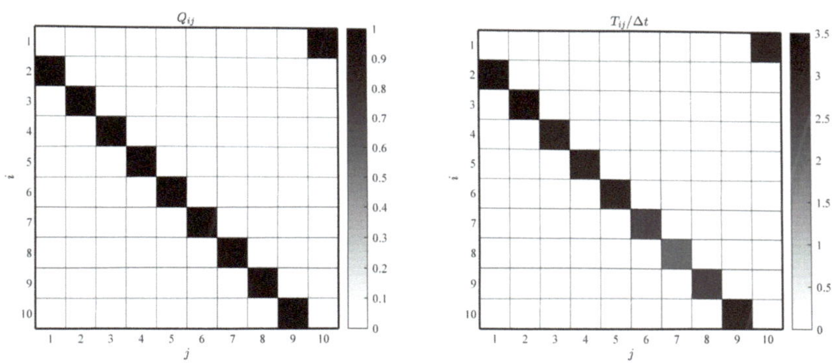

(**a**) Q_{ij} (**b**) T_{ij}

Figure 7. The matrices for CNM direct transition probabilities (Q_{ij}) and times (T_{ij}).

Notably, the most likely transitions align with the subsequent clusters in the sequence, confirming the presence of a limit cycle in the system dynamics. Furthermore, by examining the transition times matrix, it is possible to estimate the dominant frequency by the following:

$$f_{\text{lead}} \approx \frac{1}{T_{2,1} + T_{3,2} + \ldots + T_{1,10}} \approx 1.4\,\text{Hz}. \tag{13}$$

The versatility of the CNM−ROM approach becomes evident as it enables predictions of system behavior under various parameter and control configurations. In Figure 8, panel (a) illustrates the predictions for C_y, while panel (b) depicts the predictions for C_x when $\alpha = 6°$. These predictions are generated by training the CNM−ROM with snapshots from different α values, excluding $\alpha = 6°$. This demonstrates the CNM−ROM's ability to generalize the learned dynamics to unobserved parameter settings.

Within Figure 8, the blue and red lines correspond to $\alpha = 5°$ and $\alpha = 7°$, respectively. The black dashed lines represent the C_x and C_y coefficients obtained from the FOM simulations, and the black solid lines represent the predictions made by the CNM−ROM. Remarkably, the CNM−ROM accurately captures the amplitude and frequency of the force coefficients for varying α values, including the unseen $\alpha = 6°$. This showcases the robustness and generalization capability of the CNM−ROM, as it effectively predicts the aerodynamic forces even for parameter values that were not included in its training data.

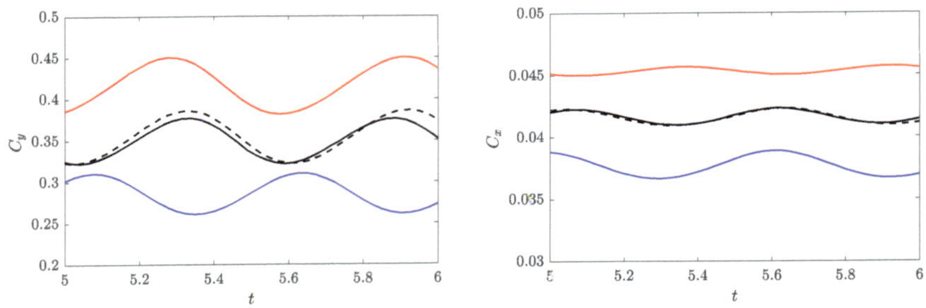

(**a**) C_y prediction (**b**) C_x prediction

Figure 8. Prediction of the effect of α using CNM−ROM. The blue and red lines correspond to $\alpha = 5°$ and $\alpha = 7°$, respectively. The black dashed lines represent the FOM simulations, while the black solid lines represent the CNM−ROM predictions.

Overall, the results presented in this section validate the effectiveness of the POD−ROM and CNM−ROM approaches in accurately capturing the flow dynamics, reproducing the flow fields, predicting force coefficients, and capturing the statistical properties of the flow. These findings highlight the potential of reduced-order modeling techniques in enhancing computational efficiency and enabling rapid design iterations for airfoil configurations.

4. Conclusions

This study employs two distinct methodologies for analyzing the flow field over a NACA0012 airfoil at a moderate Reynolds number. The first methodology, the POD−ROM, involves truncating POD expansions for the velocity, pressure, and eddy viscosity fields to a reduced number of modes. The second methodology, the CNM−ROM, utilizes a directed network model, where coarse-grained states are represented as nodes. These reduced-order modeling frameworks offer the potential to design control strategies with significantly reduced computational demands compared to the full-order approach.

Numerical simulations for the full-order model were carried out using the OpenFOAM finite-volume solver. The proper orthogonal decomposition reduced-order model was implemented through the ITHACA-FV library within the OpenFOAM framework.

The comparative analysis of POD−ROM and CNM−ROM methodologies has revealed the superior effectiveness of CNM−ROM in accurately capturing flow dynamics and statistical characteristics while demanding fewer computational resources. The CNM−ROM approach, relying on direct transition matrices and cluster analysis, demonstrated its ability to provide precise flow field predictions and aerodynamic forces.

Furthermore, the CNM−ROM model showcased its versatility by successfully predicting the effect of varying flow parameters, such as the angle of attack. Training the CNM−ROM with snapshots from different angles of attack settings accurately captured the amplitude and frequency of the lift and drag coefficients for unseen angles of attack values. This demonstrates the robustness and generalization ability of the CNM−ROM approach.

The promising results obtained from both the POD−ROM and CNM−ROM methodologies underscore their potential applicability in determining optimal flow control configurations within the reduced-order modeling framework. These methodologies provide efficient alternatives for analyzing complex flow fields and have the potential to significantly reduce computational costs in aerodynamic design and optimization processes.

Author Contributions: Conceptualization, A.C. and E.D.C.; methodology, A.C.; software, E.D.C.; validation, A.C., M.C. and L.d.L.; formal analysis, A.C., M.C. and L.d.L.; investigation, A.C., M.C. and L.d.L.; resources, M.C. and L.d.L.; data curation, A.C., E.D.C., M.C. and L.d.L.; writing—original draft preparation, A.C.; writing—review and editing, A.C., E.D.C., M.C. and L.d.L.; visualization, A.C., E.D.C., M.C. and L.d.L.; supervision, M.C. and L.d.L.; project administration, M.C. and L.d.L. All authors have read and agreed to the published version of the manuscript.

Funding: This research received no external funding.

Institutional Review Board Statement: Not applicable.

Data Availability Statement: The data presented in this study are available upon request from the corresponding author.

Conflicts of Interest: The authors declare no conflicts of interest.

Abbreviations

The following abbreviations are used in this manuscript:

URANS	unsteady Reynolds-averaged Navier–Stokes
FOM	full-order model
ROM	reduced-order model
POD	proper orthogonal decomposition
CNM	cluster-based network model

References

1. Couto, N.; Bergada, J.M. Aerodynamic Efficiency Improvement on a NACA-8412 Airfoil via Active Flow Control Implementation. *Appl. Sci.* **2022**, *12*, 4269. [CrossRef]
2. Chiatto, M.; Palumbo, A.; de Luca, L. Design approach to predict synthetic jet formation and resonance amplifications. *Exp. Therm. Fluid Sci.* **2019**, *107*, 79–87. [CrossRef]
3. Ceglia, G.; Chiatto, M.; Greco, C.S.; De Gregorio, F.; Cardone, G.; de Luca, L. Active control of separated flow over 2D back-facing ramp by an array of finite-span slotted synthetic jets. *Exp. Therm. Fluid Sci.* **2021**, *129*, 110475. [CrossRef]
4. Girfoglio, M.; Greco, C.; Chiatto, M.; de Luca, L. Modelling of efficiency of synthetic jet actuators. *Sens. Actuators A Phys.* **2015**, *233*, 512–521. [CrossRef]
5. de Luca, L.; Girfoglio, M.; Chiatto, M.; Coppola, G. Scaling properties of resonant cavities driven by piezo-electric actuators. *Sens. Actuators A Phys.* **2016**, *247*, 465–474. [CrossRef]
6. Tousi, N.; Coma, M.; Bergadà, J.; Pons-Prats, J.; Mellibovsky, F.; Bugeda, G. Active flow control optimisation on SD7003 airfoil at pre and post-stall angles of attack using synthetic jets. *Appl. Math. Model.* **2021**, *98*, 435–464. [CrossRef]
7. Tousi, N.; Bergadà, J.; Mellibovsky, F. Large Eddy Simulation of optimal Synthetic Jet Actuation on a SD7003 airfoil in post-stall conditions. *Aerosp. Sci. Technol.* **2022**, *127*, 107679. [CrossRef]
8. Lahey, R.; Cheng, L.; Drew, D.; Flaherty, J. The effect of virtual mass on the numerical stability of accelerating two-phase flows. *Int. J. Multiph. Flow* **1980**, *6*, 281–294. [CrossRef]
9. Li, Y.; Cui, W.; Jia, Q.; Li, Q.; Yang, Z.; Morzyński, M.; Noack, B.R. Explorative gradient method for active drag reduction of the fluidic pinball and slanted Ahmed body. *J. Fluid Mech.* **2022**, *932*, A7. [CrossRef]
10. Reumschüssel, J.M.; von Saldern, J.G.R.; Li, Y.; Paschereit, C.O.; Orchini, A. Gradient-Free Optimization in Thermoacoustics: Application to a Low-Order Model. *J. Eng. Gas Turbines Power* **2022**, *144*, 051004. [CrossRef]
11. Wang, T.; Yang, Y.; Chen, X.; Li, P.; Iollo, A.; Maceda, G.Y.C.; Noack, B.R. Topologically assisted optimization for rotor design. *Phys. Fluids* **2023**, *35*, 055105. [CrossRef]
12. Taira, K.; Brunton, S.L.; Dawson, S.T.M.; Rowley, C.W.; Colonius, T.; McKeon, B.J.; Schmidt, O.T.; Gordeyev, S.; Theofilis, V.; Ukeiley, L.S. Modal Analysis of Fluid Flows: An Overview. *AIAA J.* **2017**, *55*, 4013–4041. [CrossRef]
13. Rowley, C.W.; Dawson, S.T. Model Reduction for Flow Analysis and Control. *Annu. Rev. Fluid Mech.* **2017**, *49*, 387–417. [CrossRef]
14. Schmid, P. Dynamic Mode Decomposition and Its Variants. *Annu. Rev. Fluid Mech.* **2022**, *54*, 225–254. [CrossRef]
15. Herrmann, B.; Baddoo, P.J.; Semaan, R.; Brunton, S.L.; McKeon, B.J. Data-driven resolvent analysis. *J. Fluid Mech.* **2021**, *918*, A10. [CrossRef]
16. Schmidt, S.; Oberleithner, K. Global modes of variable-viscosity two-phase swirling flows and their triadic resonance. *J. Fluid Mech.* **2023**, *955*, A24. [CrossRef]

17. Stabile, G.; Hijazi, S.; Mola, A.; Lorenzi, S.; Rozza, G. POD-Galerkin reduced order methods for CFD using Finite Volume Discretisation: Vortex shedding around a circular cylinder. *Commun. Appl. Ind. Math.* **2017**, *8*, 210–236. [CrossRef]
18. Noack, B.R.; Afanasiev, K.; Morzyński, M.; Tadmor, G.; Thiele, F. A hierarchy of low-dimensional models for the transient and post-transient cylinder wake. *J. Fluid Mech.* **2003**, *497*, 335–363. [CrossRef]
19. Noack, B.; Morzynski, M.; Tadmor, G. *Reduced-Order Modelling for Flow Control*; Springer: Vienna, Austria, 2011. [CrossRef]
20. Taira, K.; Hemati, M.S.; Brunton, S.L.; Sun, Y.; Duraisamy, K.; Bagheri, S.; Dawson, S.T.M.; Yeh, C. Modal Analysis of Fluid Flows: Applications and Outlook. *AIAA J.* **2020**, *58*, 998–1022. [CrossRef]
21. Racca, A.; Magri, L. Data-driven prediction and control of extreme events in a chaotic flow. *Phys. Rev. Fluids* **2022**, *7*, 104402. [CrossRef]
22. Fernex, D.; Noack, B.R.; Semaan, R. Cluster-based network modeling—From snapshots to complex dynamical systems. *Sci. Adv.* **2021**, *7*, eabf5006. [CrossRef] [PubMed]
23. Hou, C.; Deng, N.; Noack, B.R. Trajectory-optimized cluster-based network model for the sphere wake. *Phys. Fluids* **2022**, *34*. [CrossRef]
24. Colanera, A.; Reumschüssel, J.M.; Beuth, J.P.; Chiatto, M.; de Luca, L.; Oberleithner, K. Extended Cluster-Based Network Modeling for Coherent Structures in Turbulent Flows. 2023. Available online: https://www.researchsquare.com/article/rs-3788637/v1 (accessed on 23 January 2024). [CrossRef]
25. Wang, X.; Deng, N.; Maceda, G.Y.C.; Noack, B.R. Cluster-based control for net drag reduction of the fluidic pinball. *Phys. Fluids* **2023**, *35*, 023601. [CrossRef]
26. Brunton, S.L.; Noack, B.R.; Koumoutsakos, P. Machine Learning for Fluid Mechanics. *Annu. Rev. Fluid Mech.* **2020**, *52*, 477–508. [CrossRef]
27. Saetta, E.; Tognaccini, R.; Iaccarino, G. Machine Learning to Predict Aerodynamic Stall. *Int. J. Comput. Fluid Dyn.* **2022**, *36*, 641–654. [CrossRef]
28. Stabile, G.; Rozza, G. Finite volume POD-Galerkin stabilised reduced order methods for the parametrised incompressible Navier-Stokes equations. *Comput. Fluids* **2018**, *173*, 273–284. [CrossRef]
29. Kaiser, E.; Noack, B.R.; Cordier, L.; Spohn, A.; Segond, M.; Abel, M.; Daviller, G.; Östh, J.; Krajnović, S.; Niven, R.K.; et al. Cluster-based reduced-order modelling of a mixing layer. *J. Fluid Mech.* **2014**, *754*, 365–414. [CrossRef]
30. Li, H.; Fernex, D.; Semaan, R.; Tan, J.; Morzyński, M.; Noack, B.R. Cluster-based network model. *J. Fluid Mech.* **2021**, *906*, A21. [CrossRef]
31. Di Costanzo, E.; Colanera, A.; Chiatto, M.; de Luca, L. Cluster-based network reduced order modeling for flow fields around airfoil profiles. In Proceedings of the AIAA AVIATION 2023 Forum, San Diego, CA, USA, 12–16 June 2023; American Institute of Aeronautics and Astronautics: Reston, VA, USA, 2023. [CrossRef]
32. Hijazi, S.; Stabile, G.; Mola, A.; Rozza, G. Data-driven POD-Galerkin reduced order model for turbulent flows. *J. Comput. Phys.* **2020**, *416*, 109513. [CrossRef]
33. Ballarin, F.; Manzoni, A.; Quarteroni, A.; Rozza, G. Supremizer stabilization of POD-Galerkin approximation of parametrized steady incompressible Navier-Stokes equations. *Int. J. Numer. Methods Eng.* **2015**, *102*, 1136–1161. [CrossRef]
34. MacQueen, J.B. Some methods for classification and analysis of multivariate observations **1967**, *1*, 281–297.
35. Lloyd, S.P. Least Squares Quantization in PCM. *IEEE Trans. Inf. Theory* **1982**, *28*, 129–137. [CrossRef]
36. Protas, B.; Noack, B.R.; Östh, J. Optimal nonlinear eddy viscosity in Galerkin models of turbulent flows. *J. Fluid Mech.* **2015**, *766*, 337–367. [CrossRef]
37. Deng, N.; Noack, B.R.; Morzyński, M.; Pastur, L.R. Cluster-based hierarchical network model of the fluidic pinball–Cartographing transient and post-transient, multi-frequency, multi-attractor behaviour. *J. Fluid Mech.* **2022**, *934*, A24. [CrossRef]

actuators

Article

Improvement in the Accuracy and Efficiency of Smoothed Particle Hydrodynamics: Point Generation and Adaptive Particle Refinement/Coarsening Algorithms

Jun Zhang [1], Yanchao Ding [2], Wei Wu [1], Wenjie Li [1], Zhaoming Zhang [1] and Yanmei Jiao [3,*]

1 Key Laboratory of Unsteady Aerodynamics and Flow Control, Ministry of Industry and Information Technology, Nanjing University of Aeronautics and Astronautics, Nanjing 210016, China; zhangjunrdf@nuaa.edu.cn (J.Z.); 19850078978@163.com (W.W.); jay2000@nuaa.edu.cn (W.L.); zzm603nuaa@163.com (Z.Z.)
2 China Ship Scientific Research Center, Wuxi 210084, China; dyc1119@cssrc.com.cn
3 School of Physical and Mathematical Sciences, Nanjing Tech University, Nanjing 211816, China
* Correspondence: jiaoym@njtech.edu.cn

Abstract: An adaptive particle refinement (APR) algorithm has been developed for the smoothed particle hydrodynamics (SPH) method to augment the resolution of the region of interest to achieve high accuracy and simultaneously reduce the cost of computational resources. It is widely applied in the field of fluid-controlling problems involving large interface deformations, such as the two-phase flow and fluid–structure interaction because this algorithm can capture the interface with high accuracy. Nonetheless, existing APR algorithms widely encounter computational dispersion issues at the interface of regions of different particle resolutions. Moreover, traditional shifting algorithms applied in the APR processes also have difficulties in dealing with particles with different smooth lengths. In this work, an algorithm for fast particle generation was first developed based on the accelerated ray method, which accelerates the discretization of the flow field into particles. Then, a dynamic refinement/coarsening algorithm based on the APR algorithm is proposed to solve the computational dispersion problem that occurs at the refinement/coarsening interfaces. In addition, the shifting algorithm was improved in this work to ensure the particles are always well distributed during numerical calculations and, thus, can efficiently facilitate the adaptive particle refinement/coarsening processes. Comparative analysis indicates that the robust algorithms developed for the SPH method in this work can lead to more precise and reasonable flow fields compared with the conventional SPH adaptive methods.

Keywords: smoothed particle hydrodynamics; point generation algorithm; adaptive particle refinement; dynamic refinement/coarsening; shifting algorithm

Citation: Zhang, J.; Ding, Y.; Wu, W.; Li, W.; Zhang, Z.; Jiao, Y. Improvement in the Accuracy and Efficiency of Smoothed Particle Hydrodynamics: Point Generation and Adaptive Particle Refinement/Coarsening Algorithms. *Actuators* **2024**, *13*, 174. https://doi.org/10.3390/act13050174

Academic Editor: Luigi de Luca

Received: 27 March 2024
Revised: 23 April 2024
Accepted: 1 May 2024
Published: 5 May 2024

1. Introduction

Numerous phenomena in hydrodynamic engineering, such as structure entry [1], breakwater wave dissipation [2], ship navigation [3], tidal power generation [4], etc., are usually accompanied by the large deformation of the free liquid surface and dynamic boundaries, etc. The traditional grid-based computational fluid dynamics (CFD) method can encounter issues like mesh entanglement and deformation when addressing these complexities. Smoothed particle hydrodynamics (SPH), as a kind of Lagrangian meshless method, has a natural advantage in dealing with these fluid-controlling problems by making up for the deficiencies of the mesh methods. However, uniformly distributed particles are conventionally used by SPH in numerical simulations. It, thus, generates a large number of spatial particles when the computational domain is large, which takes up a lot of the computer's memory and consumes a lot of computational time. To overcome this problem, two main types of adaptive techniques are proposed. The first one, normally referred to as the variable smooth length technique, is to set different lengths between two neighboring particles to make the particle spacing increase smoothly from the boundary to the core region of the computational domain [5]. It is similar to the practice of refining the grids of the region of interest in the grid method [6]. The variable smooth length technique can distinctly improve the accuracy of the solution. The second one is to apply dynamic refinement and coarsening algorithms to control the resolution of the distributed particles to achieve a high accuracy in the calculation of the computational domain [7].

Inspired by research in astrophysics, Monaghan introduced the concept of the spatially varied resolution, which adjusts the smoothing length according to the number of neighboring particles around the target particle [8]. Then, Nelson and Papaloizou proposed applying this variable smooth length technique to improve the SPH method [9]. In particular, particle spacing is determined by the smoothing length h, and each particle has its adaptive smoothing length h_i. By incorporating the gradient of the smoothing length in the momentum equation, an improved SPH model with variable smoothing lengths is derived, which is validated by solving the problem of a one-dimensional shock tube. Another study also improved the SPH model by modifying the SPH momentum equation so that the errors arising from the interaction of particles with different smooth lengths could be reasonably reduced. The rational simulation results of a wedge entering the water were achieved by utilizing this improved SPH model with particles densely arranged around the wedge [10]. However, this SPH method can still cause strong instability in the flow field [11]. It is brought about by the nature of the Lagrange particles in the SPH method, which inevitably induces the movement of particles over time and, thus, causes large gradients of smooth lengths between adjacent particles, ultimately leading to large errors in the kernel approximation. To avoid this problem, some studies have proposed dynamically adjusting the smooth length according to the density [12,13] or the velocity [14] of the target particles. Qiang and Gao proposed an iterative solution to the problem of fully varied smooth lengths to further improve the accuracy of physical interpolation calculations [15]. However, the efficiency of the simulation calculation is greatly affected by the need to iteratively solve the smooth lengths of the particles at each step of the calculations. To improve computational efficiency, the smooth length of the target particle was reasonably adjusted according to the variation in the number and the average smooth length of the neighboring particles [16], thus avoiding the time-consuming iterative solution method.

Some other scholars have utilized dynamic particle refinement and coarsening techniques to enhance the adaptive SPH method. Feldman and Bonet proposed a particle refinement technique that splits a parent particle into several child particles and assigns the corresponding physical properties to the child particles according to the principles of mass and energy conservation [17]. Although the dynamic particle refinement technique allows for large gradients of smooth lengths between particles, the number of particles increases significantly, thus decreasing the efficiency of numerical simulation as it proceeds. Therefore, Vacondio et al. proposed an adaptive SPH method to increase the efficiency of the numerical simulation, which dynamically distributes particles by splitting them at the target region and merging them outside the target region [18]. Wang et al. also proposed a general dynamic particle refinement strategy and, importantly, a new particle refinement criterion for two-phase flow to capture the interface more precisely in fluid–structure interaction (FSI) problems. The calculation results of the wedge entering water showed that the proposed dynamic particle refinement strategy can achieve better accuracy and efficiency [19]. It is notable that the essential disadvantage of the dynamic particle refinement technique is its high cost in computational resources, as the splitting and the merging of particles need to be performed again at the beginning of each computation step. In addition, the merging algorithm is very inefficient, and the conditions for its application are very demanding, making it difficult to be widely adopted. For this reason, Barcarolo et al. avoided using the complex merging process by keeping the split parent particles in the refinement region passively following the fluid and activating them when they left the refinement region [20]. To deal with the instability caused by the interaction of particles with different smooth lengths, Chiron et al. proposed the adaptive particle refinement (APR) technique by referring to the adaptive mesh refinement (AMR) technique from the mesh method [7]. This method makes particles of the same smooth length interact with each other only within its conservation area of uniform spatial resolution. Information between two conservation areas of different spatial resolutions can be exchanged only through the boundary of the conservation area to achieve the coupling of two different resolution regions.

In this work, systematical investigations were carried out based on the dynamic particle refinement and coarsening technique introduced above. Firstly, this work developed an initial point generation algorithm and then applied the accelerated ray method to quickly discretize the particles in the two-dimensional calculation domain. This work innovatively proposes a scheme that dynamically splits parent particles and removes child particles according to the target region based on the adaptive SPH technique proposed by Chiron [3]. We also improved the particle shifting algorithm so that it could efficiently facilitate the adaptive particle refinement and coarsening processes. Finally, the significantly improved adaptive SPH algorithm was verified through two reported cases. This work not only realizes the acceleration of the discretization of the flow field into particles through the accelerated ray method but also solves the computational dispersion problem occurring at the refinement/coarsening interfaces. In addition, the shifting algorithm is simultaneously improved in this work to significantly facilitate the adaptive particle refinement/coarsening processes.

2. Mathematical Model

In the field of computational fluid dynamics, it can be reasonably assumed that the fluid is weakly compressible. Therefore, the incompressible flow can be simulated using the weakly compressible Navier–Stokes equations [21], the Lagrangian form of which is given as follows:

$$\begin{cases} \frac{d\rho}{dt} = -\rho\nabla\cdot\boldsymbol{v} \\ \frac{d\boldsymbol{v}}{dt} = -\frac{\nabla \mathrm{P}}{\rho} + \upsilon\nabla^2\boldsymbol{v} + \boldsymbol{g} \\ \frac{d\boldsymbol{r}}{dt} = \boldsymbol{v} \\ \mathrm{P} = f(\rho) \end{cases} \tag{1}$$

where υ, ρ, $\boldsymbol{v}$, P, and $\boldsymbol{r}$ denote the kinematic viscosity, the density, the velocity, the pressure, and the position of the investigated fluid, respectively. In addition, the Tait equation [22,23] is often used to establish the relationship between pressure and density when solving hydrodynamic problems, which is provided as follows:

$$\mathrm{P} = \frac{\rho_0 c_s^2}{\gamma}\left(\left(\frac{\rho}{\rho_0}\right)^{\gamma} - 1\right) \tag{2}$$

where ρ_0 is the density of the fluid when P = 0 and c_s is the artificial speed of sound. γ is a constant that is typically set to 7 in the hydrodynamic simulations [24]. To keep the compressibility of the fluid within 1% [25], the artificial speed of sound must be set to at least 10 times the value of $u_{\max}$.

As for the SPH method, a convolution integral on the domain is used to interpolate given physical quantities based on kernel functions. The kernel approximation $\langle f(\boldsymbol{r})\rangle$ of the field function $f(\boldsymbol{r})$ at a certain spatial position r can be expressed as follows [26]:

$$\langle f(\boldsymbol{r})\rangle = \int_{\Omega} f(\boldsymbol{r}^*)W(\boldsymbol{r}-\boldsymbol{r}^*,h)dV_{\boldsymbol{r}^*} \tag{3}$$

where $W(\boldsymbol{r}-\boldsymbol{r}^*,h)$ and h represent the kernel function and the smooth length. Ω is the support domain of the kernel function, which is determined by the smooth length. Since the SPH method discretizes the flow field into particles, each of which carries the physical properties of the fluid, it is not technically possible to use Equation (3) with continuous integration to approximate the field function. Thus, the SPH scheme converts the continuous integral into a discrete summation approximation. Then, Equation (3) can be represented by the kernel particle approximation as follows:

$$\langle f(\boldsymbol{r}_i)\rangle = \sum_{j=1}^{N} f(\boldsymbol{r}_j)W(\boldsymbol{r}_i - \boldsymbol{r}_j, h)V_j \tag{4}$$

where N is the total number of particles contained in the support domain of particle i, and V_j is the virtual volume of the particle j. The smooth kernel function used in this work is the Wenland kernel function [27], with a ratio of 4 for the smooth length h compared to the particle spacing dx. There are approximately 50 particles in the kernel support domain in the two-dimensional case.

Therefore, the control equations in Equation (3) can be converted into the ones in the form of a weakly compressible SPH as follows:

$$\begin{cases} \frac{d\rho_i}{dt} = -\rho_i \sum\limits_{j=1}^{N} (\boldsymbol{v}_i - \boldsymbol{v}_j) \cdot \nabla_i W(\boldsymbol{r}_i - \boldsymbol{r}_j, h)\frac{m_j}{\rho_j} \\ \frac{d\boldsymbol{v}_i}{dt} = \frac{1}{\rho_i} \sum\limits_{j=1}^{N} (\mathrm{P}_i + \mathrm{P}_j)\nabla_i W(\boldsymbol{r}_i - \boldsymbol{r}_j, h)\frac{m_j}{\rho_j} + \boldsymbol{g} \\ \frac{d\boldsymbol{r}_i}{dt} = \boldsymbol{v}_i \\ \mathrm{P}_i = \frac{\rho_0 c_0^2}{\gamma}\left(\left(\frac{\rho_i}{\rho_0}\right)^{\gamma} - 1\right) \end{cases} \tag{5}$$

where m is the mass of the particle. The SPH scheme in antisymmetric and symmetric forms is used for the continuity and momentum equations, respectively, to reduce the errors arising from discontinuities and to ensure better conservation properties [28].

3. Numerical Methodology

3.1. Initial Point Generation

In the two-dimensional case, the generation of boundary points requires first parameterizing the physical boundary curve, which must be of first-order continuity and have good locality to ensure the smoothness of the curve, and changes in the position of a single point only affect the shape of the curve around the control point [29]. In addition, to completely represent the shape of the physical boundary, the spline interpolation function needs to proceed through every node extracted on the physical boundary. Accordingly, this work employs the Catmull–Rom cubic spline for curve parameterization, as depicted below:

$$\begin{aligned} S(u) &= S_{i-1}(-0.5u + u^2 - 0.5u^3) + S_i(1 - 2.5u^2 + 1.5u^3) \\ &+ S_{i+1}(0.5u + 2u^2 - 1.5u^3) + S_{i+2}(-0.5u^2 + 0.5u^3) \end{aligned} \tag{6}$$

where $S(u)$ is the point coordinate, and u is the parametric coordinate ranging from 0 to 1.

Once the parametric equations of the boundary curve are obtained, the discrete points of the boundary curve can be obtained, and the background point system is subsequently generated. To generate the background point system, the fluid domain needs to be determined and completely covered using the background Cartesian points. The coordinates of the background Cartesian points are calculated as follows:

$$\begin{aligned} x_{i,j} &= x_{start} + (i-1)\Delta x, (1 \le i \le N_x) \\ y_{i,j} &= y_{start} + (j-1)\Delta y, (1 \le j \le N_y) \end{aligned} \tag{7}$$

where x_{start} and y_{start} denote the coordinates of the point generated at the beginning, Δx, and Δy represent the spacings of the background Cartesian points, and N_x and N_y are the number of layers of the background Cartesian point in the x and y directions.

In the background point system, it is necessary to determine the spatial position relationship between the background Cartesian points and the boundary point system. If the background Cartesian points are in the computational domain, they are judged as valid points. The ray method is a commonly used judging algorithm in the two-dimensional case [29–31]. Due to the need to perform the judging algorithm through the ray method at least once for each background Cartesian point, when the number of background Cartesian points and boundary nodes is large, the efficiency of determining the positions of the nodes is very low. Therefore, to improve the efficiency of point generation, this work proposes an accelerated ray method, which marks the fluid domain by a set of background grids and defines the attributes of the background grids in advance. The properties of the background Cartesian points are then quickly determined according to the properties of the background grids where the background Cartesian points are located. Taking the boundary of a solid existing within a flow field as an example, the corresponding steps of the accelerated ray method are given as follows.

Firstly, the whole flow field is marked by background grids, as shown in Figure 1a. The properties of the background grids in the flow field are defined as In-D, Out-D, and On-B, which are abbreviations of In-domain, Out-of-domain, and On-boundary, respectively. Secondly, the background Cartesian points are then judged according to the properties of the background grid, as shown in Figure 1a. The properties of the background Cartesian points are set to valid, invalid, and to-be-judged according to the properties of the background grids of In-D, Out-D, and On-B. Finally, as shown in Figure 1b, the background Cartesian points whose initial properties are defined as to-be-judged are judged and defined again as valid or invalid by the ray method.

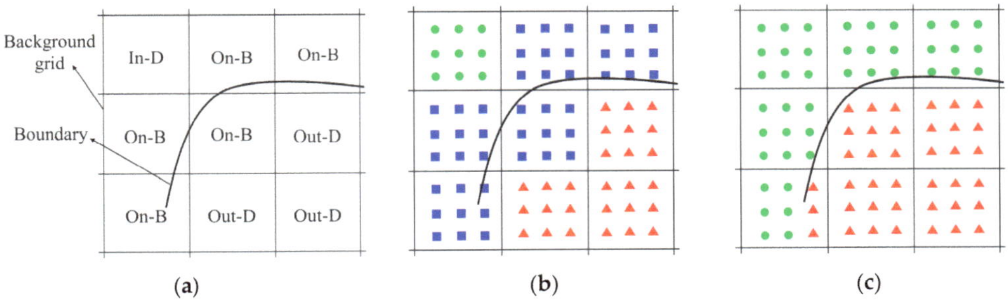

Figure 1. Schematic of the accelerated ray method: (**a**) the background grids and their properties; (**b**) the initial judgment of properties of the background Cartesian points; and (**c**) the final judgment of properties of the background Cartesian points. The abbreviations In-D, Out-D, and On-B represent the In-domain, Out-of-domain, and On-boundary, respectively. The samples •, ▲, and ■ denote valid, invalid, and to-be-judged, respectively.

3.2. SPH Adaptive Algorithm

3.2.1. Particle Refinement Algorithm

Following the initial generation of fluid particles, the refinement of particles within the region of interest needs to be performed to achieve high accuracy. The refinement algorithm applied in this work was improved based on the one reported by Feldman and Bonet [13]. It splits a parent particle requiring refinement into a number of child particles using a pre-defined square refinement pattern, as shown in Figure 2 [32]. Two parameters, the separation ratio ε and the smoothing ratio ς, are defined to determine the distance between two child particles and the ratio of the smooth length of the child particles to that of the parent particles, respectively. Then, the spacing Δx_c and the smooth length h_c of the child particles can be calculated according to Equations (8) and (9):

$$\Delta x_c = \varepsilon \Delta x_p \tag{8}$$

$$h_c = \varsigma h_p \tag{9}$$

where the subscripts c and p represent the child particle and the parent particle, respectively. The ranges of both the separation ratio ε and the smooth length ratio ς are from 0 to 1. When ε and ς equal 0, it means that the distance between two child particles and the smooth length of the child particles is 0. When ε and ς equal 1, it means that the distance between the two child particles and the smooth length of the child particles are the same as those of the parent particle.

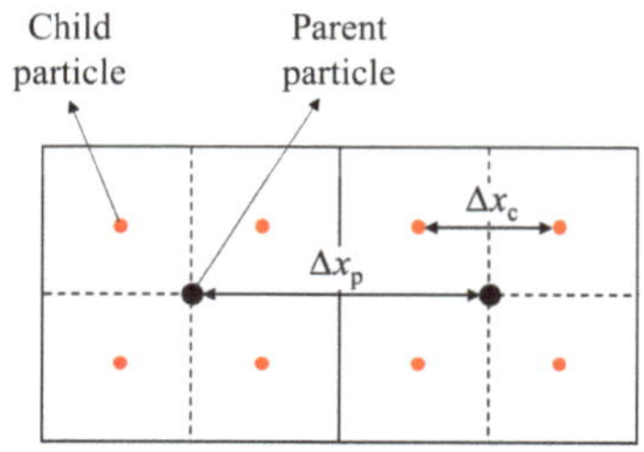

Figure 2. Schematic of square refinement pattern. The black and the red disks represent the parent and the child particles, respectively.

The distribution of the child particles approximates a uniform Cartesian lattice distribution when $\varepsilon = 0.5$. Then, as for the child particles, the ratio of the smooth length to the spacing can be derived as follows:

$$\frac{h_c}{\Delta x_c} = \frac{\varsigma h_p}{0.5\Delta x_p} = 2\varsigma\frac{h_p}{\Delta x_p} \tag{10}$$

This ratio determines the number of particles in the support domain. It is reported that these two refinement parameters need to meet certain restrictive conditions to ensure a minimized error and good stability; that is, ς should change from 0.6 to 0.65 when ε is set as 0.5 [33]. However, this significantly increases the computational load and also runs counter to the goals of the adaptive technique since the number of generated child particles in the support domain is 73% to 120% more than that of the parent particles. This work utilizes a more advanced APR technique, which directly sets two refinement parameters as $\varepsilon = 0.5$ and $\varsigma = 0.5$ to ensure the same number of neighboring child particles before and after refinement.

In addition to the refinement pattern, the physical properties of the child particles also need to be determined. The physical properties of the child particle, including the mass m_c and the velocity u_c, are derived from the parent particle below:

$$m_c = \lambda_c m_p \tag{11}$$

$$u_c = u_p \tag{12}$$

where λ_c is the refinement coefficient related to the refinement pattern. It is set as 0.25 for the square refinement pattern utilized in this work.

The refinement process should satisfy conservations of mass, energy, and momentum, which are provided as Equations (13)–(16), according to Equations (11) and (12).

$$\text{Conservation of mass}: \ m_p = \sum_{c=1}^{4} \lambda_c m_c \tag{13}$$

$$\text{Conservation of energy}: \ \frac{1}{2} m_p \boldsymbol{u}_p{}^2 = \sum_{c=1}^{4} m_c \boldsymbol{u}_c{}^2 \tag{14}$$

$$\text{Conservation of linear momentum}: \ m_{\mathrm{p}}\boldsymbol{u}_{\mathrm{p}} = \sum_{c=1}^{4} m_{\mathrm{c}}\boldsymbol{u}_{\mathrm{c}} \tag{15}$$

$$\text{Conservation of angular momentum}: \ \boldsymbol{r}_{\mathrm{p}} \times m_{\mathrm{p}}\boldsymbol{u}_{\mathrm{p}} = \sum_{c=1}^{4} \boldsymbol{r}_{\mathrm{c}} \times m_{\mathrm{c}}\boldsymbol{u}_{\mathrm{c}} \tag{16}$$

The conservation of mass, energy, and linear momentum are naturally satisfied, and the conservation of angular momentum is automatically satisfied as for the refinement pattern applied in this work. However, the conservation of angular momentum cannot be satisfied for the asymmetric refinement pattern.

3.2.2. Particle Coarsening Algorithm

The efficiency of the SPH adaptive algorithm can be significantly enhanced through the implementation of coarsening algorithms, which dynamically remove the child particles that move out of the target region. Vacondio et al. proposed a particle coarsening algorithm that allowed two child particles in pairs to be merged into a new parent particle [18]. The mass of the new parent particle was the sum of the masses of the two child particles; the position of the new parent particle was located at the center of the mass of the system composed of the two child particles, and the velocity of the new parent particle was obtained as the velocity at the center of mass of the system composed of the two child particles. Although this particle coarsening algorithm can ensure all the conservation equations above, it comes with a substantial computational cost and requires a more intricate implementation process.

Another particle coarsening algorithm is to retain and inactivate the parent particles after refinement so that they passively follow the flow within the refinement zone and activate upon exiting the refinement zone [20]. Child particles are generated once the parent particle enters the refinement zone and are removed after leaving the refinement zone. Therefore, the statuses of particles over the whole calculation domain are divided into two categories: activated and inactivated. A variable γ is defined to indicate these statuses as follows:

$$\gamma = \begin{cases} 1, & \text{Activated} \\ 0, & \text{Inactivated} \end{cases} \tag{17}$$

Introducing γ as a weighting function into the kernel particle approximation.

$$\langle f(\boldsymbol{r}_i)\rangle = \sum_{j=1}^{N} f(\boldsymbol{r}_j)W(\boldsymbol{r}_i - \boldsymbol{r}_j, h)\frac{m_j}{\rho_j}\gamma_j \tag{18}$$

$$\langle \nabla \cdot f(\boldsymbol{r}_i)\rangle = \sum_{j=1}^{N} f(\boldsymbol{r}_j) \cdot \nabla_i W(\boldsymbol{r}_i - \boldsymbol{r}_j, h)\frac{m_j}{\rho_j}\gamma_j \tag{19}$$

In this case, both the parent and the child particles are involved in the summation of the particles, and the field functions of the inactivated particles are also calculated. However, an irrational state, as shown in Figure 3, may occur when the parent particles enter or leave the refinement region [20]. When a parent particle, represented by the red disk, leaves the refinement region with an activated status, its two child particles, represented by the blue disks, are still kept inside the refinement region, also with an activated status. It causes the total mass of the system to increase transiently. To ensure a smooth transition between the two statutes of activated and inactivated, Barcarolo et al. proposed the concept of a transition zone in which γ varies linearly within the transition zone [20]. When a parent particle leaves the refinement zone and crosses the transition zone, the parameter γ increases linearly from 0 to 1. In contrast, the parameter γ decreases linearly from 1 to 0 as the child particle leaves the refinement zone and crosses the transition zone.

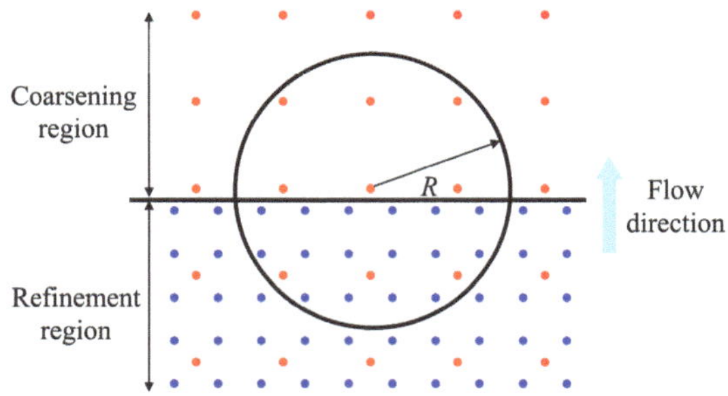

Figure 3. Illustration of the irrational state related to the particle coarsening procedure [20]. The black horizontal line is the boundary between the refinement and the coarsening regions; the red and the blue disks are the parent particles and the child particles, respectively.

Instead of using the linearly varied γ, this work proposes a new refinement-coarsening technique, as shown in Figure 4, to effectively enhance the accuracy and efficiency of the adaptive SPH method in a more effective way. Generally, the parent particles are represented by red disks in the coarsening region to indicate their activated status and by a red circle in the refinement region to indicate their inactivated status. The child particles are represented by blue disks. When a parent particle flows across the boundary from the coarsening region to the refinement region, as shown in Figure 4a, it is initially activated before the boundary at the time of t'. Once the parent particle enters the refinement region at the time of $t' + \Delta t$, it is inactivated and split into four regular child particles simultaneously. The inactivated parent particles only move with the flow field and are not involved in any calculations. When a parent particle flows across the boundary from the refinement region to the coarsening region, as shown in Figure 4b, it is initially inactivated before the boundary at the time of t''. Once the parent particle enters the coarsening region at the time of $t'' + \Delta t$, it is activated again, and the corresponding four child particles are removed from the computational domain simultaneously. The proposed refinement-coarsening technique can effectively avoid sudden changes in the mass of the system and shows great advantage in fulfilling the requirements of all the aforementioned conservation equations.

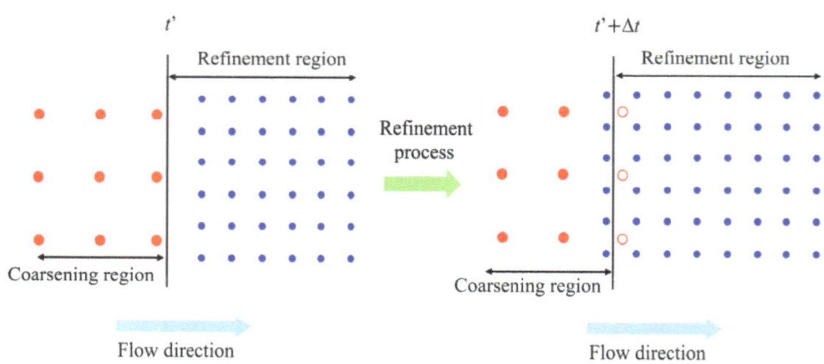

(**a**)

Figure 4. *Cont.*

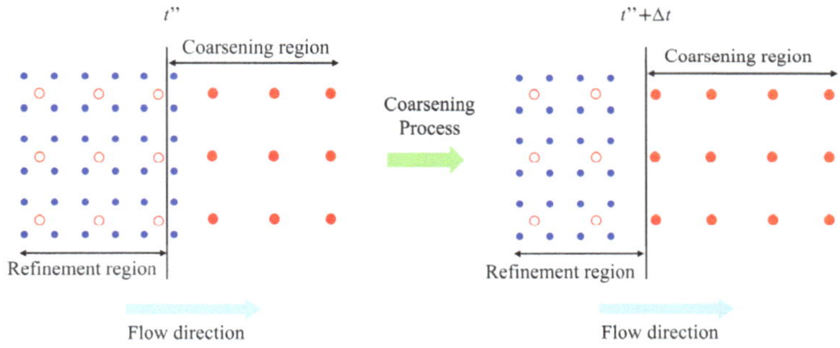

(b)

Figure 4. Illustration of the proposed refinement-coarsening technique: (**a**) the refinement process; (**b**) the coarsening process. The black vertical line is the boundary between the refinement and the coarsening regions; the red disks and the red circles represent the activated and inactivated parent particles, respectively; the blue disks represent the child particles.

3.2.3. APR Technology

According to the algorithm of the APR technique, particles possessing identical smoothing lengths within a region of consistent spatial resolution can interact directly, whereas particles with distinct smoothing lengths in two separate regions cannot directly interact. Information exchange occurs exclusively through the guard area, which serves to connect the two regions of varying resolutions. As shown in Figure 5, the guard area acts as a boundary between two regions with different levels of refinement.

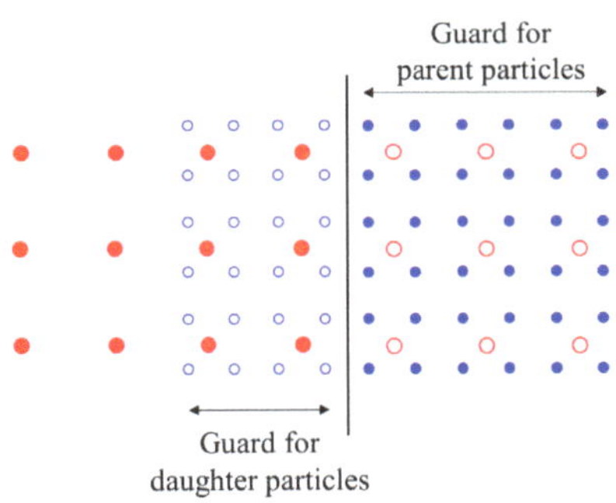

Figure 5. Schematic of the guard area for APR. The black vertical line is the boundary between the refinement and the coarsening regions; the red disks and the red circles represent the activated and inactivated parent particles, respectively; the blue disks and the blue circles represent the activated and inactivated child particles, respectively. All the solid disks are involved in SPH calculations, while all the circles are not. The circles only passively follow the fluid.

In SPH calculations, two neighboring regions of different resolutions indirectly interact with each other through particles in the guard area. The field function f of the particles in the guard area is interpolated from the fluid as follows:

$$\langle f(\boldsymbol{r})\rangle_{\text{guard}} = \frac{\sum\limits_{j=1}^{N} f(\boldsymbol{r}_j)W(\boldsymbol{r}_i-\boldsymbol{r}_j,h)}{\sum\limits_{j=1}^{N} W(\boldsymbol{r}_i-\boldsymbol{r}_j,h)} \quad (20)$$

where $\langle f(\boldsymbol{r})\rangle_{\text{guard}}$ denotes the approximation of the field function f for particles in the guard area. Interpolations are carried out at the start of each time step, and values of the field functions of the neighboring particles are assigned equally to the particles in the guard area through Equation (20). As shown in Figure 5, the field functions of the particles in the guard area are obtained via interpolations of the field functions of the activated particles so that two guard areas with different refinement levels are generated besides the boundary. Since only particles with the same refinement level can be calculated interactively, computational errors induced by the gradient of smooth length are avoided.

Since the traditional shifting algorithm can only deal with the flow where the particles are uniformly distributed, it causes erroneous displacement corrections for particles with different smooth lengths. In this work, the shifting algorithm is improved by adjusting the displacement correction equation as follows:

$$\delta \boldsymbol{r}_s = \begin{cases} -D_F \nabla C_i \Delta t, & i \notin \text{guard area} \\ 0, & i \in \text{guard area} \end{cases} \tag{21}$$

where ∇C_i is the concentration gradient of the particle, D_F is the diffusion coefficient, and Δt is the time step. When the displacement correction is applied to the particles in the guard area, the lack of a complete support domain for the protected particles leads to the incorrect calculation of the concentration gradient of the particle. It causes the particles in the guard area to spread into the fluid, which significantly disrupts the distribution of the particles in the guard area and, thus, affects the calculation of the flow field in the refinement region. Therefore, the displacement correction proposed in this work is only applied to the particles in the non-guard area. In addition, the particles in the guard area only passively move with the fluid and are regenerated at every time step to maintain a uniform distribution.

4. Results and Discussion

4.1. A Viscous Flow Passing around a Rectangular Cylinder

In this section, a case study of a viscous flow passing around a rectangular cylinder, representing a kind of blunt-headed body, is conducted to test the accuracy of the improved adaptive SPH model in this work. The geometry in this case study involves sharp corners, and the conventional SPH model commonly produces stretching instabilities when dealing with flow fields near these sharp corners. Therefore, this work demonstrates the superiority of the proposed adaptive SPH algorithm, first by figuring out the flow field of the viscous flow passing around a rectangular cylinder. The calculation domain set in this work, as shown in Figure 6, is consistent with the one reported in Ref. [34]. The gravity force is ignored, and the Reynolds number Re = ud/v is set as 200. An initial flow field with uniformly distributed particles is generated by applying the accelerated ray method. Along with the proceeding simulation, the whole calculation domain is divided into three regions with three particle resolutions, $d/\Delta x = 25$, $d/\Delta x = 50$, and $d/\Delta x = 100$, through the adaptive SPH method.

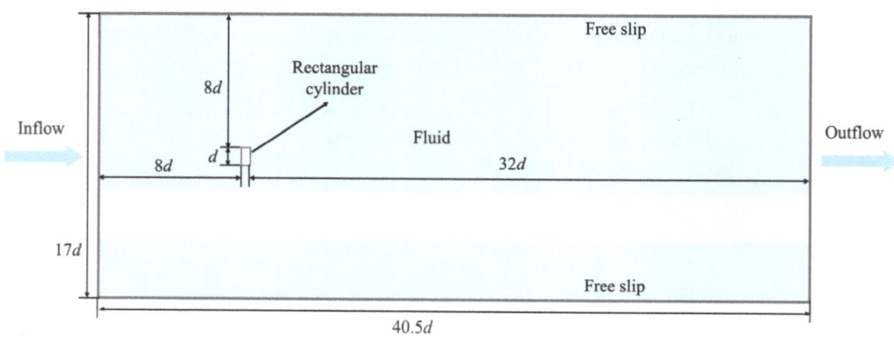

Figure 6. Schematic of the calculation domain of a viscous flow passing around a rectangular cylinder.

Usually, there is a wide zone of negative pressure behind an object when a viscous flow passes around the object. As shown in Figure 7a, the negative pressure induces significant tensile instabilities [35] and unphysical cavitation behind the object [36] by applying the conventional adaptive algorithms of the SPH method. Another problem is that the traditionally applied shifting algorithm of the SPH method normally results in diffusions of particles between regions with different resolutions, which blurs the boundaries between different regions, as shown in Figure 7b.

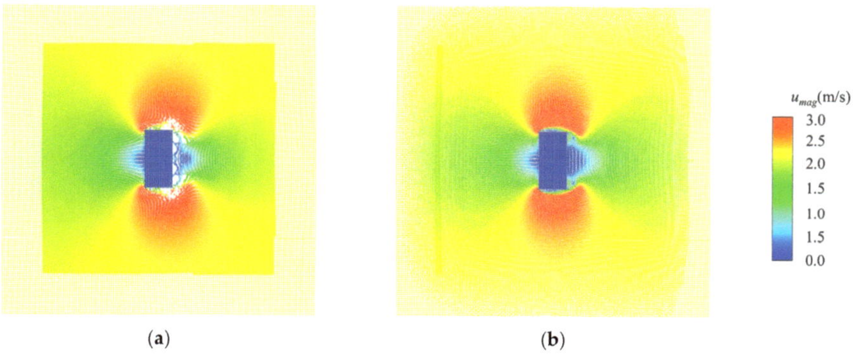

Figure 7. Common problems caused by the traditionally applied SPH method are as follows: (**a**) tensile instability and (**b**) diffusion of particles between regions with different resolutions.

All these problems can be effectively figured out by applying the adaptive algorithms improved in this work, as shown in Figure 8. It displays contour plots of the pressure and the velocity when the lift reaches the maximum value. The boundaries between regions with different resolutions are highlighted by dashed boxes. As shown in this figure, reasonable pressure and velocity are achieved, and they vary rationally and smoothly over regions with different resolutions. Most importantly, no tensile instabilities or diffusions of particles between regions with different resolutions occur.

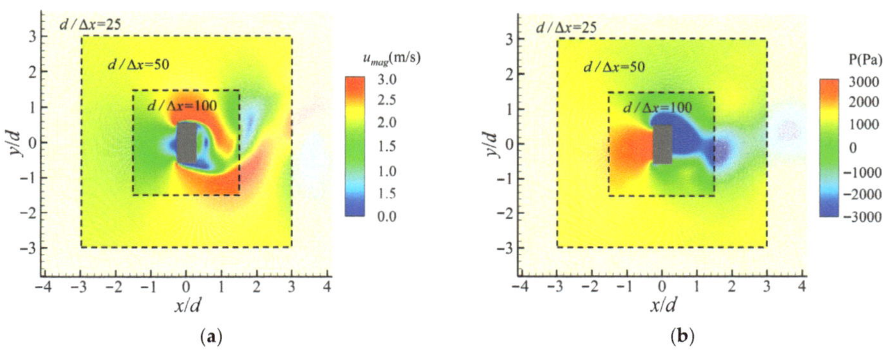

Figure 8. Simulation results of a viscous flow passing around a rectangular cylinder by applying adaptive algorithms improved in this work: (**a**) the contour plot of the pressure and (**b**) the contour plot of velocity. Dashed boxes are added to highlight the boundaries between regions with different resolutions.

The vortex and the flow field of the viscous flow passing around a rectangular cylinder at Re = 200 are calculated and given in Figure 9. It can be seen that after the flow field fully developed, a periodic vortex-shedding phenomenon occurred behind the square column, forming a Karmen vortex street. The corresponding time histories of the lift and drag coefficients were derived and are compared with that achieved using the Finite Volume Method (FVM) [34], as shown in Figure 10. The satisfactory agreements of the curves of the lift and drag coefficients in the periodic regime demonstrate that the improved adaptive algorithms can effectively capture the transportation features in this case study. Furthermore, Table 1 also compares the results of the mean of the drag coefficient $C_{d \cdot mean}$ and the root mean square of the lift coefficient $C_{l \cdot rms}$ calculated from this work and the reported work in Ref. [34]. As can be seen from the table, the errors between this work and the reported work are only 2.39% and 9.25%, corresponding to $C_{d \cdot mean}$ and $C_{l \cdot rms}$. It proves that the improved adaptive SPH model in this work can stably and accurately deal with problems such as the viscous flow passing around blunt-headed bodies.

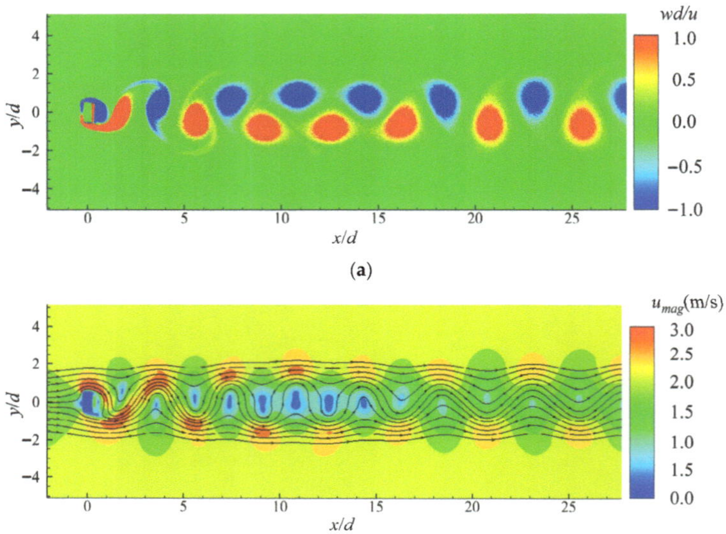

Figure 9. (**a**) The vortex and (**b**) the flow field of the viscous flow passing around a rectangular cylinder at Re = 200.

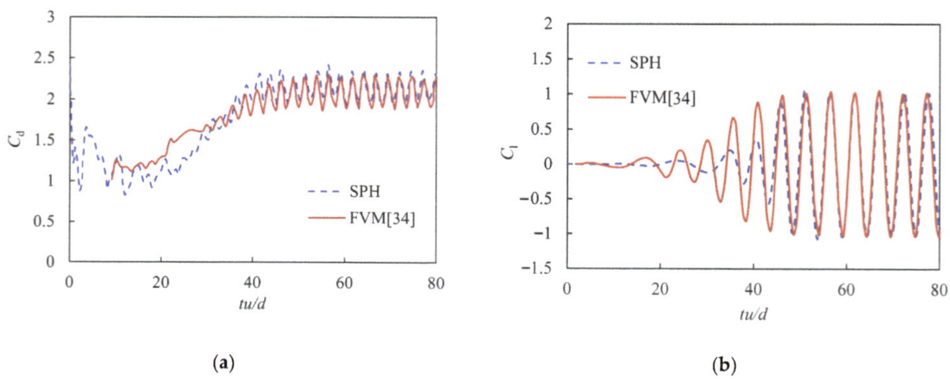

Figure 10. Comparison of the time histories of (**a**) the drag coefficient Cd and (**b**) the lift coefficient Cl between the improved SPH method and the reported FVM.

Table 1. Comparison of the mean of the drag coefficient, $C_{d.mean}$, and the root mean square of the lift coefficient, $C_{l.rms}$, calculated from this work and the reported work in Ref. [34].

	$C_{d.mean}$	$C_{l.rms}$
FVM [34]	2.090	0.724
Improved SPH	2.140	0.657
Error	2.39%	9.25%

4.2. A Free Body Sinking in a Fluid

In this section, a case study of a free body sinking in a fluid with a moving refinement domain was conducted to validate the effectiveness of the dynamic adaptive SPH method developed in this work. As shown in Figure 11, a rectangular block with dimensions of 0.5 m in height H and 1 m in width L was submerged in a fluid-filled tank with dimensions of 10 m in height H_t and 4 m in width L_t. The origin of these coordinates is located at the centroid of the rectangular block and the center of gravity of the rectangular block (x_{co}, y_{co}) is located at (0.25 m, 0 m). The densities of the fluid and the rectangular block are set as ρ_f = 1.0 kg/m^3 and ρ_b = 2.0 kg/m^3, respectively. The gravity g is 1.0 m^2/s, the initial velocity of the rectangular block is 0 m/s, and the kinematic viscosity of the fluid ν is 0.002 m^2/s. In this validation case, the variations in the center of gravity and the roll angle $\theta(z)$ of the rectangular block were analyzed.

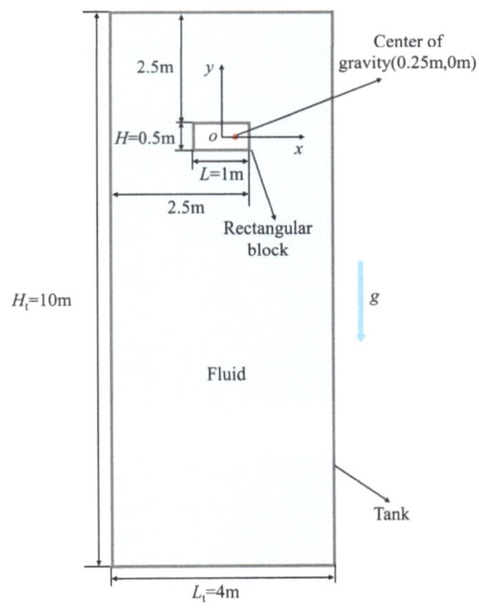

Figure 11. Schematic of a free body sinking in a fluid.

A particle-independent test was carried out first based on three different particle resolutions, $L/\Delta x$ = 50, $L/\Delta x$ = 100, and $L/\Delta x$ = 200, which was also used to verify the capability and assess the accuracy of the adaptive algorithm with the dynamic refinement and coarsening of the particles. Two refinement-coarsening sets were provided, including a refinement region with a particle resolution of $L/\Delta x$ = 100 surrounded by a coarsening region with a particle resolution of $L/\Delta x$ = 50, i.e., $L/\Delta x$ = 100–50 and a refinement region with a particle resolution of $L/\Delta x$ = 200 surrounded by a coarsening region with a particle resolution of $L/\Delta x$ = 100, i.e., $L/\Delta x$ = 200–100.

Figure 12 shows the evolution of the roll angle and the center of gravity with the time from t = 0.0 s to t = 5.0 s. First, the calculations converged gradually as the particle resolution increased from $L/\Delta x$ = 50 to $L/\Delta x$ = 200. Second, the evolution trends obtained from the refinement-coarsening sets showed growing agreement with the ones achieved from the cases of uniform particle resolutions when increasing particle resolutions. Third, the evolution trends obtained from the refinement-coarsening set of $L/\Delta x$ = 200–100 displayed a satisfied consistency with the one achieved from the work reported in Ref. [20]. The corresponding quantitative comparison of the roll angle $\theta(z)$ and the displacement of the center of gravity of the rectangular block, $x_{co}-x_c$, and $y_{co}-y_c$, at t = 5.0 s between $L/\Delta x$ = 200–100 and the Ref. [20] is provided in Table 2. It shows that the errors of $\theta(z)$, $x_{co}-x_c$, and $y_{co}-y_c$ are relatively small at 4.46%, 2.44%, and 2.69%, respectively, indicating that the adaptive algorithm developed in this work can accurately figure out the problems with moving boundaries.

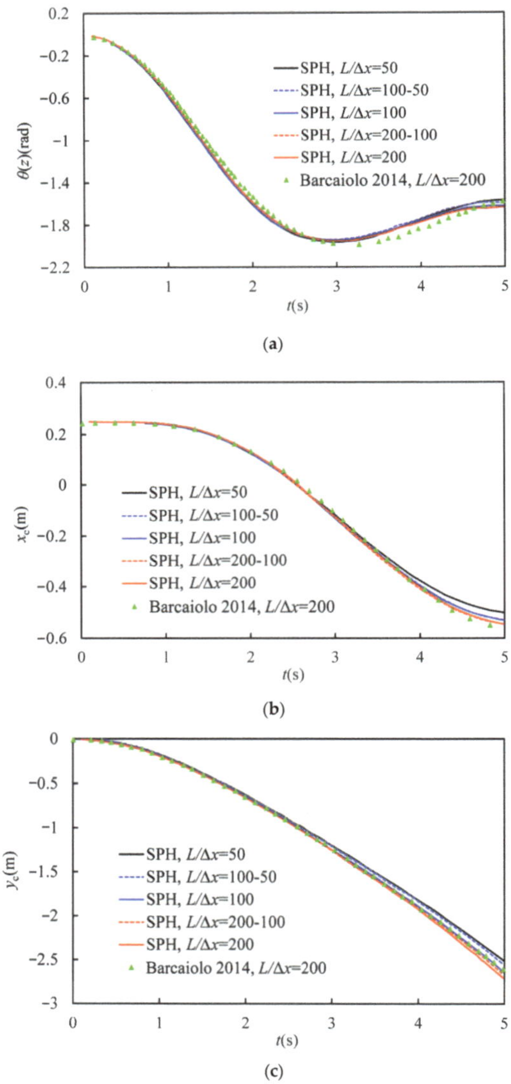

Figure 12. Evolution of selected parameters of the rectangular block sinking in a fluid: (**a**) the roll angle $\theta(z)$, (**b**) the horizontal coordinate of the center of gravity of the rectangular block x_c, and (**c**) the vertical coordinate of the center of gravity of the rectangular block y_c [20].

Table 2. Comparison of the roll angle and the displacement of the center of gravity of the rectangular block at t = 5.0 s between $L/\Delta x$ = 200–100 and the Ref. [20].

	$\theta(z)$ (rad)	$x_{co} - x_c$ (m)	$y_{co} - y_c$ (m)
Exp. [20]	−1.57	0.82	2.60
SPH, $L/\Delta x$ = 200–100	−1.64	0.80	2.67
Error	4.46%	2.44%	2.69%

Figures 13 and 14 illustrate the evolution of the pressure and the velocity when the rectangular block sinks into the fluid, respectively. Since vortexes shed from the rectangular block as it sinks, the flow field at the upper and the lower regions of the rectangular block varies more significantly and spreads more widely than that at the left and right regions. Therefore, this work applies a rectangular refinement region, with the longer edge being placed in the vertical direction in the simulation domain. As depicted in the figures, reasonable contour plots of pressure and velocity were obtained and then distributed rationally and smoothly in the dynamically varied refinement and coarsening regions.

Figure 13. Evolution of the pressure when the rectangular block sinks in the fluid.

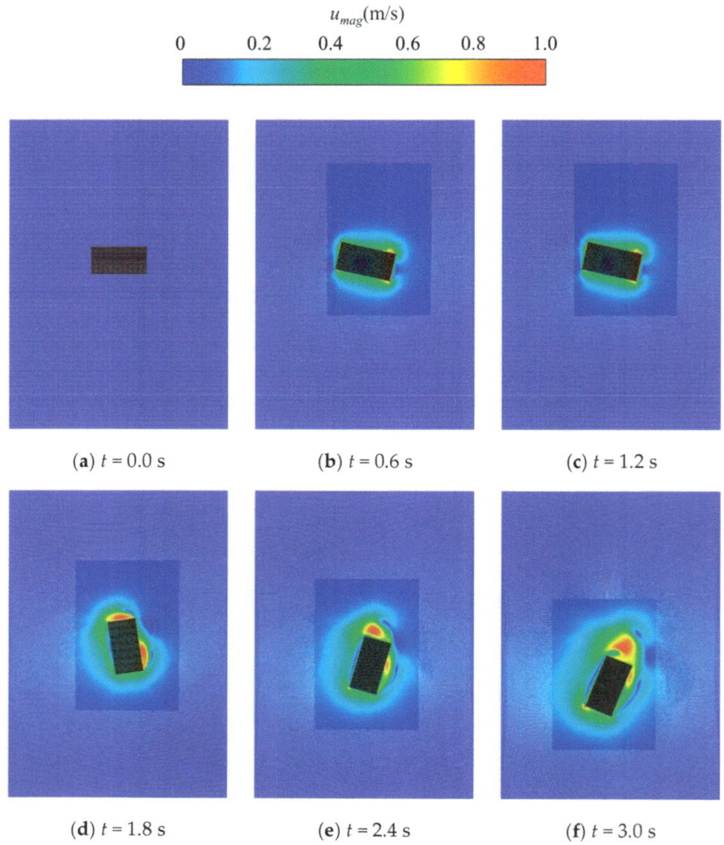

Figure 14. Evolution of the velocity when the rectangular block sinks into the fluid.

In addition, conventional adaptive techniques normally induce unreasonable pressure peaks at the junctions between the refinement and the coarsening regions [37] due to the interaction of particles in different regions with different particle resolutions. It is still impossible to eliminate the discontinuity and distributions of the pressure between two different regions, even with the use of the buffer method proposed by Barcarolo et al. [20]. However, the upgraded adaptive algorithm reestablished in this work improves the dynamic refinement and coarsening of particles using the concept of guard areas to avoid the interaction of particles with different smooth lengths. This novel approach successfully eliminates pressure perturbations at the boundary of different refinement regions, and thus, stable, and continuous pressure and velocity fields are achieved, as shown in Figures 13 and 14.

5. Conclusions

In summary, this work systematically investigates and reestablishes the SPH adaptive algorithm, the details of which are listed below:

(1) A novel point generation algorithm based on the accelerated ray method is proposed, where the boundary is parameterized using Catmull–Rom cubic splines, and background Cartesian points are composed of particles within the flow field. The new point generation algorithm accelerates the discretization of the flow field into particles.

(2) An improved dynamic particle refinement/coarsening algorithm based on the APR technique has been developed to solve the computational dispersion problem at the boundary between regions with different particle resolutions.
(3) The shifting algorithm was improved in this work to ensure the particles are always well distributed during numerical calculations and, thus, efficiently facilitate the adaptive particle refinement/coarsening processes.

Two case studies were conducted to validate the high capabilities of the reestablished SPH adaptive algorithm to deal with the problems of viscous flows passing around blunt-headed bodies and the flows within dynamically moving refinement regions. The comparative analysis between the results obtained from this study and the literature demonstrates that the improved robust SPH adaptive algorithm in this study offers distinct advantages in generating a more precise flow field when compared to traditional SPH adaptive methods.

Author Contributions: Conceptualization, J.Z.; methodology, J.Z. and Y.D.; software, J.Z., Y.D. and Y.J.; validation, Y.D. and Y.J.; investigation, W.W. and W.L.; data curation, W.W. and Z.Z.; writing—original draft preparation, W.W. and Y.J.; writing—review and editing, Z.Z. and W.L.; visualization, W.L.; supervision, Z.Z. All authors have read and agreed to the published version of the manuscript.

Funding: This research was funded by the State Key Laboratory of Hydrodynamics [grant numbers 61422030302]; the National Natural Science Foundation of China [grant numbers 52106246]; and the Natural Science Foundation of Jiangsu Province [grant numbers BK20200687]. And the APC was funded by the State Key Laboratory of Hydrodynamics [grant numbers 61422030302].

Data Availability Statement: Data will be made available on request.

Acknowledgments: This work was supported by the State Key Laboratory of Hydrodynamics [grant numbers 61422030302]; the National Natural Science Foundation of China [grant numbers 52106246]; and the Natural Science Foundation of Jiangsu Province [grant numbers BK20200687].

Conflicts of Interest: The authors declare that they have no known competing financial interests or personal relationships that could have appeared to influence the work reported in this paper.

References

1. Dong, L.; Wei, Z.; Zhou, H.; Yao, B.; Lian, L. Numerical Study on the Water Entry of a Freely Falling Unmanned Aerial-Underwater Vehicle. *J. Mar. Sci. Eng.* **2023**, *11*, 552. [CrossRef]
2. Zang, Z.; Fang, Z.; Qiao, K.; Zhao, L.; Zhou, T. Numerical Study on Wave Dissipation and Mooring Force of a Horizontal Multi-Cylinder Floating Breakwater. *J. Mar. Sci. Eng.* **2024**, *12*, 449. [CrossRef]
3. Díaz Ojeda, H.R.; Oyuela, S.; Sosa, R.; Otero, A.D.; Pérez Arribas, F. Fishing Vessel Bulbous Bow Hydrodynamics—A Numerical Reverse Design Approach. *J. Mar. Sci. Eng.* **2024**, *12*, 436. [CrossRef]
4. Kim, J.W.; Woo, S.B. A numerical approach to the treatment of submerged water exchange processes through the sluice gates of a tidal power plant. *Renew. Energy* **2023**, *219*, 119408. [CrossRef]
5. Børve, S.; Omang, M.; Trulsen, J. Regularized smoothed particle hydrodynamics with improved multi-resolution handling. *J. Comput. Phys.* **2005**, *208*, 345–367. [CrossRef]
6. Li, H.S. SPH Liquid Simulation Method Based on Adaptive Smooth Length. Master's Thesis, Guangdong University of Technology, Guangzhou, China, 2020. [CrossRef]
7. Chiron, L.; Oger, G.; Leffe, M.D.; Touzé, D.L. Analysis and improvements of Adaptive Particle Refinement (APR) through CPU time, accuracy and robustness considerations. *J. Comput. Phys.* **2018**, *354*, 552–575. [CrossRef]
8. Monaghan, J.J. Smoothed particle hydrodynamics. *Annu. Rev. Astron. Astrophys.* **1992**, *30*, 543–574. [CrossRef]
9. Nelson, R.P.; Papaloizou, J.C. Variable smoothing lengths and energy conservation in smoothed particle hydrodynamics. *Mon. Not. R. Astron. Soc.* **1994**, *270*, 1–20. [CrossRef]
10. Oger, G.; Doring, M.; Alessandrini, B.; Ferrant, P. Two-dimensional SPH simulations of wedge water entries. *J. Comput. Phys.* **2006**, *213*, 803–822. [CrossRef]
11. Marsh, A.; Oger, G.; Touzé, D.L.; Guibert, D. Validation of a conservative variable-resolution SPH scheme including ∇h terms. In Proceedings of the 6th international SPHERIC Workshop (SPHERIC 2011), Hambourg, Germany, 8–10 June 2011.
12. Attwood, R.E.; Goodwin, S.P.; Whitworth, A.P. Adaptive smoothing lengths in SPH. *Astron. Astrophys.* **2007**, *464*, 447–450. [CrossRef]
13. Kitsionas, S.; Whitworth, A.P. Smoothed Particle Hydrodynamics with particle splitting, applied to self-gravitating collapse. *Mon. Not. R. Astron. Soc.* **2002**, *330*, 129–136. [CrossRef]

14. Olejnik, M.; Szewc, K.; Pozorski, J. SPH with dynamical smoothing length adjustment based on the local flow kinematics. *J. Comput. Phys.* **2017**, *348*, 23–44. [CrossRef]
15. Qiang, H.F.; Gao, W.R. SPH method with fully variable smoothing lengths and implementation. *Chin. J. Comput. Phys.* **2008**, *5*, 569–575. [CrossRef]
16. Yang, X.F.; Kong, S.C. Adaptive resolution for multiphase smoothed particle hydrodynamics. *Comput. Phys. Commun.* **2019**, *239*, 112–125. [CrossRef]
17. Feldman, J.; Bonet, J. Dynamic refinement and boundary contact forces in SPH with applications in fluid flow problems. *Int. J. Numer. Meth. Eng.* **2007**, *72*, 295–324. [CrossRef]
18. Vacondio, R.; Rogers, B.D.; Stansby, P.K.; Mignosa, P.; Feldman, J. Variable resolution for SPH: A dynamic particle coalescing and splitting scheme. *Comput. Method. Appl. M.* **2013**, *256*, 132–148. [CrossRef]
19. Wang, L.; Xu, F.; Yang, Y.; Wang, J.Y. A dynamic particle refinement strategy in Smoothed Particle Hydrodynamics for Fluid-Structure Interaction problems. *Eng. Anal. Bound. Elem.* **2019**, *100*, 140–149. [CrossRef]
20. Barcarolo, D.A.; Touzé, D.L.; Oger, G.; Vuyst, F.D. Adaptive particle refinement and derefinement applied to the smoothed particle hydrodynamics method. *J. Comput. Phys.* **2014**, *273*, 640–657. [CrossRef]
21. Yang, Q.Z.; Xu, F.; Yang, Y.; Dai, Z.; Wang, J.Y. A GPU-accelerated adaptive particle refinement for multi-phase flow and fluid-structure coupling SPH. *Ocean Eng.* **2023**, *279*, 114514. [CrossRef]
22. Li, L.; Shen, L.; Nguyen, G.D.; El-Zein, A.; Maggi, F. A smoothed particle hydrodynamics framework for modelling multiphase interactions at meso-scale. *Comput. Mech.* **2018**, *62*, 1071–1085. [CrossRef]
23. Monaghan, J.J. Simulating Free Surface Flows with SPH. *J. Comput. Phys.* **1994**, *110*, 399–406. [CrossRef]
24. Ren, J.; Ouyang, J.; Jiang, T. Simulation of complex filling process based on the generalized Newtonian fluid model using a corrected SPH scheme. *Comput. Mech.* **2012**, *49*, 643–665. [CrossRef]
25. Xue, B.; Wang, S.P.; Peng, Y.X.; Zhang, A.M. A novel coupled Riemann SPH–RKPM model for the simulation of weakly compressible fluid–structure interaction problems. *Ocean Eng.* **2022**, *266*, 112447. [CrossRef]
26. Fürstenau, J.P.; Weißenfels, C.; Wriggers, P. Free surface tension in incompressible smoothed particle hydrodynamcis (ISPH). *Comput. Mech.* **2020**, *65*, 487–502. [CrossRef]
27. Jiang, T.; Lu, L.G.; Lu, W.G. The numerical investigation of spreading process of two viscoelastic droplets impact problem by using an improved SPH scheme. *Comput. Mech.* **2014**, *53*, 977–999. [CrossRef]
28. Colagrossi, A.; Antuono, M.; Touzé, D.L. Theoretical considerations on the free-surface role in the smoothed-particle-hydrodynamics model. *Phys. Rev. E* **2009**, *79*, 056701. [CrossRef]
29. Wang, C. Physically Based Modeling and Simulation of 1D Flexible Parts Deformation with Constraints. Master's Thesis, Zhejiang University, Hangzhou, China, 2008.
30. Wang, L.Q. Research on Algorithms of Computational Geometry in GIS. Master's Thesis, Central South University, Changsha, China, 2008.
31. Wang, H.J. 3D Solid Modeling and Boolean Operation Modeling Technique. Master's Thesis, Shandong University of Science and Technology, Qingdao, China, 2007.
32. López, Y.R.; Roose, D. Dynamic refinement for fluid flow simulations with SPH. In Proceedings of the II International Conference on Particle-Based Methods: Fundamentals and Applications, Barcelona, Spain, 26–28 October 2011.
33. López, Y.R.; Roose, D.; Morfa, C.R. Dynamic particle refinement in SPH: Application to free surface flow and non-cohesive soil simulations. *Comput. Mech.* **2013**, *51*, 731–741. [CrossRef]
34. Rossi, E.; Colagrossi, A.; Durante, D.; Graziani, G. Simulating 2D viscous flow around geometries with vertices through the Diffused Vortex Hydrodynamics method. *Comput. Method. Appl. Mech. Eng.* **2016**, *302*, 147–169. [CrossRef]
35. Swegle, J.W.; Hicks, D.L.; Attaway, S.W. Smoothed Particle Hydrodynamics Stability Analysis. *J. Comput. Phys.* **1995**, *116*, 123–134. [CrossRef]
36. Marrone, S.; Colagrossi, A.; Antuono, M.; Colicchio, G.; Graziani, G. An accurate SPH modeling of viscous flows around bodies at low and moderate Reynolds numbers. *J. Comput. Phys.* **2013**, *245*, 456–475. [CrossRef]
37. Barcarolo, D.A. Improvement of the Precision and the Efficiency of the SPH Method: Theoretical and Numerical Study. Ph.D. Thesis, Ecole Centrale de Nantes, Nantes, France, 2013.

actuators

MDPI

Article

Investigation on the Reduced-Order Model for the Hydrofoil of the Blended-Wing-Body Underwater Glider Flow Control with Steady-Stream Suction and Jets Based on the POD Method

Huan Wang, Xiaoxu Du * and Yuli Hu

School of Marine Science and Technology, Northwestern Polytechnical University, Xi'an 710072, China; wanghuan@nwpu.edu.cn (H.W.); zx670821@nwpu.edu.cn (Y.H.)
* Correspondence: nwpudu@163.com

Abstract: The rapid acquisition of flow field characterization information is crucial for closed-loop active flow control. The proper orthogonal decomposition (POD) method is a widely used flow field downscaling modeling method to obtain flow characteristics effectively. Based on the POD method, a flow field reduced-order model (ROM) is constructed in this paper for the flow field control of a hydrofoil of a blended-wing-body underwater glider (BWB-UG) with stabilized suction and blowing forces. Compared with the computational fluid dynamics (CFD) simulation, the computational time required to predict the target flow field using the established POD-ROM is only about 0.1 s, which is significantly less than the CFD simulation time. The average relative error of the predicted surface pressure is not more than 6.9%. These results confirm the accuracy and efficiency of the POD-ROM in reconstructing flow characteristics. The timeliness problem of fast flow field prediction in BWB-UG active flow control is solved by establishing a fast prediction model in an innovative way.

Keywords: proper orthogonal decomposition; active flow control; reduced-order model

Citation: Wang, H.; Du, X.; Hu, Y. Investigation on the Reduced-Order Model for the Hydrofoil of the Blended-Wing-Body Underwater Glider Flow Control with Steady-Stream Suction and Jets Based on the POD Method. *Actuators* **2024**, *13*, 194. https://doi.org/10.3390/act13060194

Academic Editor: Luigi de Luca

Received: 26 March 2024
Revised: 18 May 2024
Accepted: 18 May 2024
Published: 21 May 2024

1. Introduction

The blended-wing-body underwater glider (BWB-UG) is a new type of underwater glider with a flying wing layout. Compared with traditional underwater vehicles, BWB-UGs have advantages such as long range, long operation time, and good economic performance [1]. Various types of underwater gliders have been widely used in fields such as marine resource development and environmental monitoring. The lift–drag ratio is an important factor to measure the performance of underwater gliders. The larger the lift–drag ratio, the greater the range under a single gliding cycle and the higher the gliding efficiency. Therefore, increasing the lift–drag ratio is the key to improving the overall performance of gliders. By optimizing the shape of the underwater glider, the lift–drag ratio can be improved to some extent, but relying solely on shape optimization is limited by flow separation and reduced internal space, which severely limits the detection capability of underwater gliders [2].

Active flow control (AFC) technology is an important method for improving the lift–drag performance of underwater gliders. It can be used to improve the flow field and suppress noise by applying appropriate local perturbations to the flow environment, coupled with the original flow field, to achieve local or global flow changes with low energy consumption [3]. Common AFC techniques include blowing and suction [4], synthetic jets [5], plasma [6], electromagnetic [7], micro-electro-mechanical systems, and smart materials such as adaptive structures [8]. These methods are used in the design and aerodynamic layout of aircraft to significantly improve the maneuverability and flight efficiency. For example, constant-jet active flow control technology is widely used in aeronautical applications such as increasing the lift–drag ratio of wind turbines, controlling the flow separation of axial-pressure gases in aircraft engines, and improving the aerodynamic performance of aircraft wings [9–11]. Based on the published research results, it can be concluded that blowing and suction, synthetic jets, and electromagnetic jets can all be used in the AFC of the BWB-UG. Although plasma, micro-electro-mechanical systems, etc., have not been used in the field of underwater vehicles, they can provide new ideas for further research. Currently, research on AFC techniques is dominated by passive and open-loop active flow control, while the investigation on closed-loop active flow control is still in its infancy. In previous research, a detailed analysis of the flow field and hydrodynamic characteristics of the BWB-UG equipped with steady blowing suction active flow control has been conducted [12]. On this basis, a reduced-order model (ROM) investigation for the hydrofoil of the BWB-UG flow control with steady-stream suction and jets based on the POD method was carried out in this paper. The key to achieving closed-loop AFC is to solve the problem of real-time calculation, and the establishment of a reduced-order model of the flow field is one of the effective methods. That is, a low-order flow model is built to simulate the original high-order model. Solving the low-order model can significantly reduce the dimensionality of the computation, the amount of computation, and the computation time [13]. As a result, rapid acquisition of flow field characteristic information is achieved, so that the control strategy can be dynamically adjusted to achieve the optimum control effect.

The proper orthogonal decomposition (POD) method has become one of the most popular and widely used order reduction methods (ROMs) due to its ability to efficiently reduce the order of a nonlinear system with very small error [14]. The POD method provides the most optimal orthogonal basis, so that the projection of the sample data onto these standard orthogonal basis decreases rapidly. By intercepting the first few orders of modes with larger projections (containing higher energies), an approximate description of higher-order data with fewer modal expansions can be achieved. That is, by extracting the main feature information of the flow field to establish a ROM, the information in the flow field can be reconstructed [15].

Since Lumely [16] applied the POD method to the study of turbulence, scholars have used the POD method for flow field analysis and conducted a large number of studies. Kidambi et al. [17] studied a nonlinear control method based on POD, which effectively achieved the asymptotic adjustment of fluid velocity in a specified spatial domain. Wang et al. [18] developed a low-dimensional ordinary differential model for the control of static N-S equations based on the POD method. Min et al. [19] conducted a modal analysis of parallel double cylinders using two data-driven methods: appropriate orthogonal decomposition (POD) and dynamic mode decomposition (DMD). Sun et al. [20] proposed a ROM of wind turbine blade flow fields, coupling an air–fluid reduced-order model with wind turbine blades for fluid–solid coupling calculations, and it was shown that the ROM has good accuracy and validity. Combining the raw data from wind tunnel experiments, Wang et al. [21] applied POD combined with the interpolation method to predict the wind pressure field. A nonlinear Galerkin method using POD for the low-dimensional modeling of complex hydrodynamic systems was proposed by Kang et al. [22].

It has been shown that AFC technology has excellent potential for application in improving the performance of underwater gliders. However, from the public research literature, some preliminary research results have been achieved on the technical application of open-loop AFC for BWB-UGs, there are relatively few studies on ROMs of the flow field for closed-loop AFC, most of public papers are applied in aerospace and astronautics, and there are very few studies applied in underwater gliders. Therefore, the investigation of the reduced-order model for the hydrofoil of the BWB-UG flow control based on the POD method is of great theoretical significance and application value for solving the important problems in the closed-loop AFC of BWB-UGs.

In this paper, the hydrofoil of a BWB-UG under active flow control is investigated. Section 2 introduces the physical model, the numerical computational model, and POD method. Section 3 deals with the discussion and analysis of the results. In this section, the flow field information both before and after the application of AFC is sampled, the

ROM is calculated based on simulated data and using the POD method, then the flow field prediction is achieved using interpolated predicted modal coefficients. After comparing the results with the CFD simulations, conclusions are drawn in Section 4.

2. Model and Method

2.1. Physical Model

The hydrofoil model selected for this paper comes from the BWB-UG, which is a correction of the NACA0012 standard hydrofoil as shown in Figure 1. It has a chord length of 100 mm and a maximum thickness of 0.12 c at 0.3 c from the leading edge of the hydrofoil, which is slightly thicker near the leading and trailing edges compared to the NACA0012. The effects of parameters such as the angle of attack of the incoming flow (α), the constant suction/jet angle ($\theta_{suction}/\theta_{jet}$), and the velocity ($V_{suction}/V_{jet}$) are mainly considered in the active control of the hydrofoil flow field. The constant suction position is located at 0.7 c from the leading edge of the hydrofoil on the upper surface, with a suction opening width h. The jet position is set at 0.5 c from the leading edge of the hydrofoil on the lower surface, with a jet opening width h. As shown in Figure 2, h is defined as the width of the jet opening along the direction of the chord length with a magnitude of 0.01 c.

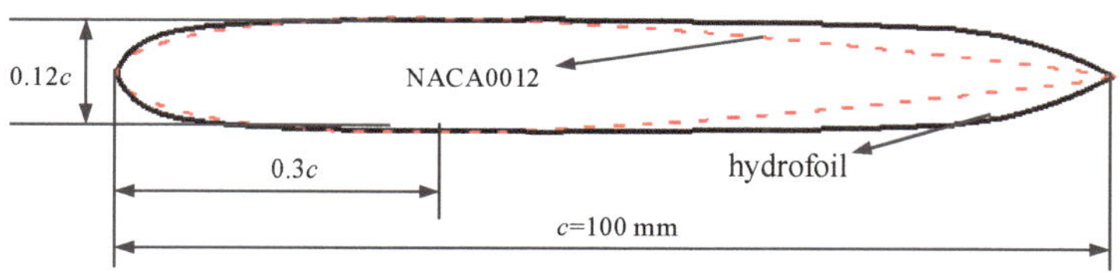

Figure 1. Hydrofoil profile geometry model.

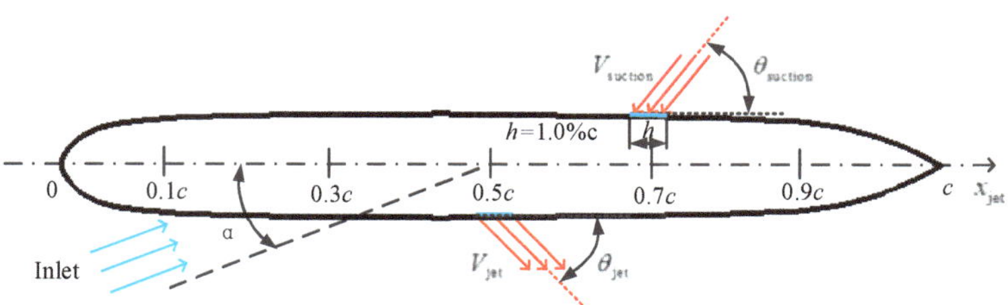

Figure 2. Hydrofoil flow control with steady-stream suction and jets.

2.2. Numerical Sampling Method

In this paper, the flow field in the hydrofoil of a BWB-UG is investigated based on the two-dimensional Reynolds-averaged Navier–Stokes equations (RANS) [23], the fluid medium is incompressible seawater, and the governing equations are as follows:

$$\frac{\partial \rho}{\partial x} + \frac{\partial \rho}{\partial y} = 0 \tag{1}$$

$$\rho\left(\frac{\partial \overline{u}}{\partial t} + \overline{u}\frac{\partial \overline{u}}{\partial x} + \overline{v}\frac{\partial \overline{u}}{\partial y} + \overline{w}\frac{\partial \overline{u}}{\partial z}\right) = \mu\left(\frac{\partial^2 \overline{u}}{\partial x^2} + \frac{\partial^2 \overline{u}}{\partial y^2}\right) + \left(\frac{\partial\left(-\rho\overline{u'u'}\right)}{\partial x} + \frac{\partial\left(-\rho\overline{u'v'}\right)}{\partial y}\right) - \frac{\partial p}{\partial x} \tag{2}$$

$$\rho\left(\frac{\partial \overline{v}}{\partial t} + \overline{u}\frac{\partial \overline{v}}{\partial x} + \overline{v}\frac{\partial \overline{v}}{\partial y}\right) = \mu\left(\frac{\partial^2 \overline{v}}{\partial x^2} + \frac{\partial^2 \overline{v}}{\partial y^2}\right) + \left(\frac{\partial\left(-\rho\overline{v'u'}\right)}{\partial x} + \frac{\partial\left(-\rho\overline{v'v'}\right)}{\partial y}\right) - \frac{\partial p}{\partial y} \tag{3}$$

where p is the pressure; μ is the dynamic viscosity coefficient; $\overline{u}$, $\overline{v}$, and $\overline{w}$ are the time-averaged velocity components of the fluid in the x, y, z directions; and u', v', w' are the velocity fluctuations of the fluid in the x, y, z directions.

A semicircular and rectangular plane region is adopted as the computational domain, considering hydrofoil flow characteristics with AFC. The semicircular section has a radius of 10 c, with the center of the circle located at the trailing edge of the hydrofoil, and the rectangular section is 20 c × 15 c. The 2D structured grid has been meshed. The inlet boundary is set to the inlet velocity, the specified velocity $V_{inlet} = 0.514$ m/s, and the velocity direction is determined by the angle of attack. The jet inlet boundary is also set as a velocity inlet, with the velocity magnitude determined by V_{jet} and $V_{suction}$, and the velocity direction is determined by $\theta_{suction}$ and θ_{jet}. The outlet boundary is set as a pressure outlet, the hydrofoil surface is a non-slip surface, and the specific computational setup and mesh setup are shown in Figures 3 and 4. The enhanced wall treatment is used. The first grid height of the boundary layer is 5×10^{-5} m, the number of boundary layers is 15, and the thickness of the boundary layer is 1.8×10^{-3} m. The gird parameters are set to ensure that the maximum $y^{+} = 0.96$ for the hydrofoil surface.

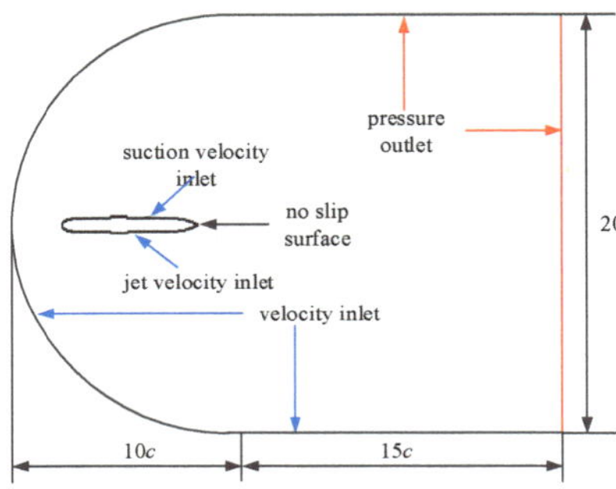

Figure 3. Computational domain and boundary conditions.

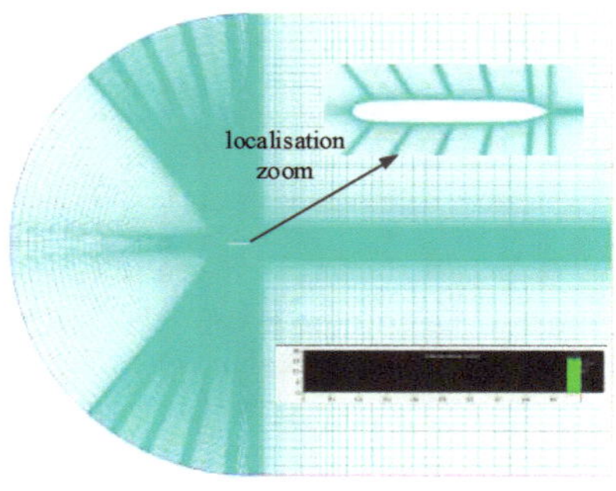

Figure 4. Computational domain meshing.

The SIMPLEC algorithm coupled with pressure and velocity in ANSYS FLUENT was chosen to solve the equations for numerical calculations. The k-ω SST is chosen as the turbulence model, and the transport equations of k and w are expressed as the following [24]:

$$\begin{cases} \frac{\partial(\rho k)}{\partial t} + \frac{\partial(\rho \overline{U}_i k)}{\partial x_i} = P_k + D_k + \frac{\partial}{\partial x_i}\left[(\mu + \sigma_k \mu_t)\frac{\partial k}{\partial x_i}\right] \\ \frac{\partial(\rho \omega)}{\partial t} + \frac{\partial(\rho \overline{U}_i \omega)}{\partial x_i} = P_\omega + D_\omega + \frac{\partial}{\partial x_i}\left[(\mu + \sigma_\omega \mu_t)\frac{\partial \omega}{\partial x_i}\right] \end{cases} \quad (4)$$

where $\overline{U}_i$ is the component of the flow velocity along the direction x_i $(i = 1,2,3)$; μ is the dynamic viscous coefficient of the fluid; σ_k and σ_ω are the modeling coefficients; P_k and P_ω are the generating terms; D_k and D_ω are the dissipative terms; and μ_t is the turbulent viscous coefficient.

The spatial discretization of the variables is chosen to be in the second-order upwind format, the finite volume method is chosen as the solver, and the residual convergence criterion is set to 10^6. The specific fluid medium is set to be incompressible seawater at 20° Celsius with a density of $\rho = 1.024 \times 10^{-3}$ kg/m^3, the dynamic viscosity coefficient is set to be a constant, $\mu = 1.0797 \times 10^{-3}$kg/(ms^{-1}), and the Reynolds number based on the hydrofoil chord length is $Re\_c = 5 \times 10^5$. The calculations are performed using the pressure solver in ANSYS FLUENT.

In order to ensure that the flow field calculations have sufficient computational accuracy and efficiency, grid independence validation is carried out to select the appropriate number of meshes. Constant suction active flow control is imposed on the hydrofoil profile by constantly adjusting the global mesh number. The incoming flow velocity is 0.514 m/s. The suction velocity coefficient is 1. The suction deflection angle is 30°. The angle of attack is 2°/6°/10°. The values of lift coefficients for different mesh sizes are shown in the Table 1.

Table 1. Grid independence validation.

Mesh Number	Lift Coefficient		
	AOA = 2°	AOA = 6°	AOA = 10°
76000	0.2683	0.5895	0.5384
105000	0.2639	0.5886	0.5378
135000	0.2612	0.5861	0.5352
164000	0.2612	0.586	0.535

2.3. POD Method

The core idea of the POD method is to find the optimal standard orthogonal basis functions in the mean-square sense from a set of time-series spatial data, based on the specified information of the sampled flow field data, which usually come from the experimental or numerical simulation results. As a result, fewer orthogonal basis expansions are used to approximate the description of the higher-order data, ultimately enabling the reconstruction and prediction of the flow field [25,26].

Considering the flow field characteristics of the hydrofoil, the incoming flow angle of attack and jet parameters are taken as state parameters, and the snapshot data obtained from the numerical simulation of the flow field are taken as sampling samples. As an example, with the pressure field $\{P(x)\} = \{P_{ij}(x) : 1 \le i \le M, 1 \le j \le N, x \in \mathbf{\Omega}\}$ under the variation of the incoming flow angle of attack α in a certain range, where M is the number of grid nodes, N is the number of appropriately selected samples, and $\mathbf{\Omega}$ is the flow field domain, the pressure field can be expressed as follows:

$$\{P_{ij}(x)\} = \begin{pmatrix} P_{11}(x) & P_{12}(x) & \cdots & P_{1N}(x) \\ P_{21}(x) & P_{22}(x) & \cdots & P_{2N}(x) \\ \vdots & \vdots & \vdots & \vdots \\ P_{M1}(x) & P_{M2}(x) & \cdots & P_{MN}(x) \end{pmatrix} \tag{5}$$

The average of N sets of pressure field snapshot data is denoted as follows:

$$\left\{\overline{P_i(x)}\right\} = \frac{1}{N}\sum_{j=1}^{N} P_{ij}(x), 1 \le i \le M \tag{6}$$

The pulsation value of the pressure field can be expressed as follows:

$$\{P_{ij}{}'(x)\} = P_{ij}(x) - \overline{P_i(x)}, 1 \le i \le M, 1 \le j \le N \tag{7}$$

The key to the POD method is to construct a set of optimal standard orthogonal basis functions $\varphi(x)$ that minimize the error value between the sampled samples and the constructed term in the least-squares sense, so that the projection of the basis functions $\varphi(x)$ onto the sampled samples achieves the maximum value,

$$\begin{cases} \max < \frac{1}{N}\sum\limits_{j=1}^{N}\left|P_{ij}{}'(x),\varphi(x)\right|^2 \\ (\varphi(x),\varphi(x)) = \|\varphi(x)\|^2 = 1 \end{cases} \tag{8}$$

$(\varphi(x),\varphi(x))$ is the inner product and $\|\varphi(x)\|^2$ is the norm over the domain $\mathbf{\Omega}$. $\varphi(x)$ is a set of standard orthogonal bases of the vector set $\{P_{ij}{}'(x)\}$, then

$$\varphi(x) = \sum_{j=1}^{N} a_i P_{ij}{}'(x) \tag{9}$$

The maximum problem of the above equation can be solved by defining a kernel function and an operator between this kernel function and the desired basis, where the kernel function W and the operator R can be defined as follows:

$$W(x,x') = (P_{ij}{}'(x), P_{ij}{}'(x')) \tag{10}$$

$$R\varphi(x) = \int_{\Omega} W(x,x')\varphi(x')\mathrm{d}x' \tag{11}$$

Suppose the kernel function in the above is defined as follows:

$$W(x,x') = \frac{1}{N}\sum_{j=1}^{N} P_{ij}{}'(x)P_{ij}{}'(x') \tag{12}$$

Bring this into Equation (11) and make an inner product of $\varphi(x)$ and $R\varphi(x)$:

$$\begin{aligned} &(\varphi(x),R\varphi(x)) \\ &= \textstyle\int_{\Omega} R\varphi(x)\varphi(x)\mathrm{d}x \\ &= \textstyle\int_{\Omega}\int_{\Omega} W(x,x')\varphi(x')\mathrm{d}x'\varphi(x)\mathrm{d}x \\ &= \frac{1}{N}\sum\limits_{j=1}^{N}\textstyle\int_{\Omega}\int_{\Omega} P_{ij}{}'(x')P_{ij}{}'(x)\varphi(x')\mathrm{d}x'\varphi(x)\mathrm{d}x \\ &= \frac{1}{N}\sum\limits_{j=1}^{N}\left|(P_{ij}{}'(x),\varphi(x))\right|^2 \end{aligned} \tag{13}$$

R is a non-negative symmetric operator. From Equation (12), we can see that the problem of finding the maximum value of Equation (8) is equivalent to the problem of finding the maximum eigenvalue of the operator R:

$$R\varphi(x) = \lambda\varphi(x), \|\varphi(x)\|^2 = 1 \tag{14}$$

Substituting Equations (9), (11) and (12) into Equation (14), we obtain

$$\sum_{j=1}^{N}\left[\sum_{k=1}^{N}\left(\frac{1}{N}\int_{\Omega} P_{ij}{}'(x')P_{ik}{}'(x')\mathrm{d}x'\right)a_k\right]P_{ij}{}'(x) = \sum_{j=1}^{N}\lambda a_j P_{ij}{}'(x) \tag{15}$$

where $P_{ij}{}'(x)$ is a set of linearly independent pressure fields, which can be simplified to the following:

$$\mathbf{C}_{jk}A^l = \lambda A^l \tag{16}$$

The autocorrelation matrix is $C_{jk} = \frac{1}{N}\int_\Omega P_{ij}'(x)P_{ik}'(x)\mathrm{d}x$, and the eigenvector $A^l = \left[a_1^l, a_2^l, \cdots, a_N^l\right]^{\mathrm{T}}$. The matrix C_{jk} is a non-negative Hermitian matrix, usually of low order, so the Jacobi method or singular value decomposition (SVD) can be chosen to solve the eigenvalues and eigenvectors of the matrix.

In this paper, the SVD method is used to solve Equation (16) to obtain the matrix C_{jk} corresponding to the eigenvectors A^l, whose eigenvalues are ordered from smallest to largest as $\lambda_1 \geq \lambda_2 \geq \cdots \geq \lambda_l > 0$, and the standard orthogonal basis function (POD basis) can be obtained as follows:

$$\varphi_l(x) = \sum_{j=1}^{N} a_j^l P_{ij}'(x) \tag{17}$$

Any flow field in the sampled flow field can be projected onto the standard orthogonal basis functions $\varphi(x)$ to obtain the corresponding modal coefficients b_j^l, and the reconstructed flow field can be expressed as the following:

$$P_{POD}^l(x) = \sum_{l=1}^{N} b_j^l \varphi_l(x) \tag{18}$$

With the dimensionality reduction property of the POD method, after solving for the eigenvalues obtained, a suitable truncation order d (modal order) can be selected for dimensionality reduction, where $d << N$, where the magnitude of the eigenvalue represents its ability to capture the generalized flow field data E in the flow field:

$$E = \int \langle P_i'(x), P_i'(x)\rangle \mathrm{d}x = \int W_{ii}(x,x)\mathrm{d}x = \sum_n \lambda_n \tag{19}$$

The flow field is reconstructed by determining the order and modal coefficients corresponding to the modes using the POD method analysis approach. The modal coefficients are determined using an interpolated solution, and an efficient approximation to the original data space is achieved by small-scale computations of the modal orders and corresponding coefficients. This reduces the computational effort required for flow field simulation. The original sampled flow field is reconstructed with a low-order POD-ROM, and the reconstructed flow field can be expressed as the following:

$$P_{POD}^l(x) = \sum_{l=1}^{d} b_j^l \varphi_l(x) \tag{20}$$

3. Discussion and Analysis of Results

3.1. Flow Field Information Sampling

3.1.1. Flow Field without AFC

Typically, a BWB-UG glides at a speed of about $1\mathrm{Kn}(0.514\ \mathrm{m/s})$ with a small angle of attack. Therefore, in this paper, the numerical simulation is carried out for the base flow field with the inlet velocity $V_{inlet} = 0.514\ \mathrm{m/s}$ and the angle of attack α is set to $0° \sim 12°$ (with an interval of $0.6°$). After the calculation, the pressure distributions of the hydrofoil surfaces under 21 operating conditions were used as sampling data to establish the reduced-order model, as shown in Figure 5.

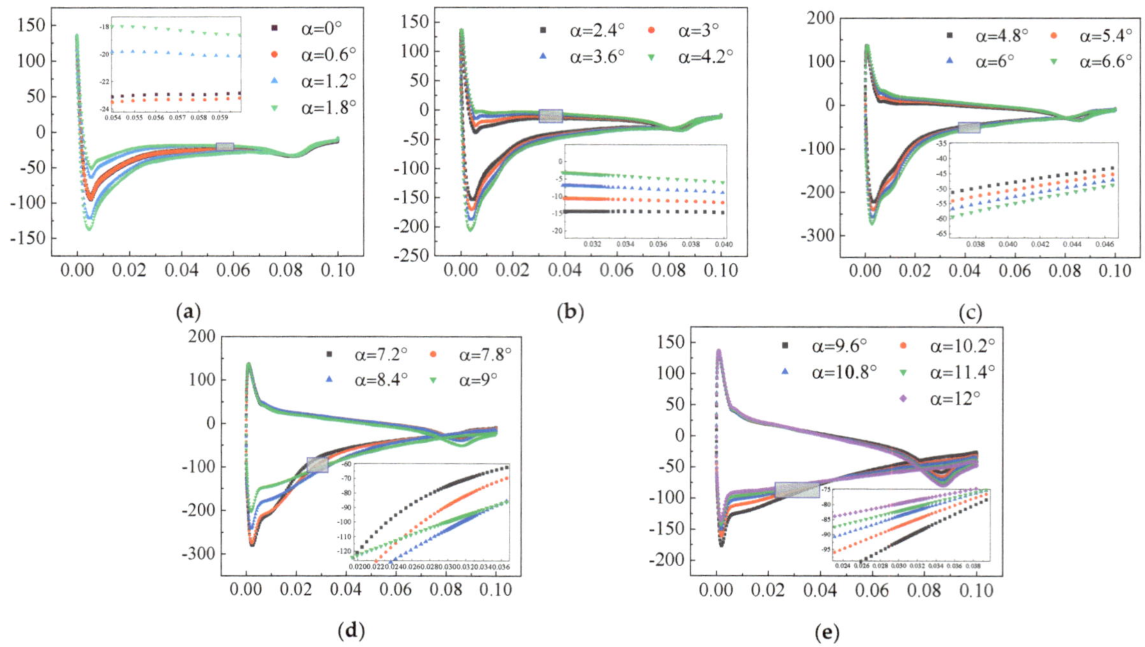

Figure 5. Surface pressure distribution of hydrofoil at different angles of attack in the base state; (**a**) $\alpha = 0°\sim1.8°$; (**b**) $\alpha = 2.4°\sim4.2°$; (**c**) $\alpha = 4.8°\sim6.6°$; (**d**) $\alpha = 7.2°\sim9.0°$; (**e**) $\alpha = 9.6°\sim12.0°$.

3.1.2. Constant Suction Control Flow Field

Firstly, the ROM of the constant suction AFC flow field in the hydrofoil is carried out, and the influence of suction velocity and suction angle on the control effect of the hydrofoil at different angles of attack is considered based on the CFD simulation. The magnitude of velocity is expressed by the suction velocity coefficient, $R_{suction} = V_{suction}/V_{inlet}$, with a range of values from 0.2 to 1.0 with an interval of 0.2. The suction deflection angle $\theta_{suction} = 30° \sim 150°$ at 30° intervals. The hydrofoil inlet velocity $V_{inlet} = 0.514$ m/s and the angle of attack α varied from 0° to 12° with an interval of 2°. The cross-combined sampling method is used to calculate the flow field. When the angle of attack and suction velocity coefficient are changed, the suction deflection angle is set to a fixed value of $\theta_{suction} = 90°$, and when the angle of attack and suction deflection are changed, the suction velocity coefficient is a fixed value of $R_{suction} = 1.0$. Each combination corresponds to 35 groups of calculation conditions, and the hydrofoil surface pressure distribution under each condition is used as the sampling data to establish the reduced-order model of the constant suction active flow control flow field, and the specific condition adjustment is shown in Figure 6. With $\theta_{\text{suction}} = 90°$, the surface pressure distribution of the hydrofoil at different suction velocities and angles of attack is shown in Figure 7.

3.1.3. Constant Jet Control Flow Field

Secondly, the ROM of the constant jet AFC flow field in the hydrofoil is carried out. The effect of jet velocity and jet angle on the control effect of the hydrofoil at different angles of attack is considered. The magnitude of the velocity is expressed by the jet velocity coefficient $R_{jet} = V_{jet}/V_{inlet}$, which takes values from 0.2 to 1.0 at 0.2 intervals. The jet deflection angle $\theta_{jet} = 30° \sim 150°$ at 30° intervals. The hydrofoil inlet velocity is $V_{inlet} = 0.514$ m/s and the angle of attack α varied from 0° to 12° at 2° intervals. Similar to Section 3.1.2, when the angle of attack and the jet velocity coefficient are changed, the jet deflection angle is set to a fixed value of $\theta_{jet} = 90°$. Each combination corresponds to 35 sets of computational conditions, and the hydrofoil surface pressure distribution is used as the sampling data for building a ROM of the jet-controlled flow field. With $\theta_{\text{jet}} = 90^{\circ}$, surface pressure distributions of the hydrofoil at different constant jet velocities and different angles of attack are shown in Figure 8.

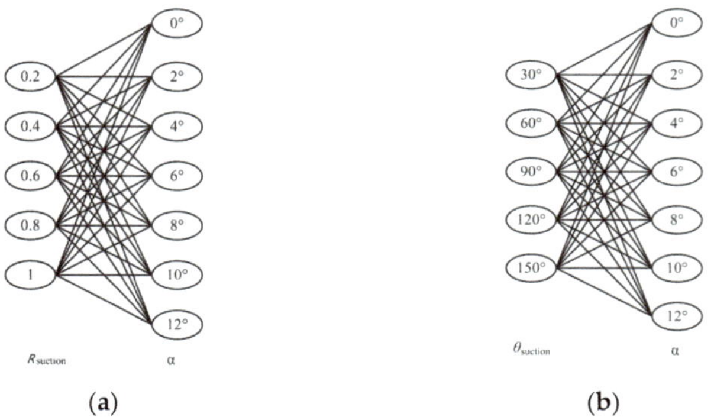

Figure 6. Hydrofoil flow field conditions with different variable combinations for suction flow control. (**a**) Suction velocity coefficient + angle of attack of incoming flow; (**b**) Suction flow deflection angle + incoming flow angle of attack.

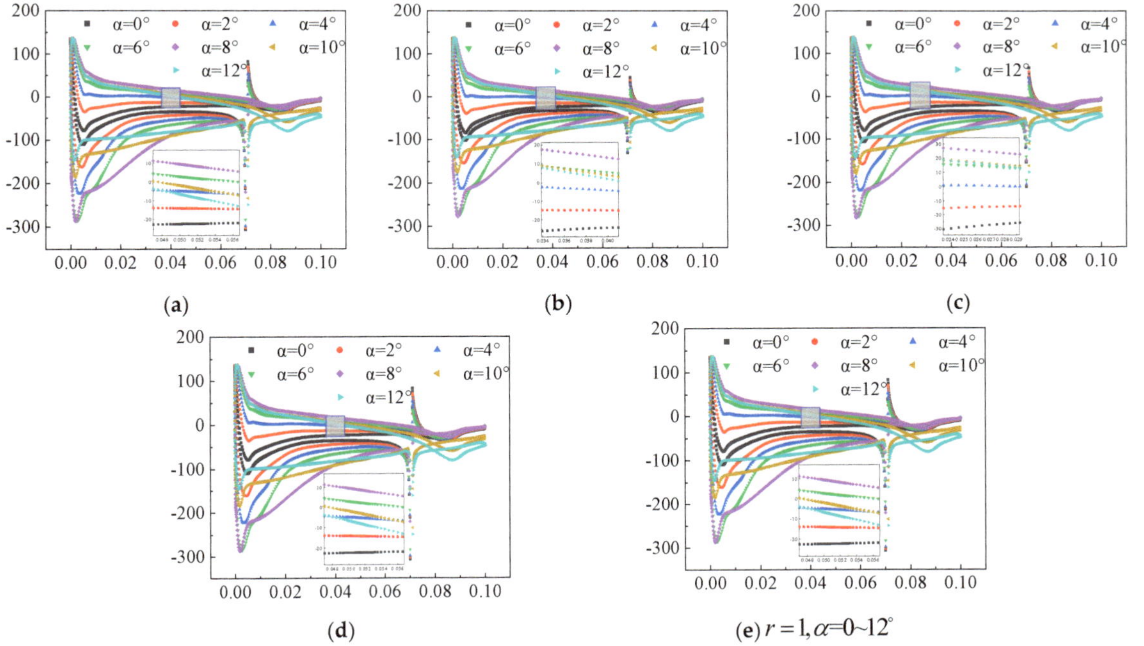

Figure 7. Surface pressure distribution of the hydrofoil at different suction velocities and angles of attack; (**a**) $R_{suction} = 0.2, \alpha = 0 \sim 12^{\circ}$; (**b**) $R_{suction} = 0.4, \alpha = 0 \sim 12^{\circ}$; (**c**) $R_{suction} = 0.6, \alpha = 0 \sim 12^{\circ}$; (**d**) $R_{suction} = 0.8, \alpha = 0 \sim 12^{\circ}$; (**e**) $R_{suction} = 1, \alpha = 0 \sim 12^{\circ}$.

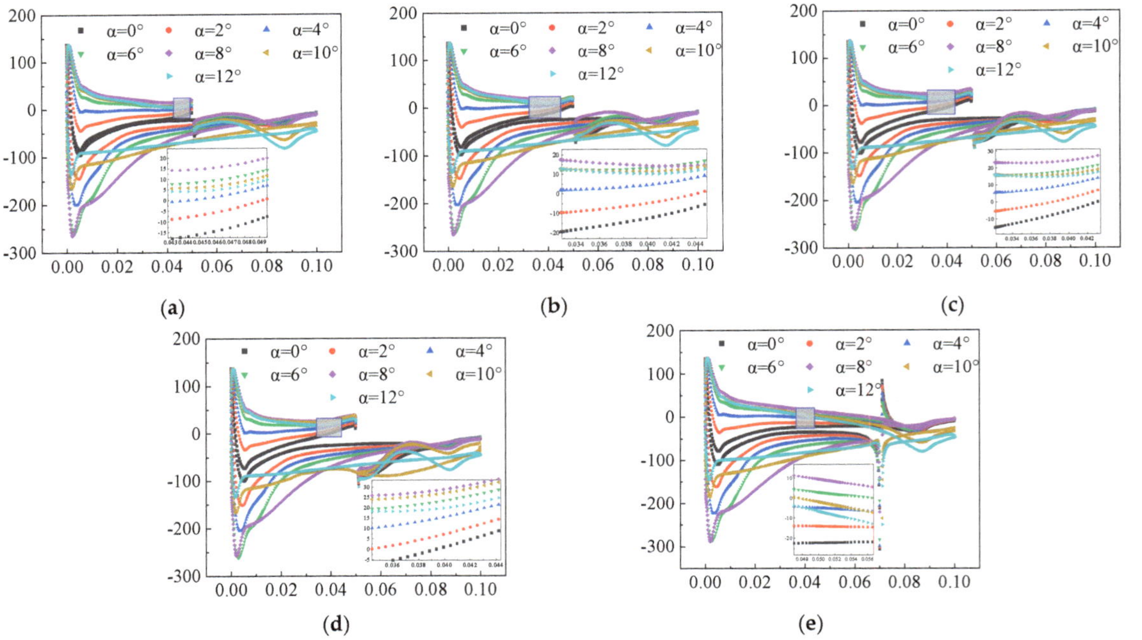

Figure 8. Surface pressure distribution of the hydrofoil at different constant jet velocities and different angles of attack; (**a**) $R_{jet} = 0.2, \alpha = 0 \sim 12°$; (**b**) $R_{jet} = 0.4, \alpha = 0 \sim 12°$; (**c**) $R_{jet} = 0.6, \alpha = 0 \sim 12°$; (**d**) $R_{jet} = 0.8, \alpha = 0 \sim 12°$; (**e**) $R_{jet} = 1, \alpha = 0 \sim 12°$.

3.2. Reconstruction of Flow Field Based on the POD Method

Based on the POD method, the modes are classified according to the difference in generalized energy levels based on the distribution characteristics of the eigenvalues of the flow field in the hydrofoil, and the order with a generalized energy share greater than 99% is taken as the truncation order *d*. The generated hydrofoil flow field data are subjected to the POD method, and the fundamental mode of the flow field and the corresponding eigenvalues can be obtained. The ratio of the sum of the eigenvalues corresponding to the first *d*-order fundamental modes to the sum of all eigenvalues is defined as the generalized energy share of the first *d*-order fundamental modes, which features characterize the amount of flow field information contained in the first *d*-order dominating modes. Mean relative error (MRE) is used as a measure of the accuracy of the reconstructed flow field data, and the MRE of the hydrofoil surface pressure field can be expressed as the following:

$$\varepsilon_P = \frac{1}{M}\sum_{m=1}^{M}\left(\left|\frac{P_{POD}^{(m)} - P_{CFD}^{(m)}}{P_{CFD}^{(m)}}\right| \times 100\%\right) \tag{21}$$

where M is the number of mesh nodes, $P_{POD}^{(m)}$ denotes the pressure value at the m-th grid node obtained from the reconstruction, and $P_{CFD}^{(m)}$ denotes the pressure value at the m-th grid node based on the CFD numerical simulation.

3.2.1. Flow Field without AFC

Based on the hydrofoil surface pressure data from the flow field without AFC, the incident angle of attack α is varied from $0° \sim 12°$ at 0.6° intervals, forming a sample size of $N = 21$, and the number of hydrofoil surface mesh nodes in each set of conditions is $M = 728$. The pressure distribution data at the mesh nodes on the surface of the 21 hydrofoils are selected as snapshot data, and the "pulsation amount" is used as a sample to construct an $M \times N$ sample matrix. The eigenvalues and eigenvectors of this matrix are solved, and the distribution of the eigenvalues is shown in Figure 9. The magnitude of the captured generalized energy is given in Table 2, where the generalized energy equal to 99% corresponds to the truncation order $d = 3$. Therefore, the truncation order 3 is used as the modal order for the ROM of the pressure distribution on the hydrofoil surface.

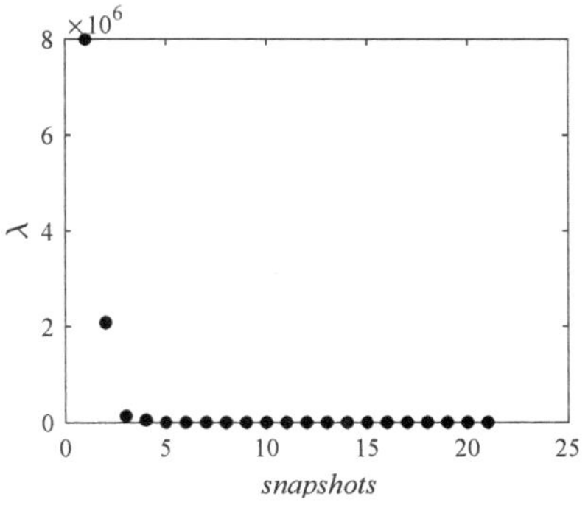

Figure 9. Distribution of eigenvalues of the base flow field.

Table 2. Comparison of the generalized energy captured cumulatively by the main eigenvalues of the base flow field.

Ordinal	Eigenvalue	Cumulative Captured Energy (%)
1	7.99×10^6	77.82
2	2.08×10^6	98.09
3	1.40×10^5	99.45

On this basis, the d-order POD basis is solved for the basal flow field in the hydrofoil after downscaling.

$$\varphi_{base}(x) = \sum_{j=1}^{3} a_j^l P_{ij}{}'(x) \tag{22}$$

where α_j^l is the coefficient obtained by solving for the autocorrelation matrix.

Based on the POD method, the corresponding modal coefficients b_j^l can be obtained by projecting any flow field in the sampled flow field onto the POD method to realize the reconstruction of the flow field.

$$P_{POD}^l(x) = \sum_{l=1}^{N} b_j^l \varphi_{base}(x) \tag{23}$$

Comparing the hydrofoil surface pressure distribution obtained from the POD step-down reconstruction with the CFD calculation data, it can be seen that the data are very similar, as shown in Figure 10. Comparing the average relative error under each working condition, it can be seen that when the hydrofoil is in the state of a small angle of attack $\alpha \leq 4.2°$, the average relative error ε_P is smaller, less than 2%; it increases and shows a fluctuating change as α increases, and the maximum value is still less than 5%, as shown in Figure 11. The reason for this is that at high angles of attack, the separation zone near the trailing edge of the upper hydrofoil surface increases significantly, which changes the generalized energy distribution of the flow modes of the flow field and affects the surface pressure distribution, forming a localized low-pressure zone in the region of the trailing edge, with a significant change in the pressure gradient on the upper surface.

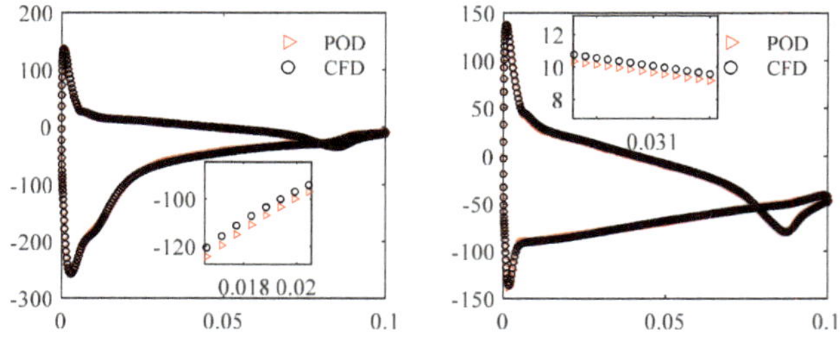

Figure 10. Comparison of hydrofoil surface pressure under base flow field; (**a**) α = 6°; (**b**) α = 12°.

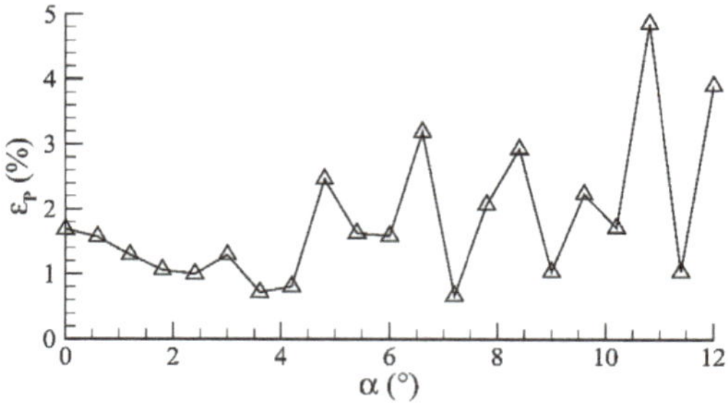

Figure 11. Variation in the mean relative error of hydrofoil surface pressure at different incoming flow angles of attack.

3.2.2. Constant Suction Control Flow Field

For the constant suction AFC field, the suction deflection angle and suction velocity are cross-combined with the angle of attack of the incoming flow, respectively, to establish a ROM of the AFC field. The hydrofoil surface pressure distribution is used with the sampling data with the number of grid nodes being $M = 719$. The suction deflection angle is $\theta_{suction} = 30° \sim 150°$ at 30° intervals, the angle of attack α is $0° \sim 12°$ at 2° intervals, and the suction velocity coefficient $R_{suction} = 1.0$ to form $N = 35$ sets of sampling conditions. The amount of pressure pulsation on the hydrofoil surface under each condition is used as the snapshot data to form a sampling matrix of $M \times N$. The eigenvalues and eigenvectors of this matrix are solved, and the distribution of the eigenvalues is shown in Figure 12. The magnitude of the captured energy is given in Table 3, where the energy equal to 99% corresponds to the truncation order $d = 11$. Therefore, $d = 11$ is used as the modal order for the ROM of the surface pressure distribution in the steady-state suction-controlled hydrofoil flow field.

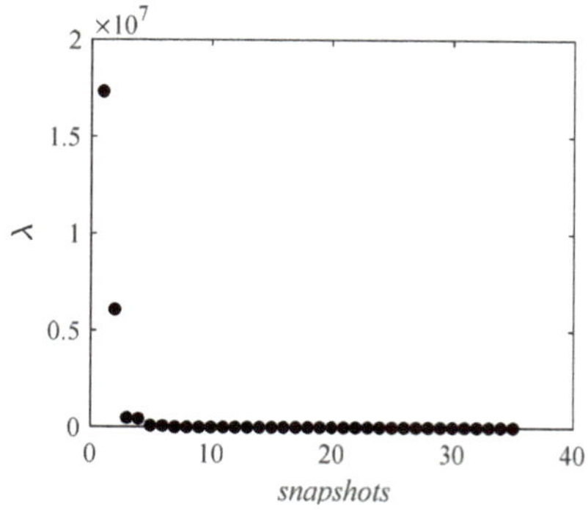

Figure 12. Distribution of the eigenvalues of the control flow field at different suction bias angles.

Table 3. Cumulative energy captured by the main eigenvalues of the flow field at different suction angle controls.

Ordinal	Eigenvalue	Cumulative Captured Energy (%)
1	1.73×10^7	70.73
2	6.09×10^6	95.58
3	4.76×10^5	97.52
...	...	...
11	8.84×10^2	99.99

Comparing the POD method with the CFD-calculated flow field data, it can be seen that the hydrofoil surface pressure distribution data are very similar in both cases, as shown in Figure 13. The average relative error ε_P of the hydrofoil surface pressure is less than 2% in most of the operating conditions, as shown in Figure 14.

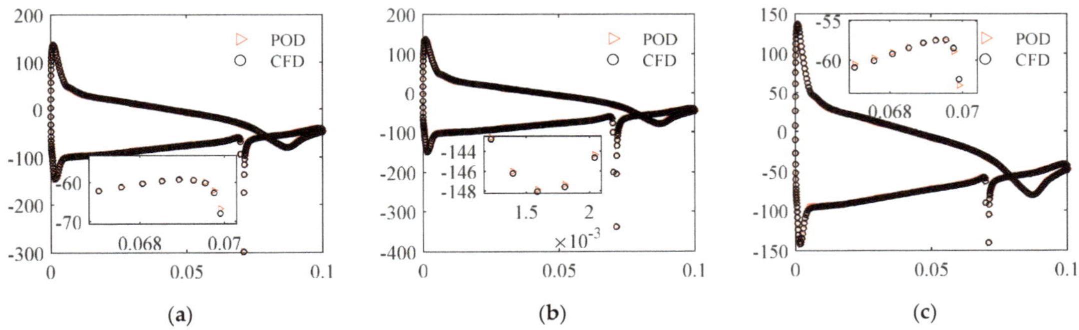

Figure 13. Comparisons of surface pressures of hydrofoil with different suction deflection angles at an angle of attack of 12°. (**a**) α = 12°, $\theta_{suction}$ = 30°; (**b**) α = 12°, $\theta_{suction}$ = 90°; (**c**) α = 12°, $\theta_{suction}$ = 150°.

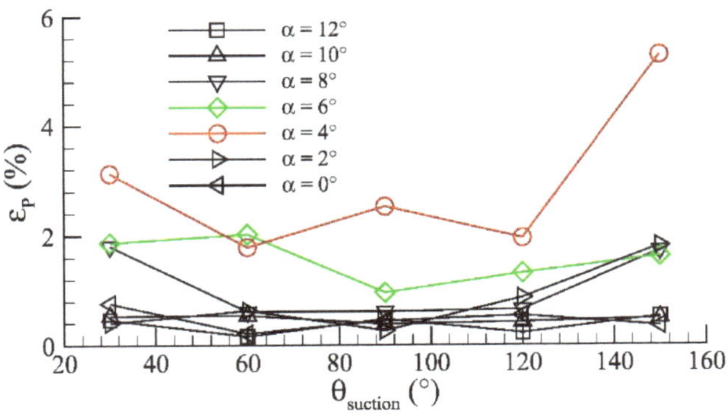

Figure 14. Variation in the mean relative error of surface pressure of suction-controlled hydrofoil at different angles of attack.

Subsequently, the suction deflection angle $\theta_{suction} = 90°$, the suction velocity coefficient $R_{suction} = 0.2$ to 1.0 at 0.2 intervals, and the angle of attack $\alpha = 0°$ to 12° at 2° intervals were investigated to form $N = 35$ sets of sampling conditions. The distribution of the eigenvalues is shown in Figure 15. The magnitude of the captur ed energy is given in Table 4. Using the same approach, $d = 10$ can be obtained as the modal order for the ROM of the surface pressure distribution in the steady-state suction-controlled hydrofoil flow field.

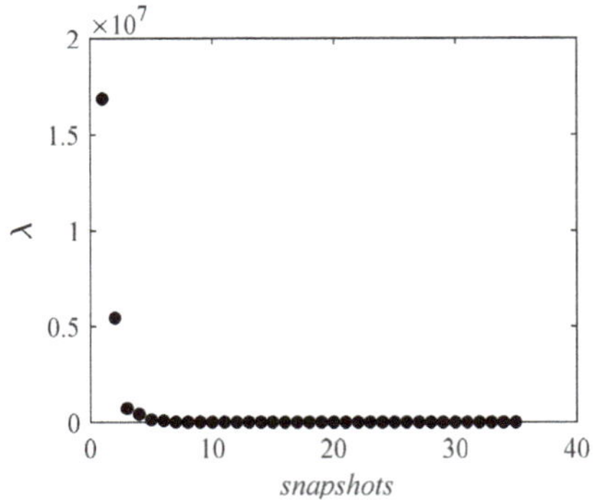

Figure 15. Distribution of the eigenvalues at different suction bias angles.

Table 4. Cumulative energy captured by the main eigenvalues of the flow field at different suc velocity coefficient controls.

Ordinal	Eigenvalue	Cumulative Captured Energy (%)
1	1.69×10^7	71.26
2	5.43×10^6	94.19
3	7.25×10^5	97.52
...	...	...
11	1.20×10^3	99.99

Similarly, when the flow field reconstructed by the POD method is compared w the flow field data calculated by CFD, it can be seen that the data of the hydrofoil surf pressure distribution are very similar in both cases, as shown in Figure 16. The aver relative error ε_P of the hydrofoil surface pressure is less than 1% for most of the work conditions, as shown in Figure 17.

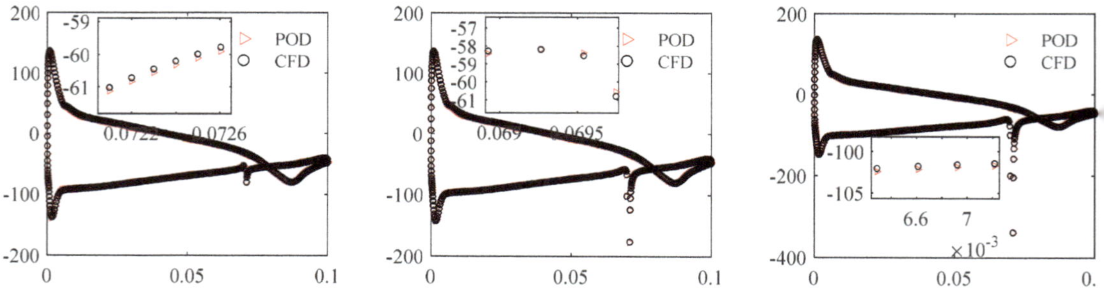

(**a**) (**b**) (**c**)

Figure 16. Comparison of surface pressures of hydrofoil with different suction deflection angle 12° angle of attack. (**a**) α = 12°, R_{suction} = 0.2; (**b**) α = 12°, R_{suction} = 0.6; (**c**) α = 12°, R_{suction} = 1.0.

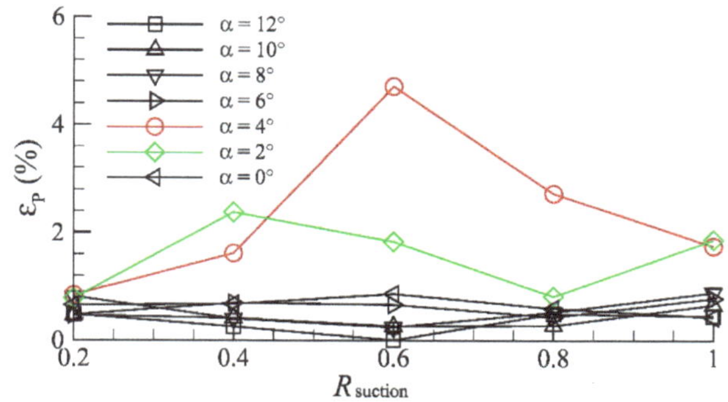

Figure 17. Variation in the mean relative error of surface pressure of suction-controlled hydrofoil at different suction velocity coefficients.

3.2.3. Constant Jet Control Flow Field

For the constant jet control flow field, the effects of the jet deflection angle and jet velocity coefficient on the control flow field are also analyzed. Using the same method, the truncation order d is determined based on the energy fraction of the flow field as well as the modes, which are 16 and 12, respectively, and the resulting eigenvalue distributions are shown in Figure 18. Comparing the POD downscaled reconstructed flow field with the CFD-calculated flow field data, it can be seen that the hydrofoil surface pressure distribution data are very similar in both cases, and the average relative error ε_P of the hydrofoil surface pressure is less than 1% for most of the working conditions, as shown in Figures 19 and 20.

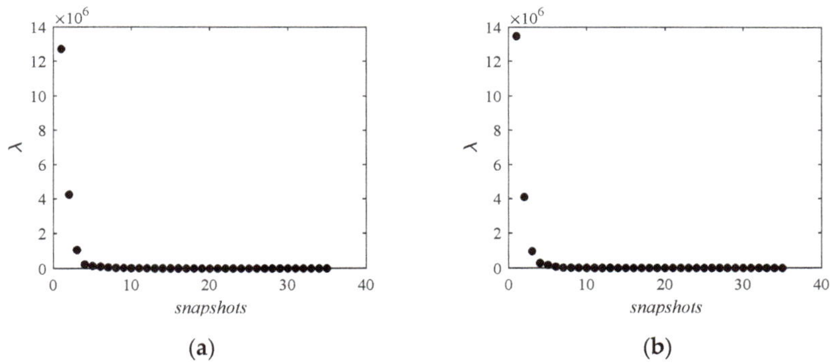

Figure 18. Distribution of eigenvalues of jet control flow field; (**a**) impact of jet deflection; (**b**) impact of jet velocity coefficients.

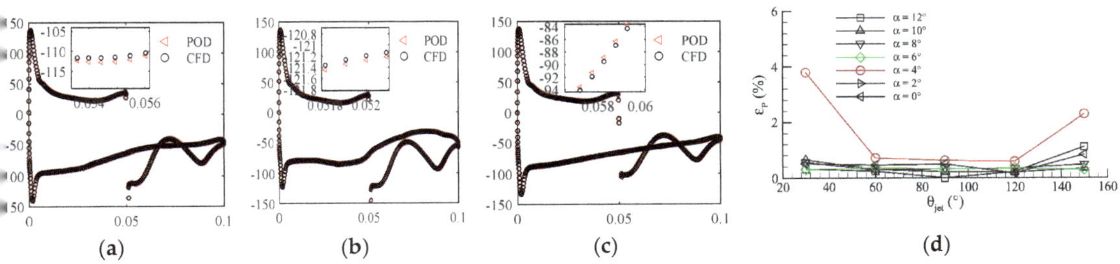

Figure 19. Comparison of surface pressure and error distribution of hydrofoil with different jet deflection angle control flow field. (**a**) α = 12°, θ_{jet} = 30°; (**b**) α = 12°, θ_{jet} = 90°; (**c**) α = 12°, θ_{jet} = 150°; (**d**) change in average relative error.

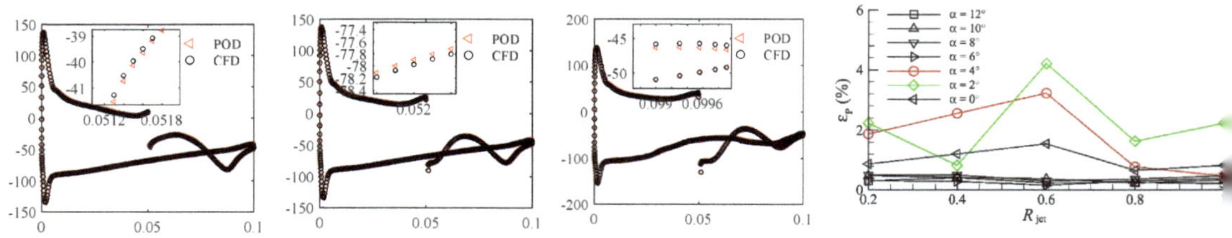

(**a**) (**b**) (**c**) (**d**)

Figure 20. Comparison of surface pressure and error distribution of hydrofoil with differen velocity coefficients to control the flow field. (**a**) $\alpha = 12°$, $R_{jet} = 0.2$; (**b**) $\alpha = 12°$, $R_{jet} = 0.6$; (**c**) $\alpha =$ $R_{jet} = 1.0$; (**d**) change in average relative error.

3.3. POD-ROM Flow Field Prediction

Based on the established POD-ROM, combined with the interpolated predictive mc coefficients, the predictive analysis of the flow field under any unknown state paramete carried out and the results are compared with those of the CFD simulations. The work condition parameters, computation time, and error analysis are shown in Table 5, wh shows that the average relative error of the hydrofoil surface pressure under the positic conditions predicted by the model is up to 6.9%, and its computation time is only in order of 0.1 s, which is much smaller than that of the CFD simulation. The current er are sufficient for engineering applications. The wall pressure distributions under differ working conditions are shown in Figure 21. The results predicted by the ROM are cl to the CFD simulation results, which can capture the main features of the flow field verify the accuracy of the POD-ROM for predicting the flow field.

Table 5. Setting and error analysis of predicted flow field conditions.

Predicted Working Conditions	Working Condition Parameters	Computational Time(s)		ε_P
		POD	CFD	
Flow field without AFC	$\alpha = 7.5°$	0.1	675	1.78
Suction-controlled flow field	$\alpha = 3°$, $\theta = 75°$ Ratio = 1.0	0.1	1516	4.27
	$\alpha = 3°$, Ratio = 0.5 $\theta = 90°$	0.1	1770	3.42
Blowing controlled flow field	$\alpha = 3°$, $\theta = 75°$ Ratio = 1.0,	0.1	575	6.90
	$\alpha = 3°$, Ratio = 0.5 $\theta = 90°$	0.1	445	1.59

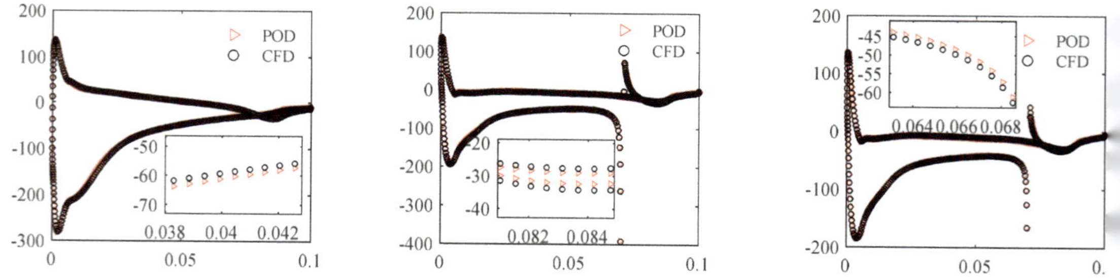

(a) (b) (c)

Figure 21. *Cont.*

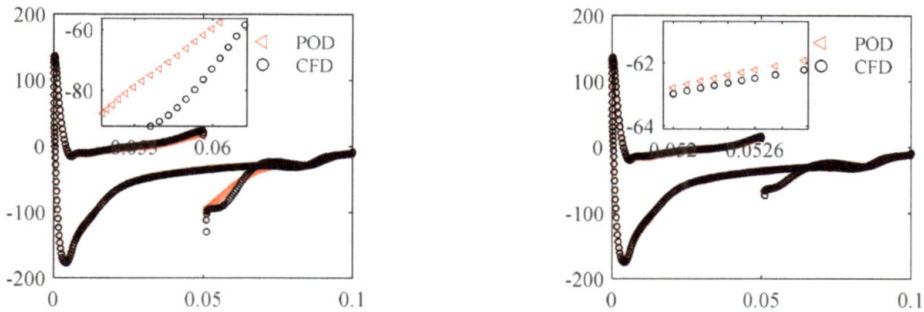

(d) (e)

Figure 21. Comparison of pressure distribution in flow field predicted by POD-ROM. (**a**) Flow field without AFC $\alpha = 7.5°$; (**b**) suction control $\alpha = 3°$, $\theta = 75°$, *Ratio* = 1.0; (**c**) suction control $\alpha = 3°$, *Ratio* = 0.5, $\theta = 90°$; (**d**) jet control $\alpha = 3°$, $\theta = 75°$, *Ratio* = 1.0; (**e**) jet control $\alpha = 3°$, *Ratio* = 0.5, $\theta = 90°$.

4. Conclusions

In this paper, based on the POD method, the ROM for the hydrofoil of the BWB-UG flow control with steady-stream suction and jets is investigated. The innovative POD method is used to solve the challenge of rapid flow field prediction in BWB-UG closed-loop active flow control. The established reduced-order model can effectively solve the timeliness problem of flow field calculation in closed-loop AFC, while the error of the model is sufficient to meet the needs of engineering applications. The main conclusions are as follows:

1. A POD-ROM prediction for the hydrofoil of the BWB-UG flow control with steady suction and blowing is successfully developed.
2. Compared to the CFD simulation, the computational time required to predict the target flow field using the established POD-ROM is only about 0.1 s, which is much less than the CFD simulation time.
3. The average relative error in predicting the surface pressure of the target using the established POD-ROM is less than 6.9% compared to CFD simulations. The established prediction model is valid.

The most difficult problem of closed-loop AFC is the timeliness of the flow field calculation, i.e., how to give relatively correct results in the shortest possible time. Therefore, in this paper, the investigation is carried out for a single variable. In future research, in-depth studies can be carried out for the reduced-order modeling containing multi-factors and high-dimensional data. The focus is on the nonlinear and strongly coupled relationship between multiple factors. This will provide more accurate guidance for closed-loop active flow control.

Author Contributions: Conceptualization, X.D. and H.W.; methodology, X.D. and Y.H.; software, H.W.; validation, X.D. and Y.H.; investigation, H.W.; resources, X.D. and Y.H.; data curation, H.W.; writing—original draft preparation, H.W.; writing—review and editing, H.W., X.D.; supervision, X.D. and Y.H.; project administration, X.D.; funding acquisition, X.D. All authors have read and agreed to the published version of the manuscript.

Funding: The author(s) disclosed receipt of the following financial support for the research, authorship, and publication of this article. This work is supported by the National Natural Science Foundation of China (No. U2341217).

Data Availability Statement: Data are contained within the article.

Conflicts of Interest: The authors declare no conflicts of interest.

References

1. Sun, C.Y.; Song, B.W.; Wang, P. Shape design and hydrodynamic characteristics analysis of the blended-wing-body underw glider. *J. Ship Sci. Technol.* **2016**, *38*, 78–83. (In Chinese)
2. Du, X.X.; Zhang, L.Y. Numerical study on the steady suction active flow control of hydrofoil in the profile of the blended-w body underwater glider. *J. Northwestern Polytech. Univ.* **2021**, *39*, 801–809. (In Chinese) [CrossRef]
3. Zhang, X.L.; Du, X.X.; Ran, X.Z. Numerical Research on Active Flow Control with Synthetic Jet for Wing Section of Underw Blended-wing-body Glider. *Shipbuild. China* **2022**, *63*, 95–106. (In Chinese)
4. Ren, F.; Wang, C.; Tang, H. Bluff body uses deep-reinforcement-learning trained active flow control to achieve hydrodyna stealth. *Phys. Fluids* **2021**, *33*, 093602. [CrossRef]
5. Lu, T.; Hu, H.; Song, J.; Ren, F. Lattice Boltzmann modeling of backward-facing step flow controlled by a synthetic jet. *J. Hydro* **2023**, *35*, 757–769. [CrossRef]
6. Dong, H.; Geng, X.; Shi, Z.; Cheng, K.; Cui, Y.D.; Khoo, B.C. On evolution of flow structures induced by nanosecond p discharge inside a plasma synthetic jet actuator. *Jpn. J. Appl. Phys.* **2019**, *58*, 028002. [CrossRef]
7. Zhang, H.; Feng, Y. Review of Flow Control by Lorentz Force. *Aerodyn. Res. Exp.* **2023**, *1*, 20–29. (In Chinese)
8. Du, X.X.; Liu, X.; Song, D. Coupled physics analysis of blended-wing-body underwater glider equipped with electromagr active flow control. *Ocean. Eng.* **2023**, *278*, 114402. [CrossRef]
9. Zhang, L.; Gao, S.Q.; Zhao, J.X.; Tang, C.; Xie, H.L. Numerical simulation of influence of steady inhalation on aerodyna performance of wind turbine. *Acta Energiae Solaris Sin.* **2018**, *39*, 2155–2162. (In Chinese)
10. Zhou, X.B.; Zhou, S.; Hou, A.P.; Zheng, X.Q. Numerical Investigation on Controlling Separated Cascade Flows by Using Jets. *Aeronaut. Et Astronaut. Sinica* **2007**, *28S*, 7–13. (In Chinese)
11. Wang, B.X.; Yang, Z.G.; Zhu, H.; Li, Y.L. Drag Reducition by Active Control with Steady Blowing/Suction Methods. *J. Tongji U (Nat. Sci.)* **2017**, *45*, 1383–1389. (In Chinese)
12. Du, X.; Liu, X.; Song, Y. Analysis of the Steady-Stream Active Flow Control for the Blended-Winged-Body Underwater Glid *Mar. Sci. Eng.* **2023**, *11*, 1344. [CrossRef]
13. Chen, G.; Li, Y.M. Advances and prospects of the reduced-order for unsteady flow and its application. *Adv. Mech.* **2011** 686–701.
14. Ren, F.; Gao, C.Q.; Tang, H. Machine learning for flow control: Applications and development trends. *Acta Aeronaut. Et Astror Sin.* **2021**, *42*, 524686.
15. Ye, K.; Wu, J.; Ye, Z.Y.; Qu, Z. Analysis Circular Cylinder Flow Using Dynamic Mode and Proper Orthogonal Decompositic *Northwestern Polytech. Univ.* **2017**, *35*, 599–607.
16. Lumley, J.L. Rational approach to relations between motions of differing scales in turbulent flows. *Phys. Fluids* **1967**, *10*, 1405–1 [CrossRef]
17. Kidambi, K.B.; Mackunis, W.; Jayaprakash, A.K. Limit Cycle Oscillation Suppression Using a Closed-Loop Nonlinear Active F Control Technique. In Proceedings of the 59th IEEE Conference on Decision and Control (CDC), Jeju, Republic of Korea, 14 December 2020. [CrossRef]
18. Axia, W.; Yichen, M.; Wenjing, Y. The proper orthogonal decomposition method for the Navier-Stokes equations. *Acad. J. X Jiaotong Univ.* **2008**, *20*, 141–148.
19. Min, G.; Jiang, N. Data-driven identification and pressure fields prediction for parallel twin cylinders based on POD and D method. *Phys. Fluids* **2024**, *36*, 023614. [CrossRef]
20. Sun, F.J.; Zhu, D.H.; Zhang, D.M. Reduced-order model for fluid-structure coupled calculation of wind turbine blades. *J. Shock.* **2021**, *40*, 175–181+215. (In Chinese)
21. Wang, Y.G.; Li, Z.N.; Gong, B.; Li, Q.S. Reconstruction & prediction of wind pressure on heliostat. *Acta Aerodyn. Sin.* **2009** 586–591. (In Chinese)
22. Kang, W.; Zhang, J.Z.; Li, K.L. Nonlinear Galerkin Method for Dimension Reduction Using Proper Orthogonal Decompositic *Xi'an Jiaotong Univ.* **2011**, *45*, 58–62+67. (In Chinese)
23. Anderson, J.D.; Wendt, J. *Computational Fluid Dymanics*; McGraw-Hill Companies: New, York, NY, USA, 1995.
24. Menter, F.R. Two-equation eddy-viscosity turbulence models for engineering applications. *AIAA J.* **1994**, *32*, 1598–1605. [Cross
25. Eiximeno, B.; Miró, A.; Rodríguez, I.; Lehmkuhl, O. Toward the Usage of Deep Learning Surrogate Models in Ground Veh Aerodynamics. *Mathematics* **2024**, *12*, 998. [CrossRef]
26. Lu, K.; Jin, Y.; Chen, Y.; Yang, Y.; Hou, L.; Zhang, Z.; Li, Z.; Fu, C. Review for order reduction based on proper orthog decomposition and outlooks of applications in mechanical systems. *Mech. Syst. Signal Process.* **2019**, *123*, 264–297. [CrossRe

actuators

MDPI

cle

Steady-Pressure Control Method for Emulsion Pump Station ısed on Online Updating of Optimal Flow Rate

g Xu [1,2], Ziming Kou [1,2,*], Juan Wu [1,2], Tengyan Hou [1,2], Yanwei Peng [1,2] and Buwen Zhang [1,2]

1 College of Mechanical and Vehicle Engineering, Taiyuan University of Technology, Taiyuan 030024, China; m15135134430@163.com (P.X.); wujuanz@163.com (J.W.); houtengyan@163.com (T.H.); pengyanwei0049@link.tyut.edu.cn (Y.P.); zhangbuwen0101@link.tyut.edu.cn (B.Z.)

2 National and Local Joint Engineering Laboratory for Mining Fluid Control, Taiyuan 030024, China

* Correspondence: zmkou@163.com

Abstract: In order to solve the problem of unstable fluid supply pressure and serious impact caused by the complicated and changeable working condition of a fully mechanized mining face in coal mines and the sluggish response of the fluid supply system to the fluid demand for the hydraulic support, a control method based on online updating generalized regression neural network (GRNN) was proposed. Firstly, the simulated hydraulic support test platform and co-simulation model were built. Secondly, The optimal flow dataset of steady-pressure fluid supply under different working conditions is calculated by simulation. Furthermore, the GRNN prediction model was established by using dataset and online updating learning technology to predict the optimal fluid supply flow according to environmental parameters. Finally, the optimal flow control method of online updating GRNN was established, and numerical research and experimental verification were also carried out in different working conditions. The results indicated that the proposed control method could track the working conditions of the working face in real time and adjusted the fluid supply flow of the emulsion pump station adaptively, which effectively alleviated the pressure fluctuation and pressure shock, and the system pressure was more stable, meeting the demand of steady-pressure fluid supply on the working face.

Keywords: GRNN neural network; steady-pressure fluid supply; optional flow control; online update

ion: Xu, P.; Kou, Z.; Wu, J.; Hou, ng, Y.; Zhang, B. A Steady- ure Control Method for sion Pump Station Based on ıe Updating of Optimal Flow Actuators **2024**, *13*, 247. ://doi.org/10.3390/act13070247

emic Editors: Hui Tang, Ven and Feng Ren

ved: 16 May 2024
ed: 20 June 2024
pted: 25 June 2024
shed: 28 June 2024

1. Introduction

The construction of intelligent mines in China is developing at a high speed. In order to meet the accelerating mining speed of the working face, higher requirements are put forward for the reliability and stability of intelligent mining of the fully mechanized working face [1,2]. As the power source of hydraulic support in a fully mechanized mining face, the emulsion pump station directly affects the working efficiency and stability of hydraulic support, which is the key equipment to ensure the safety and high efficiency of coal mine production [3,4]. Due to the coupling relationship between the operating speed, action type, working resistance, fluid supply flow and system pressure of the hydraulic support system [5], the pressure-transient characteristics of the hydraulic support system are obvious. Meanwhile, with the increase in working resistance of the hydraulic support at large mining height and the increase in working face length, the problem of poor stability of fluid supply pressure becomes more obvious. Therefore, how to improve the stability of the fluid supply system as much as possible under different working conditions of hydraulic support is an important and complicated problem to be solved urgently in the fluid supply control of an emulsion pump station.

Improving the stability of the fluid supply system is the key to safe and efficient production of coal face. At present, scholars mainly strive to maintain the stable pressure of the system through two aspects: one is to optimize the structure of the hydraulic system to reduce the pressure fluctuation, and the other is to improve the fluid supply quality of the emulsion pump station by the intelligent control algorithm. Among th there were many studies on optimizing the structure of hydraulic systems by scho the system pressure was stabilized by optimizing the accumulator parameters and accumulator configuration scheme [6,7]; the transient dynamic characteristics of pipel were improved by optimization of pipeline size parameters [8,9]; the layout of the emul pump station and pipeline layout were improved to reduce pressure loss [10,11]; unloading valve was reasonably utilized to recover pressure to ensure the stability of emulsion pumping station [12]. and other methods. However, optimizing the structur the hydraulic system can reduce the pressure loss and pressure shock, but it can only p an auxiliary role in the problem of pressure stability in the system with long-distance f supply and strong time-varying load.

In order to make the operation of the equipment on the working surface more sta the key to ensure the pressure stability of the hydraulic system is to adopt the appropr intelligent control method to improve the fluid supply quality of the emulsion pu station. Tian et al. [13] studied the application of PLC variable frequency speed regula and constant pressure technology in the mine emulsion pump. The experimental res showed that the design of the variable frequency–constant voltage control system ba on PLC could effectively save energy consumption. T. Qiao et al. [14] applied the fu control and PID control method to the variable frequency and constant pressure con system of a mine emulsion pump station, and obtained good dynamic performance. et al. [15] trained field data based on Elman neural network, combined with movem time and pressure setting, and carried out pressure predictive control on an emuls pump station. Wang and Li [16] optimized the variable frequency driving mode adopted the control mode of combining frequency converter with combined switc realize constant-pressure fluid supply in coal mining face. By predicting the traction sp of the shearer, the power demand of hydraulic support is predicted, and then the intellig adjustment of pressure and flow of the shearer station is realized by using a mixed par swarm optimization algorithm. Li [17] synthesized the parameters such as mining mach position, pillar pressure, advance stroke, etc., and constructed the feature vector. Combi with the attention mechanism, the problem of long-term time prediction of fluid sup demand was solved, and the remote intelligent fluid supply control strategy of the pump station was put forward. Ma [18] proposed a new concept of overlapping time, and sol the pressure pulsation problem of the constant-flow parallel mechanical displacem micropump from the perspective of control theory by realizing an RBF neural netw combining unsupervised learning with supervised learning. Umrao and Chaturvedi put forward a load frequency fuzzy control method, and the results showed that the con method had good robustness to the system control of nonlinear and complex mathema models and avoided a large number of rules. It could be seen that the intelligent algori had a good application prospect in the field of pressure control of the pumping stat Tian et al. [20,21] predicted the running speed of the shearer, calculated the power dem of the hydraulic support based on the speed of the shearer, and then controlled the po of the emulsion pump station and optimized the flow output of the emulsion pump stat Si et al. [22] proposed an immune particle swarm optimization fuzzy neural network algorithm to realize the pressure stabilization control of the fluid supply system of a f mechanized mining face. The simulation results showed that the load interference l little influence on the controller, and its convergence was rapid, its robustness was gre improved, and it had good anti-disturbance and disturbance compensation ability.

In the above research, different degrees of success have been achieved in realizing smooth fluid supply of the hydraulic support system. However, in view of the time-vary actual working conditions of underground working face, due to the low accuracy and sp of the existing frequency conversion control, the emulsion pump station cannot prov accurate flow in time when the working conditions change, which causes the fluctua in fluid supply pressure and flow and affects the control accuracy of hydraulic suppo working face. Therefore, the research on intelligent control of the emulsion pump is suitable for actual production without considering the variable operating conditions, and there are few reports on this kind of related research.

In this paper, a control method based on online updating generalized regression neural network (GRNN) is proposed for the actual working conditions of coal face. The method tracks the state of the system in real time through online updating learning technology and adaptively adjusts the steady-pressure fluid supply flow. Firstly, the experimental platform and co-simulation model of the simulated hydraulic support are built, and the consistency and effectiveness of the experimental platform and simulation model are verified. Secondly, the concept of optimal steady-pressure fluid supply flow rate is put forward by analyzing the field-measured data of working face and the process of steady-pressure fluid supply, and the optimal flow datasets of steady-pressure fluid supply under different working conditions are obtained by simulation. Then, the GRNN prediction model is established by using datasets and online updating learning technology. Finally, an optimal flow control method for updating GRNN online is established through cyclic online correction, and numerical and experimental studies are carried out under different working conditions by using the optimal pressure regulator flow control method to verify the effectiveness of the method.

2. Establishment of a Simulation Platform and Its Verification

2.1. Experiment System

In order to obtain experimental data to validate the simulation model, a hydraulic support system experimental platform was built. The experimental platform included three parts: emulsion pump station system, hydraulic support system, and control system. The experimental equipment and principle are shown in Figures 1 and 2, respectively. The emulsion pumping station consists of one emulsion tank and two emulsion pumps, each of which is driven by a frequency converter and equipped with unloading valves and accumulators. The hydraulic support system adopts three sets of cylinders to simulate the hydraulic support, the simulated support cylinder and the loading cylinder interact to realize the loading, of which two sets simulate the column cylinder and one group simulates the pushing cylinders. The pressure sensors and flow sensors were installed at the pump outlet, and displacement sensors and pressure sensors were installed in front of the cylinder for feedback. The hardware of the control system mainly consists of an industrial computer (ADVANTECH, Taiwan, China), PC (Lenovo, Shanghai, China), PCI data acquisition card (Art-Control, Beijing, China), and PLC controller (Siemens AG, Munich, Germany), and the software uses a Simulink/Real-time control system (MATLAB 2022b) and writes the control program of the voltage regulator fluid supply in Labview. In addition, a filter module is added to the control program to filter the signal collected by the sensor. The details of the experimental equipment are shown in Table 1.

Table 1. Parameters of the experimental platform.

Name	Parameter	Value	Units
Emulsion	Density	998	kg/m^3
Emulsion pump	Flow	200/80	L/min
Energy accumulator	Capacity	20	L
Loading cylinder	amount	3	/
	Cylinder/rod diameter	160/105	mm
Column cylinder	amount	2	/
	Cylinder/rod diameter	110/80	mm
Pushing cylinder	amount	1	/
	Cylinder/rod diameter	110/80	mm

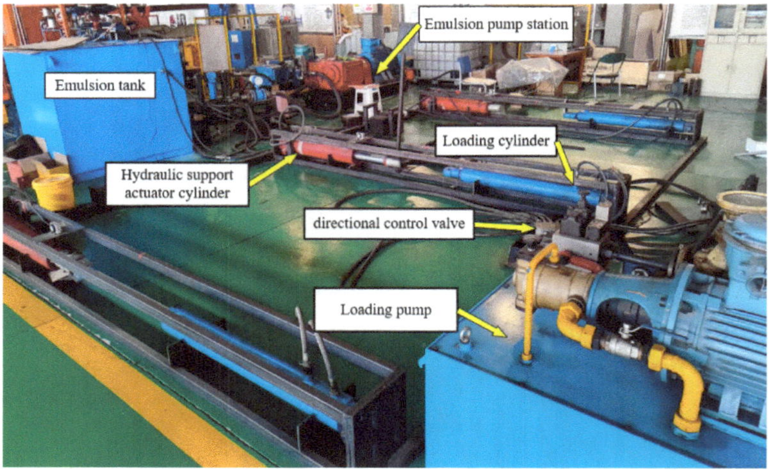

Figure 1. The experimental platform.

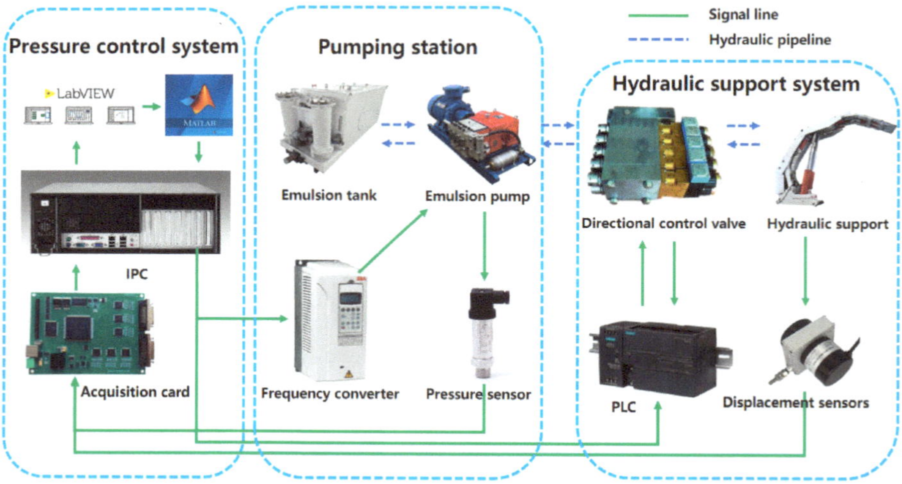

Figure 2. Diagram of the experimental platform.

2.2. Establishment and Verification of the Simulation Platform

The AMESim-Simulink co-simulation model is established according to the b experimental platform, and the parameters of each device are set according to the tual experimental equipment parameters to ensure the authenticity and reliability of calculation. The co-simulation model is shown in Figure 3.

In order to avoid the interference of redundant factors, the following simplification assumptions are made in the establishment of the simulation model: (1) the bulk modu and absolute viscosity of the emulsion are constant. (2) The emulsion is an incompressi fluid, and its density is independent of temperature. (3) Do not consider the leakag each component in the system. (4) Ignore the influence of the electromechanical convers device on system characteristics. (5) Cavitation is not considered.

Based on the established hydraulic system simulation model, the accuracy of simulation model is verified by comparing the simulation results with the test results un the rated fluid supply condition. Start emulsion pumps (3) and (4) before the operating The speed of the converter is adjusted, and the two emulsion pumps operate at the fl supply flow of 200 L/min and 80 L/min, respectively. When the emulsion pump runs stably, the rated fluid supply flow of the pumping station is 280 L/min. The system pressure data of the simulated hydraulic support were collected when the hydraulic support performed no action with leakage, and the actions of descending, pulling, raising, and pushing were performed. The comparison of experiment and simulation results under rated fluid supply conditions is shown in Figure 4.

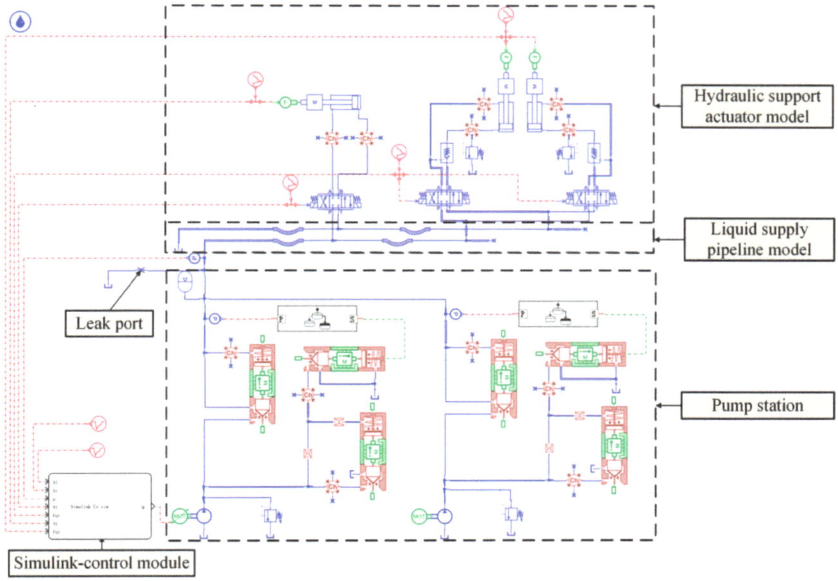

Figure 3. Co-simulation model of steady-pressure fluid supply.

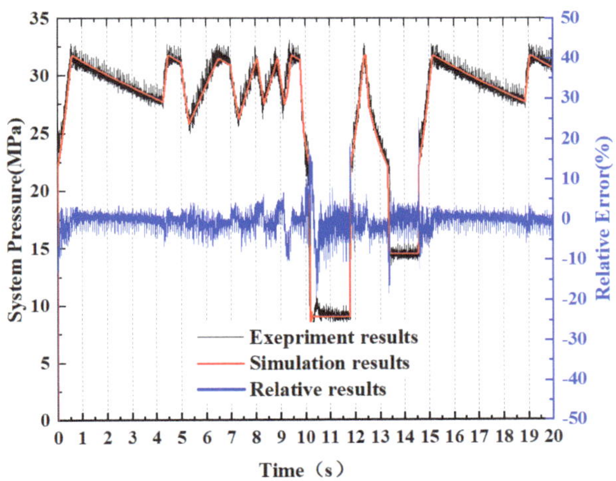

Figure 4. Comparison of system pressure between simulation and experimental results.

As shown in Figure 4, the overall change trend of system pressure measured by experiment is in good agreement with the simulation results under rated fluid supply conditions. It is divided into five stages, which are hydraulic support in no action leakage stage, descending, pulling, raising, and pushing actions. In the pulling stage, the system pressure fluctuates frequently in the operating pressure range of the unloading valve because the fluid supply flow is greater than the fluid demand of the system. On contrary, due to the large demand for fluid in the system, the fluid supply is insufficier the raising and pushing stage. The minimum system pressure drops to 8.9 MPa, and maximum pressure fluctuation is 22.6 MPa. The maximum relative error and the aver relative error between the simulation results and the experimental results are 16.70% 3.73%, respectively, and the experimental results have good consistency with the simula results, which verifies the validity of the simulation model of the hydraulic support sys

2.3. Establishment of Optimal Flow Datasets under Different Operating Conditions

2.3.1. Steady-Pressure Fluid Supply Process

Nowadays, many underground hydraulic support systems in coal mines are ra fluid supply, and the system pressure range is maintained through the unloading va in the pump station. The stability of the system pressure is an important measure of quality of fluid supply from the emulsion pumping station, while the fluid supply flow also determines the hydraulic support operating speed and the degree of system press stability. Figure 5 illustrates the measured data of hydraulic support system pressure mine, and the rated flow rate of the emulsion pump station is 400 L/min.

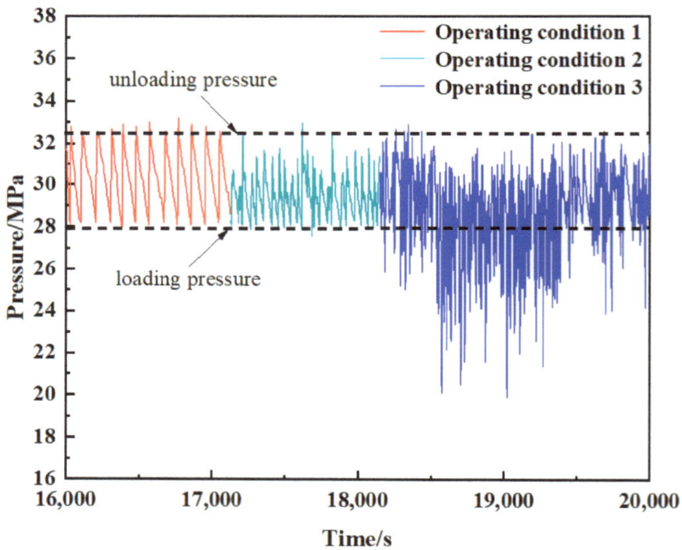

Figure 5. Curve of measured pressure data.

As can be seen from the system pressure data in Figure 5, because the diffe operating modes of the hydraulic support actuator (descending, pulling, raising, pushing actions) have great differences in the demand for fluid supply flow, and the lo are strongly time-dependent, this results in corresponding sudden changes in sys pressure. According to the classification of different amounts of fluid consumption of hydraulic support in the working face, the operation of the working face hydraulic sup is divided into three typical working conditions, and there are different degrees of press fluctuations in each working condition:

Operating Condition 1: When the hydraulic support is inoperative and there is on small amount of system leakage, the leaking fluid is supplemented by the accumulator, the pumping station does not supply fluid to the system. At this stage, the system press fluctuates regularly within the pressure limit of the unloading valve. Under this condit the demand for system fluid is very small, and excessive output flow of the pump station will lead to low system efficiency and increase the amount of system press fluctuation. Operating Condition 2: When the amount of fluid used by the hydraulic support to perform the action is less than the amount supplied by the pumping station, the system pressure continues to fluctuate frequently within the pressure limit of the unloading valve, resulting in system instability. At the same time, due to the frequent action of the unloading valve, the pump station output flow cannot fully enter the system and act on the actuator, but directly into the fuel tank through the unloading valve, which affects the movement speed of the support. Operating Condition 3: When the amount of fluid used by the hydraulic support to perform the action is greater than the amount of fluid supplied by the pump station, the system pressure is greatly reduced until it rises after the operation of the hydraulic support is completed, which may cause the cylinder pressure of the column to fail to reach the ideal initial support force, and the hydraulic support moves slowly due to the low system pressure in the whole process.

For this reason, a steady-pressure fluid supply method adapted to the flow requirements of different working conditions of the hydraulic support was proposed. Reasonable fluid supply flow was output through the emulsion pump station to ensure that the system pressure remained relatively stable within a high-level range. The ideal steady-pressure fluid supply process pressure curve is shown in Figure 6. The process is divided into four stages: the system slowly leaks when the pump station is not replenished (stage a), the system slowly leaks when the pump station is quickly replenished (stage b), the support action during the accumulator supply (stage c), the support action during the pump station supply (stage d). Through this pressure process curve, analyzing the support system of the ideal steady-pressure fluid supply of the four phases of the hydraulic characteristics of the theoretical analysis is specified as follows.

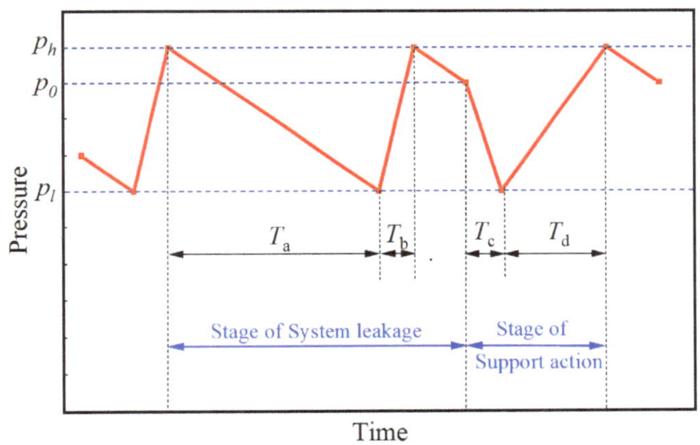

Figure 6. Pressure curve of steady-pressure fluid supply process.

(a) Stage of pump station not replenishing when the system is leaking slowly

In this stage, the hydraulic support does not perform any action, and the system slowly leaks fluid, because the system pressure is higher than the unloading valve loading pressure p_l, the pumping station is in the unloading state, the pumping station does not replenish the fluid to the system, and the leaking fluid is replenished by the accumulator, which leads to a slow reduction of the system pressure from the unloading valve unloading pressure p_h to p_l. The duration of this stage:

$$T_a = \frac{V_h - V_l}{Q_{out}} \tag{1}$$

where T_a is the duration of stage a; V_h is the volume of the accumulator when the system pressure is p_h; V_l is the volume of the accumulator when the system pressure is p_l. Q_{out} is the leakage volume of the system.

The relationship of accumulator pressure and volume:

$$p_x \cdot (V_e - V_x) = p_e \cdot V_e$$

where p_e is the rated pressure of the accumulator; V_e is the rated volume of the accu lator; p_x is the current pressure of the accumulator; and V_x is the current volume of accumulator.

(b) Stage of pumping station rapid replenishment when the system is leaking slo

In this stage, the hydraulic support does not perform any action, the system le slowly, because the system pressure is higher than p_l, the pumping station is in the load state, the pumping station replenishes the fluid to the system quickly. Due to the sh duration of this stage, the leakage of the system can be ignored, the output flow of pumping station is regarded as all the fluid charging to the accumulator, and the syst pressure is rising rapidly from p_l to p_h. The duration of this stage is:

$$T_b = \frac{V_h - V_l}{Q_p}$$

where Q_p is the flow rate of fluid supplied from the pumping station.

(c) Stage of accumulator fluid supply during the hydraulic support operation

In this stage, the hydraulic support began to act, due to the system pressure be higher than p_l, the pumping station is in the unloading state, the pumping station did supply fluid to the system, the hydraulic support action power source for the accumula the accumulator output pressure fluid to push the hydraulic support hydraulic cylin resulting in a rapid decline in system pressure to p_l, at this time the hydraulic supp hydraulic cylinder to the initial speed v_l into the pumping station supplying the fluid st

$$v_l = \frac{Q_e - Q_{out}}{A_{in}}$$

where Q_e is the flow rate of fluid supply to the accumulator, Q_{out} is the fluid leakage of system; A_{in} is the action area of the hydraulic support feeding cylinder.

(d) Stage of pumping station fluid supply during hydraulic support operation

At the beginning of this stage, the hydraulic cylinder of the hydraulic support mo at the initial speed v_l, the system pressure is the loading pressure of the unloading va p_l, and the pump station enters the loading state to supply fluid to the system. Ic steady-pressure fluid supply situation system pressure rises uniformly, when the hydra cylinder completes the current action at the same time, the system pressure rises to Hydraulic cylinder force to meet the formula:

$$pA_{in} = F_q + \theta v A_{out} + ma$$

where p is the real-time pressure, A_{in} and A_{out} are the area of the hydraulic cylinder in and outlet chambers, respectively, F_q is the load on the hydraulic cylinder, θ is the resista coefficient of the hydraulic cylinder in relation to the speed of action, m is the mass of hydraulic cylinder, and a is the acceleration of the hydraulic cylinder action. The hydra cylinder action stroke is:

$$\Delta S = v_l T_d + \frac{1}{2} a T_3^2$$

According to the principle of volume balance within the hydraulic system, the en sion pumping station supply flow rate is the sum of the flow rate required for the actio the hydraulic cylinder of the hydraulic support, the flow rate required for the charging of accumulator, and the leakage flow rate of the system, which satisfies the following form

$$Q_p T_d = \Delta L A_{in} + Q_{out} T_d + V_h - V_l$$

According to Formulas (1)~(7), when the rated capacity of the accumulator and the amount of fluid leakage in the system are constant, the flow rate of the ideal steady-pressure fluid supply changes with the action area, action stroke, and load size of the hydraulic cylinder in the execution of the hydraulic support.

2.3.2. The Concept of Optimal Flow Rate

As mentioned before, in the existing research [23], the control strategy of the output flow rate of the pumping station is based on the optimal flow rate of the current actuator action, which is always a fixed and stable flow rate of the current actuator action, ignoring the influence of the actuator resistance and load. However, the downhole working conditions are complex and variable, the actuator resistance and load have strong time-varying characteristics, and the coupling of each factor with the fluid supply flow has a great influence on the system pressure. Therefore, under different working conditions, a fixed supply flow rate will limit the movement speed of the hydraulic cylinder and cause the system pressure to fluctuate.

In order to achieve the highest operating efficiency of the emulsion pump under partload conditions, the flow output control can be further relaxed so that a higher set of flow outputs can be used. In order to achieve the optimum flow output control of the emulsion pump, the objective function is defined as:

$$p_{set} = \text{Max}(Q) \tag{8}$$

The optimized constraint equation is expressed as Equation (9), which is used to maintain an acceptable system pressure.

$$P_l \leq p_{set} \leq P_h \tag{9}$$

where P_l and P_h are the lower and upper limits of system pressure, which are 28 MPa and 31.5 MPa (unloading valve operating pressure), respectively.

Since the system pressure will be increased by increasing the emulsion pump flow output, the above optimization problem can be simplified to find the maximum flow output value Q_{set}, at which the system pressure can be kept high to allow the actuator to achieve a faster speed of operation. Therefore, under different operating conditions, there will be a corresponding value of Q_{set}.

On the other hand, higher flow output values also lead to higher system pressure. When the system pressure value reaches the regulated unloading pressure of the unloading valve, the unloading valve opens, leading to unloading of the pumping station, and ultimately there are fluctuations in the system pressure due to the oversupply of fluid, which reduces the quality of the fluid supply. Therefore, in order to maintain acceptable system pressures, the fluid supply flow should be limited to the range of the best control method to be developed. In the current development, the lower limit of the system pressure P_l is set to 28 MPa, which is the closing pressure setting value of the unloading valve. The upper limit of the system pressure, P_h, is set to 31.5 MPa, which is the opening pressure setting of the unloading valve.

Therefore, when the system pressure is higher than 28 MPa, the higher system pressure accelerates the actuator's movement [24] and ensures a faster movement speed maintained by the actuator. In the process of determining the Q_{set} value, when the matched pressure value p reaches 31.5 MPa, the control method adjusts Q_{set} downwards to avoid unloading the unloading valve until p is in the range of 28–31.5 MPa.

2.3.3. Optimal Flow Dataset under Different Working Conditions

A simulation model is built to simulate a hydraulic support; the purpose is to establish an emulsion pump output flow dataset and set the Q_{set} dataset under different working conditions by artificially controlling the flow rate of the fluid supply, so as to realize the effect of the steady-pressure fluid supply as described in Section 2.3.1. Under specific working conditions, the inlet and outlet areas of different hydraulic cylinders repre different types of hydraulic support movements, and the rest of the data are obtai through experimental records. The accuracy of the sensors and sampling intervals r cause some errors in some of the data, but these errors are acceptable for the accu requirements of the flow prediction of the steady-pressure fluid supply.

The actual value of the system pressure p is limited to between 28 MPa and 31.5 M The required system pressure value is adjusted according to the difference between actual system pressure value p and the set p_{set}, thus adjusting the output flow rate of emulsion pump to cope with the actual system pressure. The emulsion pump regulates required system pressure with a fixed output flow rate and supplies fluid to the actu at this system pressure. Thereafter, the output flow rate from the experiment can be u as the desired Q_{set}, which can then be specified as the optimum output flow rate Q_{s} the emulsion pump for that particular operating condition, with the goal of maximiz system pressure and minimizing system pressure fluctuations.

Using the simulation model of the analogue hydraulic support, obtain the datase emulsion pump output flow rate for different working conditions set up as represer by the four actions of the hydraulic support including descending, pulling, raising, pushing actions, which contains 300 sets of data on the hydraulic support action t action stroke, hydraulic cylinder load, and pipeline length.

3. Establishment of a GRNN Model for Predicting the Optimal Dataset of Flow Rat

This section attempts to present a new approach aimed at providing stabiliza of emulsion pressure. The GRNN artificial neural network modelling approach has advantages of simple network architecture, fast convergence, and excellent predic results when the sample data are small and unstable, and has been widely used in fields of heat load prediction [25], hydraulic control [26], time series prediction [27], trajectory tracking control [28]. Its specific modelling approach is extensively describe the open literature [29]. It is suitable for the problem of prediction of unstable data ur such complex operating conditions as underground hydraulic support systems.

According to the optimal flow rate requirements introduced in the previous sec the optimal flow output dataset of the hydraulic support system under different work conditions was determined by simulation, and based on the relationship between inputs of the obtained dataset and the corresponding outputs, a GRNN prediction mo connecting the optimal flow output value with four environmental parameters, namely type of action of the hydraulic support, the stroke of the action, the length of the pipe and the load of the actuator, was established.

3.1. Input and Output Parameters of the GRNN Model

Previous studies [15,22] have shown that when the actuator action resistance co cient, system leakage, and unloading valve set pressure are certain, the type of actu action and action stroke, the length of the pipeline, and the actuator load are the m factors influencing the flow rate of the fluid supply, so these four parameters are speci as inputs to the GRNN model. At the same time, the optimal emulsion pump flow ra specified as the only output parameter of the GRNN model. The flow rate when the sys pressure profile is the steady-pressure fluid supply profile shown in Figure 6 is collecte the optimal steady-pressure fluid supply flow rate output.

In order to be closer to the actual working conditions in the borehole, the size of pushing cylinder is selected according to the ZY9200/25/50D shielded hydraulic sup configuration cylinder. The action stroke range of descending, pulling, raising, and pusl actions is determined according to the actual working conditions of top removal, the load range of the actuator is calculated by the size of the pushing cylinder and allowable range of system pressure. The optimal flow combination design scheme is sh in Figure 7.

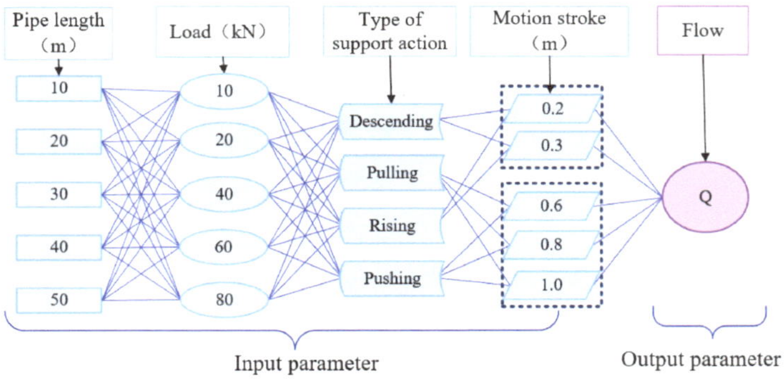

Figure 7. Combined design scheme of optimal flow.

3.2. Training and Testing of GRNN Models

The total number of available data is 300 groups, and the random division method is adopted. Considering the working characteristics of the hydraulic support, the main action is divided into four actions: descending, pulling, raising, and pushing, and the four actions have a large difference in the amount of fluid, so the experiments in this section randomly adopt 70% of the data in each action data as the training data, and the remaining 30% of the data as the test data, respectively. At the same time, the data underwent normalization processing before modelling to eliminate the effect of different magnitudes of data on the model error. The Min–Max normalization processing was used with a mapping range of [0, 1], and the transformation function is as follows:

$$x^* = \frac{x - x_{min}}{x_{max} - x_{min}} \tag{10}$$

where x^* represents the normalized data, x_{max} is the maximum value of the sample data, and x_{min} is the minimum value of the sample data.

3.2.1. Training of the GRNN Model

A GRNN neural network for predicting the optimal flow output was constructed. There are four layers in the GRNN model, an input layer, a pattern layer, a summation layer, and an output layer [30,31]. The model structure is shown in Figure 8. The input layer consists of four simple neurons, which pass the four input parameters of hydraulic bracket action type N, hydraulic cylinder action stroke ΔS, pipe length L, and actuator load $\mathrm{F_q}$ directly to the pattern layer, which contains 210 neurons, where each neuron corresponds to a different training sample, and the transfer function of the neurons in this layer is a Gaussian function [32,33]:

$$\varphi_\mathrm{i} = \exp\left(\frac{-\|\mathrm{D_i}\|^2}{2\sigma^2}\right) \quad \mathrm{i} = 1, 2, \ldots, \mathrm{n} \tag{11}$$

where φ_i is the network output of each neuron in the pattern layer, the weight function of the layer is $\|\mathrm{D_i}\|^2 = (x - x_i)^T(x - x_i)$, x is the input vector, x_i is the ith training sample corresponding to the ith neuron, σ is the smoothing factor. When σ is too large, the model will be insensitive to the changes in the details of the input data and will not learn enough, which may reduce its adaptability to new data. When σ is too small, the model will be too sensitive to the changes in the input data, resulting in overfitting, which may reduce the stability of the network.

In the summation layer, two neurons are used to perform the summation operation, a simple arithmetic summation neuron and a weighted summation neuron. Finally, the output layer outputs the predicted value, consisting of one neuron representing the opti emulsion pump flow output:

$$y = \frac{\sum_{i=1}^{n} y_i \exp\left(-\|D_i\|^2/2\sigma^2\right)}{\sum_{i=1}^{n} \exp\left(-\|D_i\|^2/2\sigma^2\right)}$$

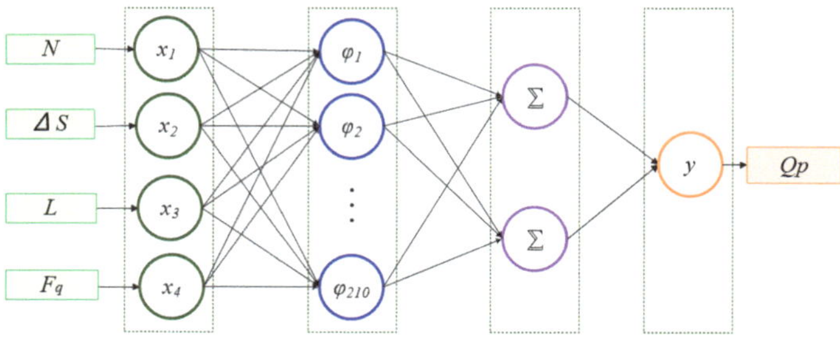

Figure 8. The structure of the GRNN model for predicting the optimal Q_p.

From Equation (12), when the GRNN model is developed using the training dat and its structure and weights are determined, the smoothing factor σ has an impor impact on the performance of the GRNN model, so in order to obtain the optimal predic results and to avoid over-smoothing of the data due to too large a value of σ or over-fit of the data due to too small a value of σ, the parameter smoothing factor obtained the cross-validation method [23] is used as parameter values, the optimal value of σ determined based on the expected error percentage EEP [25], the optimal σ value determined to be 0.06 according to Equation (13), and the minimum EEP value was 1.82

$$EEP = \frac{\sqrt{\sum_{i=1}^{210} (\hat{y}_i - y_i)^2/210}}{|y_{max}|} \times 100\%$$

where $\hat{y}_i$ is the ith predicted flow output value, y_i is the flow output value in the ith dat and y_{max} is the maximum flow output value in the 210 sets of training data.

A GRNN model has been developed which can be used to predict the optim regulated supply flow rate to be set for the emulsion pump under different opera conditions. In Figure 9, the predicted data using the GRNN model are compared with training data. It can be seen that after training and fitting, the final average relative erro the training set flow rate values was obtained to be 0.94%, and the R^2 value reached 0.9 when training the GRNN model.

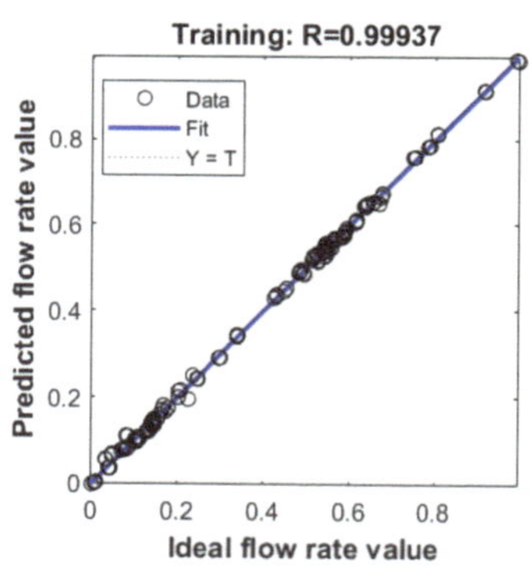

Figure 9. Comparisons between the predicted data and the training data.

3.2.2. Testing of GRNN Model

To test the learning ability of the model, the trained GRNN model was tested using the test set. Figure 10 compares the predicted supply flow values with the test supply flow values. The results show that after training and fitting, the average relative error of the final test set flow values is 1.02%, the R^2 value during the model testing reaches 0.9985, and the correlation coefficient R reaches 0.99894. Therefore, the established GRNN model can be used to predict the optimal set of fluid supply flow of the hydraulic bracket system, which can satisfy the demand of steady-pressure fluid supply with high accuracy.

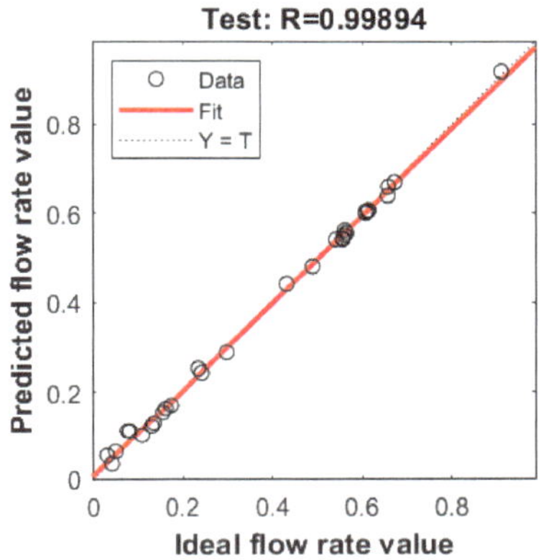

Figure 10. Comparisons between the predicted data with the testing data.

3.3. Online Updating of GRNN Model

Because of the complex and changeable working conditions of hydraulic support, it is difficult for an offline algorithm to predict new working conditions, so an online updating prediction algorithm is needed to identify the model fitting in real time to ensure continuous learning to adapt to the new working conditions. The flowchart of the online updating control approach can be summarized as shown in Figure 11.

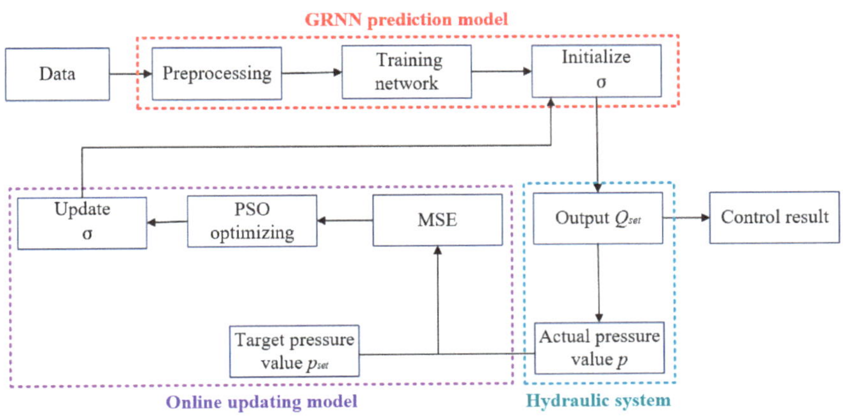

Figure 11. The flowchart of the online updating control approach.

The smoothing parameter σ of the GRNN model is updated based on the PSO optimization algorithm, and the prediction accuracy of the GRNN model is improved by adjusting σ reasonably. The MSE of the collected mean square error between the system pressure and the ideal pressure for steady-pressure fluid supply within 0.5 s before and after the end of the hydraulic support action is selected to comprehensively evaluate the performance of the current GRNN, and the training dataset is constantly updated. MSE calculation formula is as follows:

$$MSE = \frac{1}{N}\sum_{i=1}^{N}[p(t) - p_{set}(t)]^2$$

where N is the data length, p is the current pressure value, and p_{set} is the target press value of steady-pressure fluid supply.

Taking MSE as the input of the PSO optimization problem, the fitness function of is calculated, and the smoothing parameters in each time step are continuously calcula by PSO. Find the individual extreme value $\sigma_{bi} = (\sigma_{bi1}, \sigma_{bi2,\ldots,}\sigma_{bim,})$ and the optimal posi of population history $\sigma_{pb} = (\sigma_{pb1}, \sigma_{pb2,\ldots,}\sigma_{pbm,})$, and the optimal smoothing paramet obtained. Wherein, the speed and position updating formulas are, respectively:

$$V_{im}^{(k+1)} = uV_{im}^{(k)} + c_1 r_1\left(\sigma_{bim}^{(k)} - \sigma_{im}^{(k)}\right) + c_2 r_2\left(\sigma_{pbm}^{(k)} - \sigma_{im}^{(k)}\right)$$

$$\sigma_{im}^{(k+1)} = \sigma_{im}^{(k)} + V_{im}^{(k+1)}$$

where k is the number of iterations, ${V_{im}}^{(k)}$ is the mth dimension component of the velo vector of the ith particle in the kth iteration, m = 1,2, . . ., M, u are inertia weights, c_1 an are acceleration weights, and r_1 and r_2 are random numbers in the interval (0,1).

$$u^{(k)} = u_{max} - k\left(\frac{u_{max} - u_{min}}{iter_{max}}\right)$$

where: $inter_{max}$ is the maximum number of iterations.

4. A Numerical Study on the Development of Steady-Pressure Fluid Supply Metho and Its Numerical Implementation

In the numerical research introduced in this section, the GRNN model is used to pre the optimal steady-pressure fluid supply flow rate, the GRNN model and AMESim/Simu co-simulation model of the hydraulic support system are established, and the opti control method of steady-pressure fluid supply is realized numerically. The self-adap steady-pressure fluid supply control simulation is carried out for single-cycle cons load and variable load conditions, respectively. In addition, the online updating of ste pressure fluid supply control simulation of the unexpected working condition of sys leakage, and the pressure fluctuation of the system, is obtained.

4.1. Single-Cycle Constant Load Steady-Pressure Fluid Supply Control

The action process of single-cycle hydraulic support with constant load is as follows: the system is stabilized for 5 s, four actions of descending, pulling, raising, and pushing executed in sequence, with a waiting time of 0.5 s before each action, and the load is cons Under different working conditions of hydraulic support, the constant fluid supply f will cause the system pressure to decrease in a large range or fluctuate frequently du the imbalance between supply and demand of flow. As a contrast, the rated fluid sup mode and the steady-pressure fluid supply mode are taken as the research objects. In simulation model, the rated fluid supply flow rate of 280 L/min and the optimal ste pressure fluid supply flow rate mentioned above are used to supply fluid to the sys respectively. The pressure curves of the system are shown in Figures 12 and 13.

As can be seen from Figure 12, in the traditional rated fluid supply scheme, when hydraulic support carries out the action of descending and pulling, the system press continues to fluctuate frequently within the pressure limit of the unloading valve du the amount of fluid supplied by the pumping station being larger than that require the actuator; when the hydraulic support carries out the action of raising and pushing, system pressure decreases greatly due to the fluid supplied by the pumping station being smaller than that required by the actuator until it rises after the completion of the action.

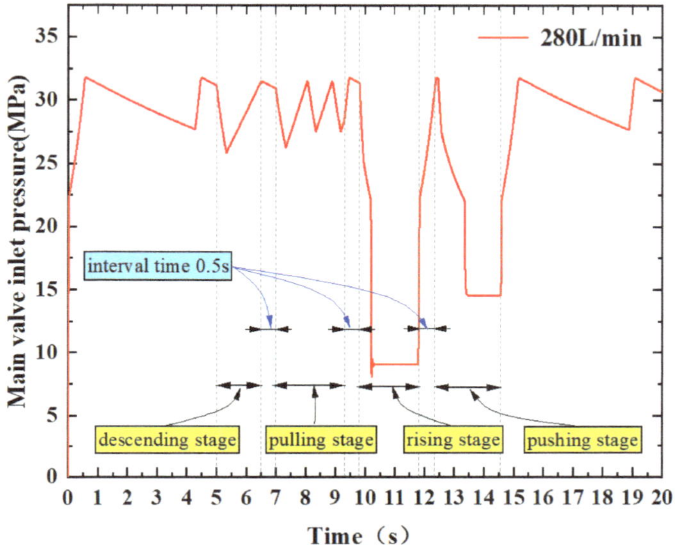

Figure 12. Pressure fluctuation curve of rated fluid supply scheme.

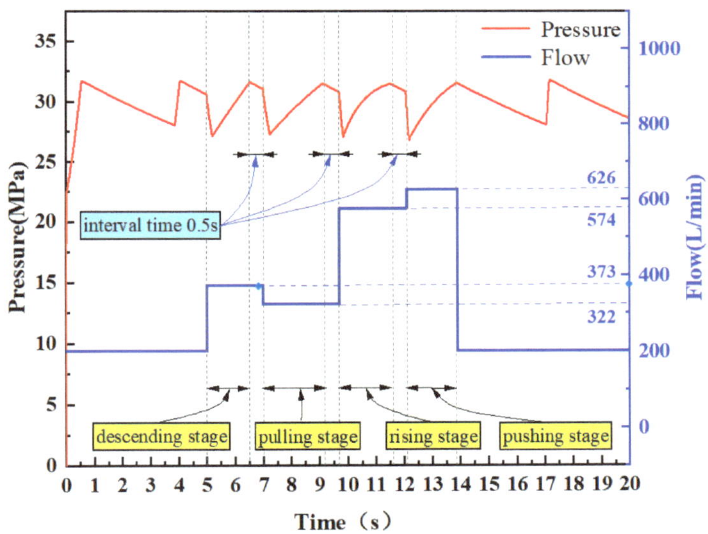

Figure 13. Pressure curve of steady-pressure fluid supply scheme.

As can be seen from Figure 13, the established method of steady-pressure fluid supply can significantly improve the hydraulic system pressure condition, effectively reduce the number of pressure fluctuations during the action of descending and pulling action as well as the amplitude of pressure fluctuations during the action of raising and pushing, and slow down the loss of hydraulic components caused by pressure fluctuations. At the same time, the system pressure is maintained at a higher level, which improves the movement speed of hydraulic support and reduces the cycle time of overall action from 9.61 s to 8.8 saving 7.8% of time.

4.2. Online Updating of Steady-Pressure Fluid Supply Control for Variable Load Conditions

The actual working condition of underground working face is very complicated, load force on the actuator cannot be constant, the operation process of the hydraulic supp follower will be affected by the load characteristics, and different loads have a large imp on the system pressure and fluid supply efficiency. It is difficult to adapt to the real-ti demand of the working surface by only considering the constant load condition for fl supply; therefore, for the variable load condition, the online update of the steady-press fluid supply control is proposed. When the load force on the actuator changes, the emuls pump output flow dataset is adjusted online, so that the next moment of the fluid sup flow more closely matches the working conditions.

In the simulation, varying loads are applied in the operation stages of the simula column raising and descending action stages, and the load signals are shown in Figure Under the variable load conditions as shown in Figure 14, GRNN neural network trained fline and GRNN neural network updated σ online were used to control the steady-press fluid supply, and the pressure curves of the simulated column raising and descend action were collected, as shown in Figure 15. As can be seen from Figure 15: When offline trained GRNN neural network is used for the steady-pressure fluid supply con due to the sudden increase in load at 5.2 s and 6.3 s, the output flow of the emulsion pu station cannot adapt to the changing working conditions in time. As a result, the syst pressure reaches the unloading pressure before the cylinder reaches the target strok 5.5 s and 6.7 s. At this time, the unloading valve is in the unloading state, and the out flow of the emulsion pump station cannot enter the system, so the flow control of the pu station is invalid. In contrast, when using the GRNN neural network of online updat σ for steady-pressure fluid supply control, in the stage of a sudden increase in load, to excessive fluid supply flow, the system pressure change rate suddenly changes, the cylinder moving speed changes, especially in the process of raising and descend action, but because the output flow is constantly updated to adapt to the load force, system pressure does not fluctuate greatly. And the actuator completes the raising descending action smoothly. The results show that the pressure change of the system more stable when the GRNN neural network is updated and trained online, and it good adaptability.

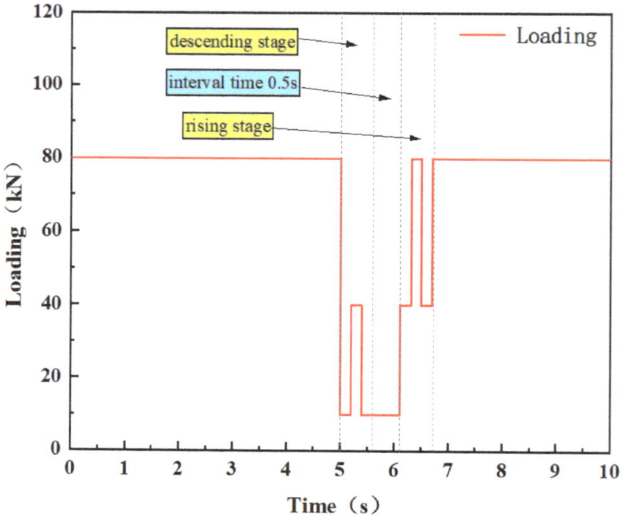

Figure 14. Load signals of the raising and descending stage.

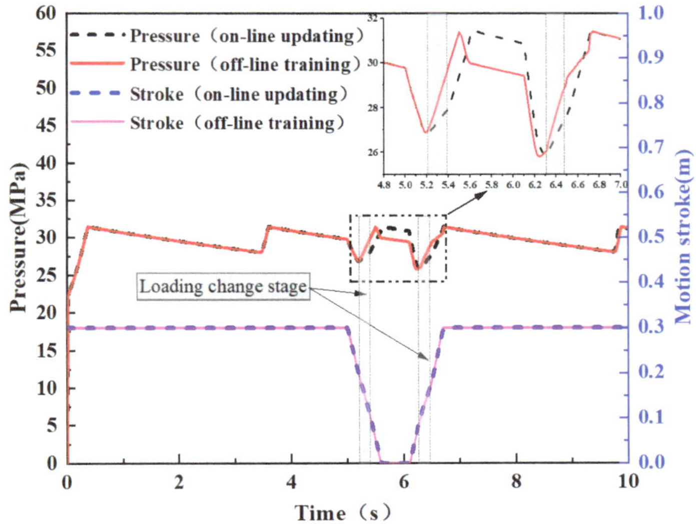

Figure 15. Online updating steady-pressure fluid supply pressure curve.

4.3. Online Updating of Steady-Pressure Fluid Supply Control

Because the downhole hydraulic support system is a high-pressure, large-flow, and long-distance fluid supply system, system leakage is often encountered in practical engineering. Under this working condition, due to system leakage, the output flow calculated by GRNN neural network trained offline in advance will be less than the steady-pressure fluid supply flow, which cannot meet the fluid demand of the actuator. Therefore, by online updating steady-pressure fluid supply flow control, the system pressure of the hydraulic support during each movement (descending, pulling, raising, and pushing) will be different from the target pressure of the steady-pressure fluid supply, and the dataset of fluid supply flow will be continuously adjusted to update the steady-pressure fluid supply flow trained under this working condition until the difference between the system pressure and the target pressure is controlled within the allowable range.

In order to simulate the system leakage condition above, a throttle port with an opening diameter of 3 mm is designed in the pipeline in the simulation model, the steady-pressure fluid supply control is carried out by online updating training, the system pressure is compared with the ideal target pressure of steady-pressure fluid supply at the end of each hydraulic support action, and the training dataset and the smoothing parameter σ are updated. The fluid supply flow during each operation cycle under five different working conditions was randomly recorded and compared with the optimal steady-pressure fluid supply flow under corresponding working conditions. The error is shown in Figure 16. It can be seen that when the system leaks, the error between the output flow calculated by the GRNN neural network trained offline in advance and the optimal steady-pressure fluid supply flow reaches a maximum of 14.8%. However, due to the online updating of σ, the error between the output flow and the optimal steady-pressure fluid supply flow decreases continuously with the adjustment of the fluid supply flow during each operation cycle until the fifth online updating control. The error is all controlled below 3.7%, which meets the requirements of pressure control of hydraulic support.

The changes in the system pressure curves under the online updating control for several times are shown in Figure 17. From the figure, it can be seen that during the first hydraulic support cycle action, the system pressure cannot rise stably due to the system leakage. At the moment of the end of the action, the difference between the system pressure and the unloading pressure is large, resulting in a sudden increase in the system pressure after the reversing valve action, causing a system pressure shock. With each online update control to adjust the amount of fluid supply, during each hydraulic support action, system pressure rise tends to stabilize until the fifth online updating control; at the of the hydraulic support action, the system pressure shock basically disappeared. T shows that the online updating GRNN neural network has good adaptability under sud changes in working conditions.

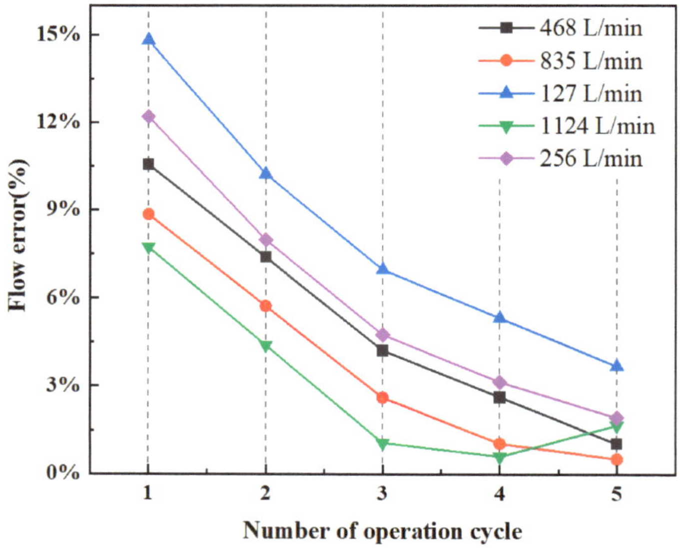

Figure 16. Fluid supply error of online updating.

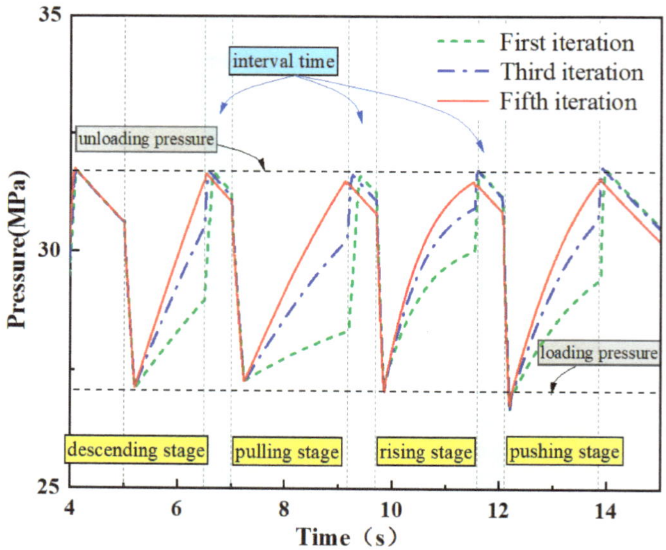

Figure 17. Pressure curve for multiple online updating.

5. Experiment

In order to further verify the executability of the control method, an experime study was carried out through the constructed experimental platform. Since the total rate of the two emulsion pumps in the experimental platform is up to 280 L/min, this pa only studies the two action processes of descending and pulling, which use less fluid. The adjusting pressure of the proportional relief valve at the outlet of the loading pump is set to 20 MPa, which can achieve the loading force loading of the actuator. The flow output of the emulsion pumping station is controlled by the online updating steady-pressure fluid supply flow of the GRNN neural network, and after several online updates, the system pressure curve during the hydraulic support action is measured by the pressure sensor, as shown in Figure 18.

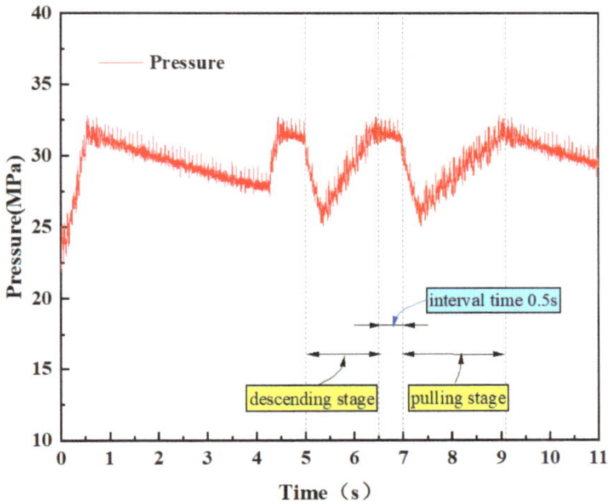

Figure 18. Pressure curve of steady-pressure fluid supply experiment.

It can be seen from the pressure curve that the trend of the system pressure curve and the ideal steady-pressure control curve is the same during the operation of the hydraulic support, which verifies the consistency of the simulation and experiment results. There is no obvious pressure shock in the whole process, and there is only a small range of pressure fluctuation during each action, so the pressure stability of the system is good, which is beneficial to improve the smooth operation of the hydraulic support system.

6. Conclusions

In this paper, on the basis of analyzing the process of steady-pressure fluid supply in the hydraulic support system of the comprehensive mining face, the optimal steady-pressure fluid supply flow prediction model based on the online updating GRNN neural network is proposed for the optimal tracking control of the fluid supply flow output of the emulsion pumping station, and simulation and experimental validation are carried out. The following conclusions can be drawn:

(1) The GRNN model established based on the theory of optimal flow rate for steady-pressure fluid supply and the dataset evaluated by the simulation platform for simulating hydraulic support system has good prediction accuracy, and different operating parameters can be set according to different working conditions.

(2) The optimal control method of steady-pressure fluid supply flow developed based on online updating of the GRNN model can realize good control of hydraulic support system pressure. Under regular actuator load change conditions and special conditions such as system leakage, the PSO optimization algorithm can be updated online in real time to identify the model fitting in real time, ensuring continuous learning to adapt to new working condition characteristics, so as to maintain stable system pressure and improve the efficiency of following the machine to move the frame, which is of good practical performance.

Author Contributions: Conceptualization, P.X. and Z.K.; methodology, J.W.; software, P.X. Y.P.; validation, P.X., Y.P. and B.Z.; formal analysis, Y.P. and T.H.; investigation, P.X.; resou Z.K.; data curation, T.H., Y.P. and B.Z.; writing—original draft preparation, P.X.; writing—rev and editing, P.X., Z.K. and J.W.; visualization, Z.K.; supervision, Z.K.; project administration, and J.W.; funding acquisition, Z.K. All authors have read and agreed to the published versio the manuscript.

Funding: This work was supported by the National Natural Science Foundation of China (U1910

Data Availability Statement: This data is provided upon request due to privacy or ethical restrict

Conflicts of Interest: The authors declare no conflicts of interest.

References

1. Li, R.; Liu, B.; Wang, D.; Chen, J.; Wang, T.; Zhou, R.; Wang, J.; Zhao, K.; Yu, Y.; Qin, H.; et al. Progress and applicatio intelligent fluid supply technology in working face. *Coal Sci. Technol.* **2022**, *50*, 247–253.
2. Ran, L. Intelligent fluid supply technology in fully-mechanized coal mining face and its development trend. *Coal. Sci. Tec* **2019**, *47*, 203–207.
3. Su, J. *Research on Constant Pressure Controlsystem in the Emulsion Pump Station*; Taiyuan University of Technology: Taiy China, 2012.
4. Wang, G.; Xu, Y.; Ren, H. Intelligent and ecological coal mining as well as clean utilization technology in China: Review prospects. *Int. J. Min. Sci. Technol.* **2019**, *29*, 161–169. [CrossRef]
5. Zhou, R.; Li, S.; Wei, W.; Zhang, L.; Wang, W. Study on coupling mechanism of pressure and flow in following hydraulic sys of mining face. *Coal. Sci. Technol.* **2020**, *48*, 129–136.
6. Mamcic, S.; Bogdevicius, M. Simulation of dynamic processes in hydraulic accumulators. *Transport* **2010**, *25*, 215–221. [Cross
7. Van de Ven, J.D. Constant pressure hydraulic energy storage through a variable area piston hydraulic accumulator. *Appl. En* **2013**, *105*, 262–270. [CrossRef]
8. Li, Z.; Guo, Y.; Lian, Z. Analysis of frequency characteristics of hydraulic support's liquid pipeline. *J. Taiyuan Univ. Technol.* **2** *43*, 203–206.
9. Li, Z. Research and analysis on main supply and return system of hydraulic support for large mining height. *Mech. Manag.* **2018**, 46–48.
10. Shu, F.; Wang, Y.; Li, J. Simulate calcultion about pressure loss in hydraulic pipe system of roof supports. *Coal. Technol.* **2009** 26–28.
11. Zhao, X. *Pressure Loss and Dynamic Characteristics of Emulsion Supply and Return Pipeline for Hydraulic Support*; Taiyuan Unive of Science and Technology: Taiyuan, China, 2018.
12. Yang, S. Simulation of the unloading pressure and recovery pressure of internal unloading valve for emulsion pump station, *Mech. Mater.* **2011**, *48–49*, 531–536.
13. Tian, C.; Su, C.; Zhao, W. Application of plc in variable frequency constant pressure control system for mine emulsion pu station. *J. Phys. Conf. Ser.* **2021**, *1982*, 012187. [CrossRef]
14. Qiao, T.; Yan, L. Fuzzy-PID control and application in constant fluid feeding system. *Coal. Eng.* **2011**, *10*, 95–97.
15. Chao, T.; Nan, Q. A pressure control method for emulsion pump station based on elman neural network. *Comput. Intell. Neu* **2015**, *5*, 1–8. [CrossRef]
16. Wang, G.; Li, Z. Optimization design of emulsion pump station using variable frequency driving method in working face high mining height. *Coal Sci. Technol.* **2014**, *42*, 66–69.
17. Li, R.; Wei, W.S.; Lai, Y.H. Long-distance intelligent liquid supply for coal mining faces based on liquid demand predictic *Phys. Conf. Ser.* **2023**, *2561*, 012018. [CrossRef]
18. Ma, C.; Xu, B.; Liu, H. The Optimization of the Constant Flow Parallel Micropump Using RBF Neural Network. In Proceeding the 2021 6th International Conference on Robotics and Automation Engineering (ICRAE), Guangzhou, China, 19–22 Noven 2021; pp. 185–195.
19. Umrao, R.; Chaturvedi, D.K. A novel fuzzy control approach for load frequency control. In *Recent Advancements in Sys Modelling Applications, Proceedings of National Systems Conference 2012*; Springer: New Delhi, India, 2013; pp. 239–247.
20. Tian, C.; Wei, X.W.; Zheng, Y. The intelligent control of emulsion pump station. *J. Phys. Conf. Ser.* **2021**, *1881*, 022044. [Cross
21. Tian, C.; Wei, X.W.; Yang, C.; Su, C. Optimization Technology of frequency conversion constant pressure control system of n emulsion pump station in electrical engineering and automation Specialty. *Adv. Civ. Eng.* **2021**, *2021*, 5569994. [CrossRef]
22. Si, M.; Wu, B.; Wang, Z. Research on large flow intelligent liquid supply system in fully mechanized working face. *J. Mine Au* **2022**, *48*, 66–72.
23. Peng, Y.; Kou, Z.; Wu, J.; Luo, J.; Liu, H.; Zhang, B. Research on a Pressure Control Method for a Liquid Supply System Base Online Updating of a Radial Basis Function Neural Network. *Processes* **2023**, *12*, 57. [CrossRef]
24. Fu, X. *Research on Intelligent Hydraulic Fluid Supply Theory and Technology for Adapting Supports Operation*; Taiyuan Universit Technology: Taiyuan, China, 2017.

Sun, Y.; Xue, H.; Wang, W.; Wu, S.; Wu, Y.; Hong, Y.; Deng, S. Development of an optimal control method of chilled water temperature for constant-speed air-cooled water chiller air conditioning systems. *Appl. Therm. Eng.* **2020**, *180*, 115802. [CrossRef]
Pławiak, P. An estimation of the state of consumption of a positive displacement pump based on dynamic pressure or vibrations using neural networks. *Neurocomputing* **2014**, *144*, 471–483. [CrossRef]
Bukhari, A.H.; Raja, M.A.Z.; Alquhayz, H.; Abdalla, M.Z.; Alhagyan, M.; Gargouri, A.; Shoaib, M. Design of intelligent hybrid NAR-GRNN paradigm for fractional order VDP chaotic system in cardiac pacemaker with relaxation oscillator. *Chaos Solitons Fractals* **2023**, *175*, 114047. [CrossRef]
Lopez-Sanchez, I.; Rossomando, F.; Pérez-Alcocer, R.; Soria, C.; Carelli, R.; Moreno-Valenzuela, J. Adaptive trajectory tracking control for quadrotors with disturbances by using generalized regression neural networks. *Neurocomputing* **2021**, *460*, 243–255. [CrossRef]
Li, C.; Bovik, A.C.; Wu, X. Blind image quality assessment using a general regression neural network. *IEEE Trans. Neural. Netw.* **2011**, *22*, 793–799. [PubMed]
Aengchuan, P.; Wiangkham, A.; Klinkaew, N.; Theinnoi, K.; Sukjit, E. Prediction of the influence of castor oil–ethanol–diesel blends on single-cylinder diesel engine characteristics using generalized regression neural networks (GRNNs). *Energy Rep.* **2022**, *8*, 38–47. [CrossRef]
Iban, M.C.; Şentürk, E. Machine learning regression models for prediction of multiple ionospheric parameters. *Adv. Space Res.* **2022**, *69*, 1319–1334. [CrossRef]
Specht, D.F. A general regression neural network. *IEEE Trans. Neural Netw.* **1991**, *2*, 568–576. [CrossRef]
Xia, C.; Lei, B.; Wang, H.; Li, J. GRNN short-term load forecasting model and virtual instrument design. *Energy Procedia* **2011**, *13*, 9150–9158.

actuators

Article

Optimization Design of a Polyimide High-Pressure Mixer Based on SSA-CNN-LSTM-WOA

Guo Yang [1], Guangzhong Hu [1,2], Xianguo Tuo [1,*], Yuedong Li [1,*] and Jing Lu [1]

1 School of Mechanical Engineering, Sichuan University of Science & Engineering, Zigong 643000, China; yg2511535580@gmail.com (G.Y.); hgzdhx@163.com (G.H.); lujing_017@live.cn (J.L.)
2 Key Laboratory of Advanced Manufacturing Technology of Panzhihua City, Panzhihua 617000, China
* Correspondence: tuoxg@cdut.edu.cn (X.T.); lyd_cdut@hotmail.com (Y.L.)

Abstract: Foam mixers are classified as low-pressure and high-pressure types. Low-pressure mi rely on agitator rotation, facing cleaning challenges and complex designs. High-pressure mi are simple and require no cleaning but struggle with uneven mixing for high-viscosity substa Traditionally, increasing the working pressure resolved this, but material quality limits it at hi pressures. To address the issues faced by high-pressure mixers when handling high-viscosity mate and to further improve the mixing performance of the mixer, this study focuses on a polyimide h pressure mixer, identifying four design variables: impingement angle, inlet and outlet diameters, impingement pressure. Using a Full Factorial Design of Experiments (DOE), the study investig the impacts of these variables on mixing unevenness. Sample points were generated using Opti Latin Hypercube Sampling—OLH. Combining the Sparrow Search Algorithm (SSA), Convoluti Neural Network (CNN), and Long Short-Term Memory Network (LSTM), the SSA-CNN-LS model was constructed for predictive analysis. The Whale Optimization Algorithm (WOA) optim the model, to find an optimal design variable combination. The Computational Fluid Dynamics (C simulation results indicate a 70% reduction in mixing unevenness through algorithmic optimiza significantly improving the mixer's performance.

Keywords: DOE; foam polyimide high-pressure mixer; foam mixers; mixing unevenness optimiza SSA-CNN-LSTM; WOA

Citation: Yang, G.; Hu, G.; Tuo, X.; Li, Y.; Lu, J. Optimization Design of a Polyimide High-Pressure Mixer Based on SSA-CNN-LSTM-WOA. *Actuators* **2024**, *13*, 303. https://doi.org/10.3390/act13080303

Academic Editor: Luigi de Luca

Received: 24 June 2024
Revised: 2 August 2024
Accepted: 5 August 2024
Published: 8 August 2024

1. Introduction

Polyimide foam is considered one of the most versatile high-performance material the new century, with excellent mechanical properties, dielectric properties, organic solv resistance, high-temperature resistance, and radiation resistance. Due to these exceptio properties, its demand is increasing in fields such as engineering materials process microelectronics, aerospace, and the military [1]. Currently, polyimide foaming machi are adapted from polyurethane foaming machines, leading to discrepancies between two materials. In actual production, the uneven mixing of raw materials can resul the poor quality of the final molded products, low yield rates, and an inability to m market demands. To improve the material's evenness, the most important compon mixer should be sufficiently considered and designed.

The mixer is the core component of the polyimide foaming machine, and its per mance directly affects the quality of the foam. To address the issue of uneven mixin the mixer, in addition to altering the operating parameters of the mixing head during mixing process [2], research on the optimization of the mixing head structure has dem strated that parameters such as the impingement angle, inlet diameter, outlet diam and pressure all influence the mixing effect of the mixer [3,4]. Yuan Xiaohui et al. [5] u the software "Ansys" (ANSYS 2022r1) to analyze the flow field of materials in the mix foaming machine, focusing on the unfolded flow channel. The shear stress was then de mined based on the velocity field, revealing that circulation in the cross-sectional direc promotes the thorough mixing of materials in the foaming machine. Trautmann et al. [6] conducted experimental research on the impingement angle and found that it influences the mixing effect of the mixing head. Guo et al. [7] analyzed the mixing mechanism of the foaming machine, confirming that high-pressure impingement mixing is superior to low-pressure mixing. They also analyzed factors affecting the mixing effect, such as the impingement pressure and feed diameter. Schneider [8] and Bayer [9] designed a feed port with an asymmetric structure, where the fluid inlet is designed as a high-pressure nozzle. Their research suggested that high-pressure mixing achieves a more uniform blend compared to low-pressure mixing. Murphy et al. [10] conducted a detailed study of the factors affecting rapid liquid–liquid mixing in impingement stream mixers, providing a theoretical basis for the optimization design of mixers. Van Horn et al. [11] designed flow channels in a Venturi shape, utilizing a reduction in diameter to increase the flow velocity of the two fluid streams.

There are various optimization methods for fluid machinery, including traditional verification methods and intelligent optimization techniques [12,13]. The former relies on extensive data and experience, while the latter employs approximate models such as the response surface method, kriging models, and artificial neural networks, utilizing intelligent algorithms like genetic algorithms and particle swarm optimization for the optimization process. Luo et al. [14] established a predictive model of centrifugal pump parameters using a backpropagation (BP) neural network and then sought the optimum value using genetic algorithms. After optimization, the impeller efficiency increased by 1.4%. Jiang et al. [15] utilized the orthogonal experimental method to analyze the impact of design parameters on mixing efficiency, and then applied the GA-BP-GA methodology for the structural optimization of the mixer. Following optimization, the mixing index decreased by 52%. Li et al. [16] employed the response surface method to create a regression predictive model for supersonic nozzles, and using genetic algorithms, determined the optimal structural parameters for the nozzle. Compared to traditional optimization methods, the approach that combines approximate models with intelligent algorithms for fluid machinery optimization can effectively enhance the efficiency of the optimization process and reduce the time cycle.

In summary, to address the issue of uneven mixing in polyimide mixers handling high-viscosity liquids, we focused on the impact of the mixing head's structure on its performance. Both domestic and international studies primarily employ fluid simulation techniques to visualize flow fields, optimizing structures and operational parameters based on the mixer's performance [17–19]. Building on the existing high-pressure foam equipment of a particular company, we assessed the impact of each structural component on mixing unevenness using full factorial DOE and factor analysis. Samples were extracted through OLH, and predictive models of mixing unevenness were established using the SSA-CNN-LSTM. Subsequently, the WOA was used to optimize structural parameters, aiming to enhance the mixing efficiency and structure of the mixing head.

2. CFD Simulation of Three-Phase Fluid Flow in a Polyimide Mixer

A high-pressure polyimide mixer is essential for producing polyimide foams. The research, development, and design of such equipment often rely on empirical formulas, leading to high uncertainty and lengthy development cycles. In some experiments, techniques like Planar Laser-Induced Fluorescence (PLIF) and Particle Image Velocimetry (PIV) visualize fluid concentration and velocity within the flow field [20,21]. For mixers synthesized from reactants, methods such as Transmission Electron Microscopy (TEM) and wavelength dispersive X-ray fluorescence spectroscopy accurately characterize nanoparticle composition and size distribution [22,23]. However, experimental research methodologies require significant time and resources. This chapter will treat the mixer as a black box, using CFD simulation techniques to visually investigate the polyimide high-pressure mixer.

2.1. Mixed Head Structure Analysis

High-pressure mixers are essential for mixing polyimide materials, making the st of fluid flow characteristics within these mixers important. This research involves a si lated study of a high-pressure mixer (Figure 1) used in a polyimide foam production During operation, material flows into the mixer through the inlet and circulates bac the feed pipe via the return duct. When the pressure reaches the set value, the mate enters the mixing chamber. Materials A and B mix and are ejected through the outlet the mold.

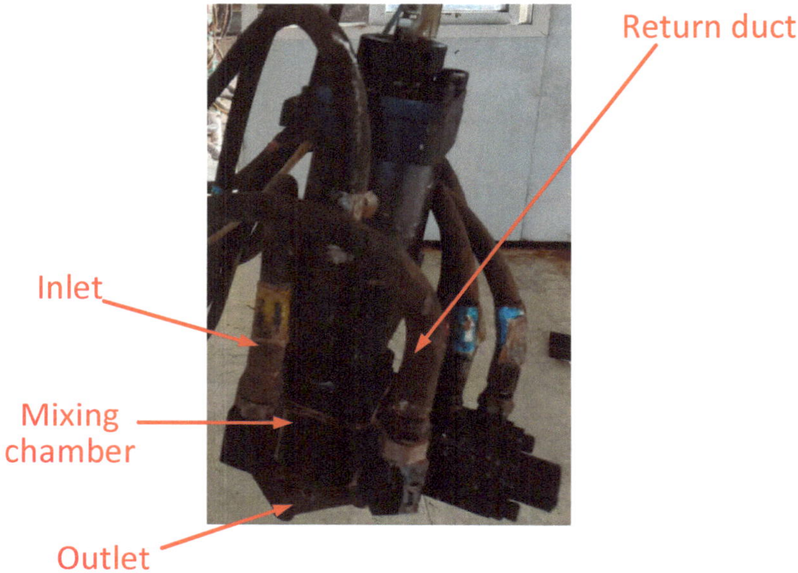

Figure 1. High-pressure mixer.

In the high-pressure mixer, materials A and B are fed into the mixing chamber thro two separate inlets at a high speed and pressure by metering pumps, achieving high-sp collision mixing. The geometric model of the polyimide high-pressure mixer was crea using 3D drawing software (Solidworks 2021) to explore the impact of the mixing hea structural parameters and optimize its structure. The model was used for the fluid anal of the high-speed collision mixing process of liquids A and B. The structure of the mix chamber is shown in Figure 2. Materials A and B enter the chamber through inlets A B. Inside, they collide and mix at high speed at a specific impingement angle determi by the two inlets. The shape of the fluid collision area is complex and irregular due to high-speed, angled collision, generating intense turbulence and vortices. Rapid chan in the fluid velocity and direction after the collision result in highly irregular, nonlin flow paths. Additionally, the significant velocity gradient within the collision area cau uneven material distribution and pronounced shear interactions between the fluids. Th factors collectively contribute to the complexity and unpredictability of fluid flow wit the mixing zone. After mixing, the materials exit the chamber through the outlet. inlet diameters for components A and B are 8 mm, the outlet diameter is 35 mm, and impingement angle is 180 degrees.

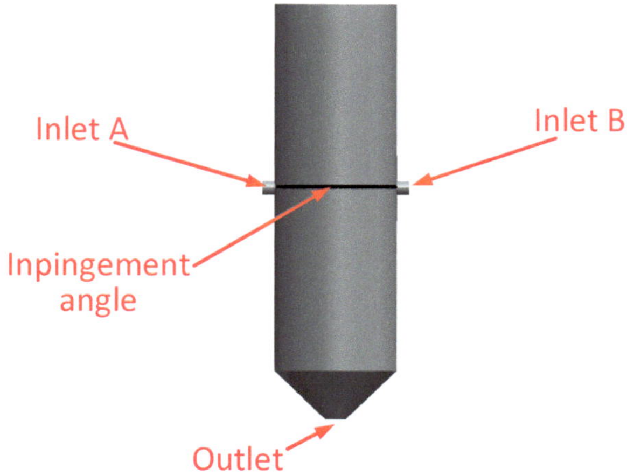

Figure 2. High-pressure-mixer three-dimensional model.

2.2. Mathematical Model

This section introduces the conservation laws of energy, mass, and momentum that fluid follows in a mixer. Based on the Navier–Stokes equation theory, a mathematical model is established. The mass and energy conservation equations are described, clarifying parameters like fluid density, velocity components, and temperature. The turbulence model, proposed by Launder and Spalding, is explained, detailing the relationship between turbulence energy, dissipation rate, and viscosity. The multiphase flow model is discussed, employing the Volume of Fluid (VOF) model to handle gas–liquid–liquid three-phase flow, introducing the continuity equations for each phase and the momentum conservation equation. These theories and equations provide a foundation for numerical simulations and analyses.

2.2.1. Reynolds Time-Averaged Control Equation

When fluid flows within a mixer, it adheres to the laws of conservation of energy, mass, and momentum. The mathematical model is established based on the theory of the Navier–Stokes control equations. The specific expression of these equations is as follows.

Mass conservation equation:

$$\frac{\partial \rho}{\partial t} + \frac{\partial(\rho\mu)}{\partial x} + \frac{\partial(\rho v)}{\partial y} + \frac{\partial(\rho w)}{\partial z} = 0 \tag{1}$$

Vector form:

$$\frac{\partial \rho}{\partial t} + \nabla \cdot (\rho v) = 0 \tag{2}$$

In the equation: ρ—fluid density, with units in Kg/m^3; u, v, w—velocity components in the x, y, z directions, respectively, in m/s.

Energy conservation equation:

$$\frac{\partial(\rho t)}{\partial t} + div(\rho \vec{u} T) = div(\frac{k}{c_p} gradT) + S_T \tag{3}$$

In the equation: C_p—specific heat capacity, with units in KJ/(kg×°C); T—temperature, in units K; k—fluid heat transfer coefficient; and S_T—viscous dissipation term.

$$\begin{aligned} &\frac{\partial(\rho u)}{\partial t} + \frac{\partial(\rho u w)}{\partial x} + \frac{\partial(\rho v w)}{\partial y} + \frac{\partial(\rho w w)}{\partial z} \\ &= \rho f_z - \frac{\partial p}{\partial z} + \frac{\partial(2\mu \frac{\partial u}{\partial z} + \bar{\lambda} div v)}{\partial z} + \frac{\partial[\mu(\frac{\partial v}{\partial z} + \frac{\partial u}{\partial x})]}{\partial x} + \frac{\partial[\mu(\frac{\partial w}{\partial y} + \frac{\partial u}{\partial z})]}{\partial y} \end{aligned}$$

$$\frac{\partial(pu)}{\partial t}+\frac{\partial(puu)}{\partial x}+\frac{\partial(\rho vu)}{\partial y}+\frac{\partial(\rho wu)}{\partial z}$$
$$=\rho f_x-\frac{\partial p}{\partial x}+\frac{\partial(2\mu\frac{\partial u}{\partial x}+\overline{\lambda}divv)}{\partial x}+\frac{\partial[\mu(\frac{\partial v}{\partial x}+\frac{\partial u}{\partial y})]}{\partial y}+\frac{\partial[\mu(\frac{\partial w}{\partial x}+\frac{\partial u}{\partial z})]}{\partial z}$$
$$\frac{\partial(\rho v)}{\partial t}+\frac{\partial(\rho uv)}{\partial x}+\frac{\partial(\rho vv)}{\partial y}+\frac{\partial(\rho wv)}{\partial z}$$
$$=\rho f_x-\frac{\partial p}{\partial y}+\frac{\partial(2\mu\frac{\partial u}{\partial y}+\overline{\lambda}divv)}{\partial y}+\frac{\partial[\mu(\frac{\partial v}{\partial x}+\frac{\partial u}{\partial y})]}{\partial x}+\frac{\partial[\mu(\frac{\partial w}{\partial x}+\frac{\partial u}{\partial z})]}{\partial z}$$

In Equation (4), u, v, w—velocity components in the x, y, z directions, respecti in m/s; ρ—fluid density, in kg/m^3; P—static pressure, in Pa; and μ—molecular visco in Pa·s.

2.2.2. Turbulence Model

The standard model, proposed by Launder and Spalding, is a predominant to engineering flow field computations due to its broad applicability, cost-effectiveness, an curacy. It is recognized as the simplest model producing effective computational outco especially in simulating a single-phase fluid flow within stirred tanks. Consequently, turbulence model is chosen to address turbulence in mixers. In this model, the relation among turbulence energy k, the rate of turbulence dissipation ε, and the turbulent visc μ_t is represented as follows:

$$\mu_t=\rho C_\mu\frac{k^2}{\varepsilon}$$

The turbulence kinetic energy k equation is:

$$\frac{\partial(\rho k)}{\partial t}+\frac{\partial(\rho k\mu_i)}{\partial x_i}=\frac{\partial[(\mu+\frac{\mu_t}{\sigma_k})\frac{\partial k}{\partial x_j}]}{\partial x_j}+G_k-\rho\varepsilon$$

The equation for the turbulence energy dissipation rate is:

$$\frac{\partial(\rho\varepsilon)}{\partial t}+\frac{\partial(\rho\varepsilon u_i)}{\partial x_i}=\frac{\partial[(\mu+\frac{\mu}{\sigma_\varepsilon})\frac{\partial\varepsilon}{\partial x_j}]}{\partial x_j}+\frac{C_{1\varepsilon}}{k}G_k-C_{2\varepsilon}\rho\frac{\varepsilon^2}{k}$$

The production term of the turbulent kinetic energy k due to the mean velocit calculated by the following expression:

$$G_k=\mu(\frac{\partial u_i}{\partial x_j}+\frac{\partial u_j}{\partial x_i})\frac{\partial u_i}{\partial x_j}$$

2.2.3. Multiphase Flow Model

Extensive numerical simulations have demonstrated [24–28] that the VOF mc exhibits an excellent performance in managing both steady-state and transient gas–liq interfaces. This study examines a gas–liquid–liquid three-phase flow, employing the V multiphase flow model. The continuity equations for each phase are as follows:

$$\rho_q\frac{\partial\alpha_q}{\partial t}+\nabla(U\vec{\alpha_q}\rho_q)=S_{\alpha_q}+\sum_{p=1}^{n}(\dot{m}_{pq}-\dot{m}_{pq})$$

In these equation, p and q denote different phases; ρ_q is the density of phase represents time; α is the volume fraction; U is the velocity at any point in the flow fi S_{α_q} indicates the source phase; and $\dot{m}_{pq}$ and $\dot{m}_{qp}$ are the mass transfer rates between two phases.

In the VOF model, all phases share a single velocity field. The momentum conservation equation is given by:

$$\frac{\partial}{\partial t}(\rho\vec{U}) + \nabla\bullet(\rho\vec{U}\vec{U}) = -\nabla P + \nabla\bullet[\mu(\nabla U + \vec{U}^T)] + \rho g + \vec{F} \tag{10}$$

2.3. Numerical Solving Method

The standard k-ε model is used to solve the flow within the mixer. The inlets for materials A and B both use pressure inlet boundary conditions, with the pressure at the inlets set to the impingement pressure. The inner walls of the mixer are treated as stationary walls. Gravity is also considered, with its direction being perpendicular to the outlet plane. The pressure–velocity coupling method utilizes the SIMPLE algorithm, with pressure discretization using PRESTO! All other discretization schemes adopt a second-order upwind scheme, with under-relaxation factors set to default values and variable residuals all set to 10^{-6}. Additionally, the VOF multiphase flow model is incorporated. According to actual working conditions, fluids A and B are selected as working substances. Fluid A has a density of 1.17 g/cm^3 and a viscosity of 0.16 Pa·s; fluid B has a density of 1.25 g/cm^3 and a viscosity of 0.36 Pa·s, with an operating pressure of 10 MPa. Because the fluid's mixing time within the mixer is brief, and the viscosity has a minimal impact on the mixing performance. To simplify the calculations, the following assumptions are made: (1) no reactions occur within the mixer; (2) temperature and pressure effects on viscosity are neglected; and (3) the flow field within the mixer is considered as a three-dimensional, steady, isothermal, incompressible viscous fluid in stable turbulent flow.

2.4. Grid Partitioning and Investigation of Grid Independence

The mixer was meshed using the grid function in Fluent software (Fluent 2022). As the study focuses on impingement mixing effects rather than flow details near the wall, the boundary layer was not considered. This simplifies the model and conserves computational resources. Due to the complexity and irregular shape of the fluid collision area, the number of grids significantly affects calculation accuracy. Too many grids increase the computational burden and prolong calculation time, while too few compromises precision. Therefore, selecting an appropriate number of grids is essential. The mesh element quantities for the five sets of plans are listed in Table 1, and the resulting pressure drop is shown in Figure 3.

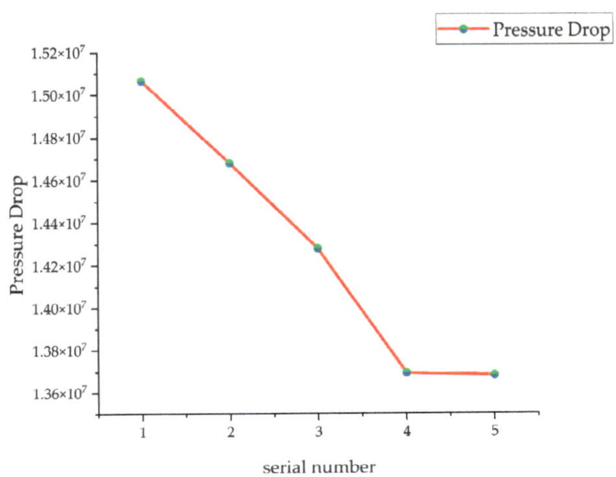

Figure 3. Mesh independence verification.

From Figure 3, it is observed that when the number of grids reaches 146,142, the p sure drop remains nearly constant. Therefore, the selected number of grids for calcula is 146,142. The results of the mesh division are shown in Figure 4. The quality of the m elements is greater than 0.444, meeting the computational and accuracy requirements.

Table 1. Mesh element quantities for five plans.

Plan	Number of Mesh Elements
Plan 1	58,574
Plan 2	63,874
Plan 3	86,824
Plan 4	146,142
Plan 5	221,348

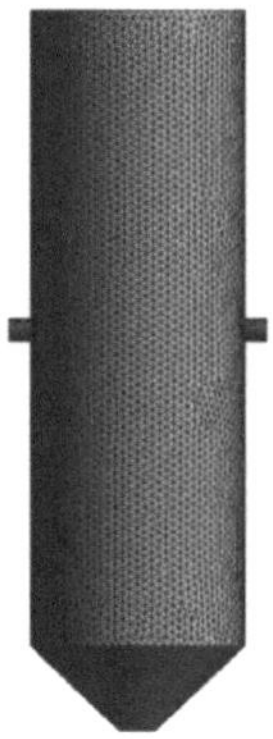

Figure 4. Mesh division results.

2.5. *Results Analysis and Discussion*

This section evaluates the mixing performance of the mixer through both qualita and quantitative analyses methods. The qualitative method includes plotting den contour maps of the outlet cross-section and observing color bands and distribution assess mixing uniformity. The quantitative method uses the root mean square deviatio the liquid density at various sampling points to measure mixing quality. The specific st involve, after completing numerical calculations, uniformly selecting liquid density val from nodes on the outlet cross-section for calculation.

2.5.1. Mixer Mixing Performance Evaluation Index

To ensure the accuracy of the conclusions, both qualitative and quantitative analy are employed to evaluate the mixing effects. The qualitative method involves plotting a c sity contour map of the outlet section to observe color bands and distribution, determir the uniformity of the mixing.

The quantitative method uses mixing unevenness, the root mean square deviat of the liquid density at various points. This metric represents the degree of differe between the liquid density at the sampling points and the average liquid density of mixing system under investigation. Thus, this article uses the root mean square deviat of the density values on the mixer's outlet cross-section to assess the mixing quality of designed mixer. The S parameter can be determined:

$$S = \sqrt{\frac{\sum_{i=1}^{n} (\rho_i - \overline{\rho})^2}{(n-1)}} / \overline{\rho}$$

In Equation (11), ρ represents the liquid density, and n stands for the number of sample points. Generally, the smaller the mixing unevenness, the smaller the difference in liquid density across the cross-section, indicating a smaller density gradient and better mixing quality.

To compute the mixing index, after the numerical calculations are completed, the liquid density values at 50 uniformly selected nodes on the outlet cross-section are taken, as shown in Figure 5.

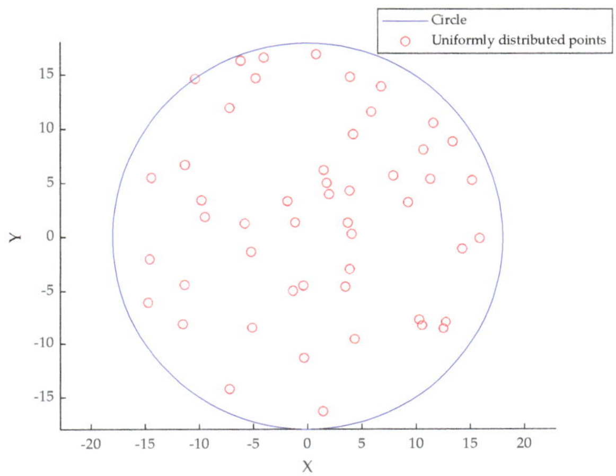

Figure 5. Distribution of sampling points.

2.5.2. Simulation Results

To improve the mixing performance of the mixer head, an internal flow field analysis is conducted. Based on the prototype mixer, the effect of the mixer head's geometric parameters on mixing efficiency is investigated by altering dimensions such as the inlet diameter, outlet diameter, impingement angle, and impingement pressure. This serves as a foundation for further explorations into enhancing mixing efficiency and optimizing the mixer's structure. After setting the boundary conditions, the flow field analysis of the mixing head is conducted. A density contour plot for the outlet section of the mixing head is generated to make the results more visually intuitive. The figures used in this document are exclusively density contour plots. The relevant density contour plot is shown in Figure 6.

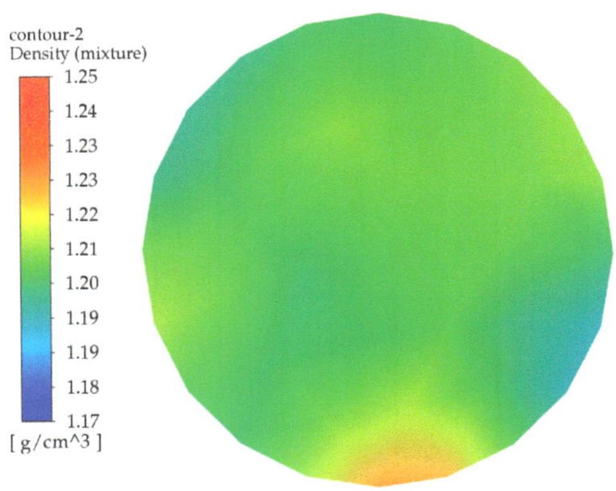

Figure 6. Original mixer head outlet section density contour.

As shown in Figure 6, the density is greater toward the middle sides and the lo middle side, with less density on the left side and the lower right side. Given the 1:1 mi ratio of the materials, thorough mixing would result in fewer color bands, and the der should be between 1.20 and 1.21 g/cm^3. In the figure, the high-density and low-der areas occupy a larger proportion, indicating that the surface mixing effect is poor further optimization is needed.

The calculated mixing unevenness of the original mixing head is 0.11, indicating a p mixing performance. The structure of the original mixing head is optimized to improve mixing effect.

3. Full Factorial DOE and Significance Analysis of Influencing Factors

In this chapter, a full factorial DOE was used to analyze the significance of parameters affecting the mixer's performance. A factor level table was established, and mixing unevenness for each experimental group was calculated. Finally, factor anal was conducted to determine the impact of each parameter on the mixer's performa laying the foundation for subsequent structural optimization.

3.1. Full Factorial DOE

Experimental design is a common statistical method, including confounding, respo surface, full factorial, and fractional factorial designs. The full factorial DOE is a cla method characterized by its comprehensive nature, typically including two or more fac each with multiple levels. This approach encompasses all combinations, providing a c plete understanding of factor interactions. When designing a full factorial DOE, researcl consider which factors could impact results and select variables accordingly. The us full factorial DOE helps researchers understand various factors in an experiment, but factors, like sample size and repeatability, must also be considered to ensure accu and reliability.

Taking mixing unevenness as the experimental impact index, four factors—m inlet diameter D_1, outlet diameter D_2, impinging angle θ, and impinging pressure P—w chosen as screening parameters. Each factor was set at two levels according to the exp mental design requirements. The two-level values for these four parameters are detaile Table 2. The mixer inlet diameter D_1 ranges from 8 mm to 16 mm, the outlet diameter L set at 25 mm and 45 mm, the impinging pressure P is set at 8 MPa and 20 MPa, and impinging angle θ is set at 60° and 180°.

Table 2. Factor-level table.

Parameters [Unit]	Value1	Value
D_1 [mm]	8	16
D_2 [mm]	25	45
θ [°]	60	180
P [MPa]	8	20

A two-level full factorial DOE was conducted on the four parameters, yield 19 sample points. For the experimental design of the mixer's structural parameters, meant obtaining 19 slightly different mixer models. Simulation analyses were separa performed on these mixer models, yielding the mixing unevenness (S) for each mode shown in Table 3.

Table 3. Full factorial DOE results.

S.N	D_1 [mm]	D_2 [mm]	P [MPa]	θ [°]	S
1	16	25	20	180	0.600989
2	16	45	20	60	0.653101
3	16	45	8	180	0.742879

Table 3. *Cont.*

S.N	D_1 [mm]	D_2 [mm]	P [MPa]	θ [°]	S
4	6	25	8	180	0.824658
5	6	25	8	60	0.795684
6	11	35	14	120	0.239548
7	16	25	8	60	0.593242
8	16	45	8	60	0.644044
9	11	35	14	120	0.238963
10	11	35	14	120	0.233965
11	6	45	8	180	0.832547
12	6	45	20	60	0.429569
13	6	45	20	180	0.826549
14	6	45	8	60	0.448426
15	16	25	8	180	0.614530
16	6	25	20	180	0.815843
17	6	25	20	60	0.786324
18	16	45	20	180	0.717372
19	16	25	20	60	0.551930

3.2. Factor Analysis

To analyze the impact of varying structural parameters on mixing unevenness, factor analysis was conducted using the simulation results of the mixing head flow field. The numerical simulation results were used as the ordinate, and the levels of its four structural parameters were used as the abscissa. The experimental data were processed with MINITAB, resulting in a pareto chart of the standardized effects of the response variables, as shown in Figure 7. From Figure 7, it is evident that the inlet diameter, outlet diameter, and impingement angle significantly impact the mixing effect, while the effect of impingement pressure is relatively small. The influence order of each structural parameter on mixing unevenness is: mixer outlet diameter > impingement angle > inlet diameter > impingement pressure. It can be observed from the figure that the impact of impingement pressure on the mixing effect is minimal and can therefore be disregarded.

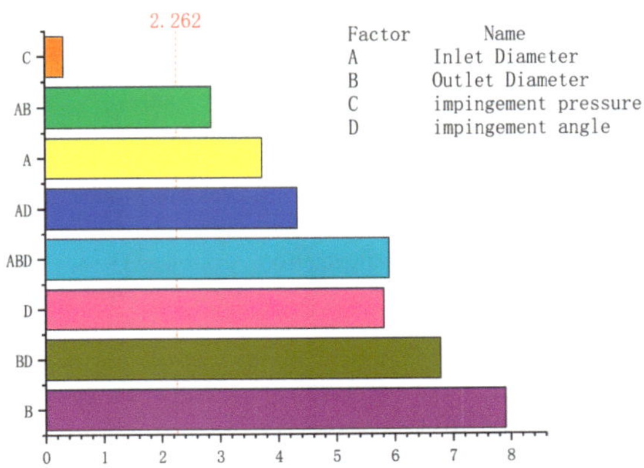

Figure 7. Pareto chart of standardization effect.

4. Optimization of Mixing Head Structure

In this chapter, the mixer's structure is optimized using a joint optimization algorithm combining SSA-CNN-LSTM and the WOA. Initially, the optimization objectives were determined, and a sample library was established using OLH. Subsequently, an SSA-CNN-LSTM prediction model was constructed and compared with CNN-LSTM and LSTM prediction models. Finally, the WOA was employed to optimize the constructed predic model to obtain the optimal structural parameters.

4.1. SSA-CNN-LSTM-WOA Optimization Process

This paper optimizes the structural parameters of the polyimide mixer to minin mixing unevenness. Four structural parameters were selected for significance analysi determine their impact and eliminate insignificant ones, thereby reducing computatic effort. The proposed optimization process combines sample extraction, approxim model establishment, and optimization algorithms. OLH was used to select samples training the approximate model, and numerical simulation results for each sample w computed. An SSA-CNN-LSTM surrogate model was constructed to fit the relations between optimization objectives and variables. The WOA was used to find the opti solution. If the optimization results meet the design requirements, the search is conclu and the results are subjected to simulation and experimental validation. If the results unsatisfactory, variables are reselected for further optimization until the requirements are Figure 8 illustrates the mixer optimization process based on the SSA-CNN-LSTM-WOA.

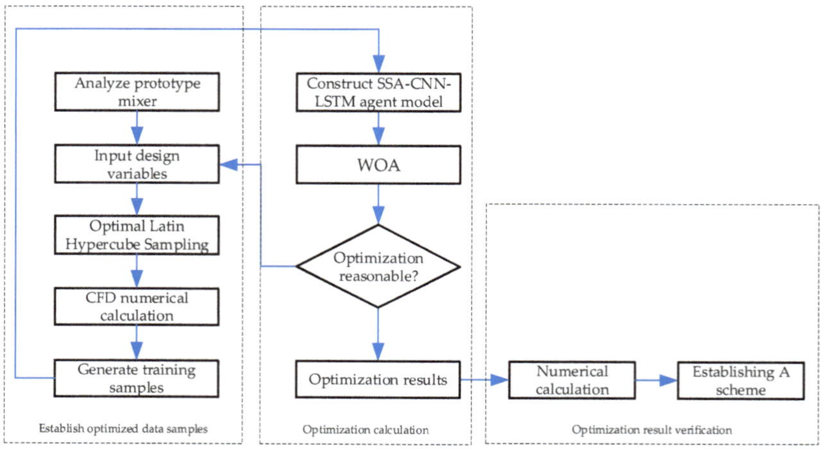

Figure 8. SSA-CNN-LSTM-WOA optimization process.

4.2. Selection of Optimization Objectives

From the full factorial DOE, it is evident that impingement pressure has a mini impact on mixing uniformity. Hence, the inlet diameter, outlet diameter, and impingem angle are chosen as the optimization variables. Thus, establishing its mathematical mo for optimization. The optimization model consists of three factors: the design object variables, and constraints. The minimum mixing unevenness of the mixing head is optimization goal. The impingement angle, inlet diameter, and outlet diameter are design variables. The constraints are the value ranges of these variables. The mathemat model is shown in Equation (12).

$$\begin{cases} W = F(S) \rightarrow \min \\ S = f(D_1, D_2, \theta) \\ 6 \leq D_1 \leq 16 \\ 25 \leq D_2 \leq 45 \\ 60 \leq \theta \leq 180 \end{cases} \quad ($$

In Equation (12):

- W—represents the optimization objective function of the mixer;

- D_1—denotes the inlet diameter of the mixer;
- D_2—denotes the outlet diameter of the mixer;
- θ—represents the impingement angle.

4.3. Construction and Validation of the SSA-CNN-LSTM Model

This section employs OLH to extract a certain number of samples, followed by the construction of an SSA-CNN-LSTM prediction model. The model is evaluated using Mean Absolute Error (MAE), Mean Absolute Percentage Error (MAPE), and Root Mean Square Error (RMSE) as performance metrics. To demonstrate the superiority of the SSA-CNN-LSTM model, comparative analyses are conducted with CNN-LSTM and LSTM models.

4.3.1. Establishment of Training Samples

Before constructing the model, multiple sets of samples must be obtained to serve as training data, ensuring they are uniformly distributed within the design space. This paper selects three optimization variables. Manually assigning samples often leads to an uneven distribution, resulting in training data that do not adequately represent the sample space. Therefore, uniform sampling and Latin Hypercube Sampling (LHS) are commonly used.

Uniform sampling considers only the distribution of samples, without ensuring they are neat and comparable. Scholars later proposed the LHS method, which divides the sample space into equidistant single spaces and randomly takes one sample from each. This method is efficient and stable, but the samples are not uniformly distributed within the design space. To better represent the sampling space, OLH was developed as an improvement to LHS. OLH is a stochastic multidimensional stratified sampling approach that divides the probability distribution function of experimental factors into N non-overlapping sub-regions based on the value ranges of the influencing factors, with independent equal-probability sampling within each sub-region. Compared to orthogonal experiments, LHS offers more flexible grading of level values, and the number of experiments can be manually controlled. However, the experimental points may still not be evenly distributed, and as the number of levels increases, some regions of the design space may be missed. OLH improves upon LHS, enabling uniform, random, and orthogonal sampling within the design space of experimental factors, allowing substantial model information to be acquired with relatively few sampling points [29]. Figure 9 shows the results of 16 samples at 2 levels of OLH, demonstrating that this method better fills the sampling space. Figure 10 shows the sampling results of LHS.

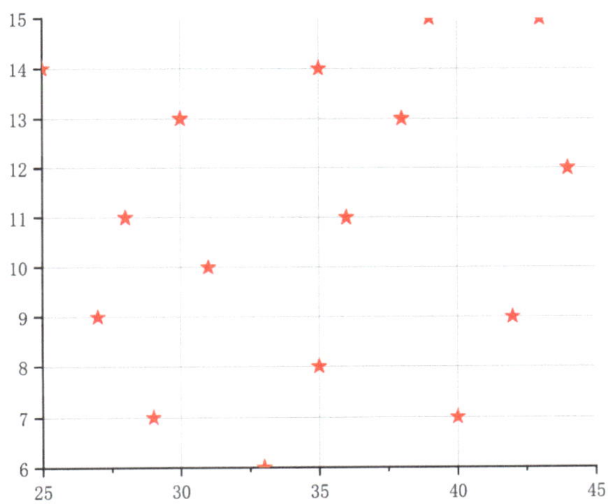

Figure 9. OLH sampling results.

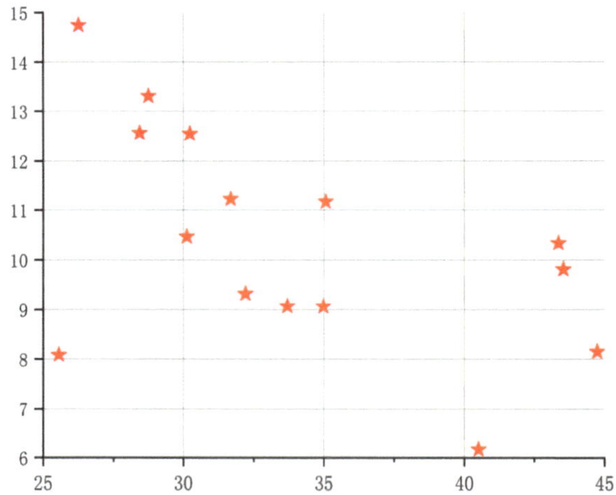

Figure 10. LHS sampling results.

The training samples should be at least ten times the number of input variables ensure credible results, this paper uses OLH to draw 60 sample points from the sam space, as depicted in Figure 11. Based on these samples, 60 mixer models are constructe 3D drawing software, resulting in 60 CFD computational models. The numerical simula method is identical to the one used for the prototype mixer discussed earlier, produc 60 sets of data for mixing unevenness corresponding to 60 sets of parameters. A sam database is established, as shown in Table 4.

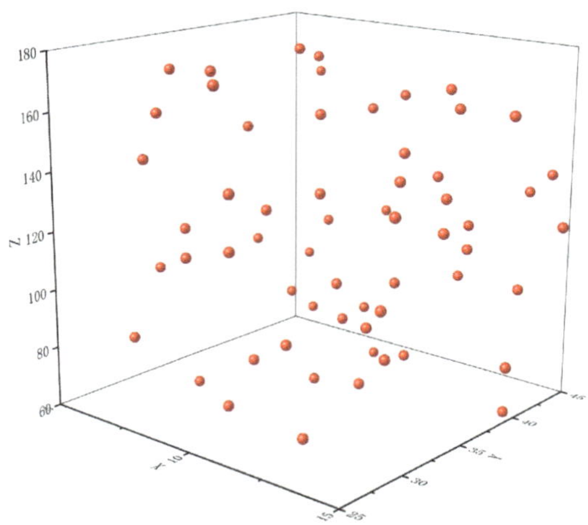

Figure 11. Sampling space.

Table 4. Sample library.

S.N	D_2 [mm]	D_1 [mm]	θ [°]	S	S.N	D_2 [mm]	D_1 [mm]	θ [°]	S
1	27	9	85	0.565601332	31	40	16	161	0.7920212
2	36	8	66	0.695298038	32	40	6	70	0.6582259
3	29	11	76	0.439453453	33	30	16	167	0.2772654
4	44	7	71	0.604243547	34	28	13	93	0.322916
5	35	9	160	0.039434946	35	34	10	129	0.388527

Table 4. *Cont.*

S.N	D_2 [mm]	D_1 [mm]	θ [°]	S	S.N	D_2 [mm]	D_1 [mm]	θ [°]	S
6	36	15	122	0.230791242	36	31	12	132	0.288808952
7	35	8	123	0.090646802	37	29	14	107	0.48610121
8	28	10	150	0.03733436	38	42	13	110	0.420938856
9	40	7	142	0.156511798	39	37	8	170	0.220676555
10	38	10	66	0.433840623	40	27	14	80	0.379446807
11	44	9	101	0.358081664	41	38	6	154	0.246376589
12	32	9	72	0.53680958	42	39	12	117	0.539765996
13	32	10	89	0.33418985	43	37	6	61	0.615474219
14	26	8	176	0.730686246	44	35	16	165	0.508707806
15	28	13	110	0.566211296	45	30	11	131	0.314845462
16	31	14	96	0.571509591	46	31	9	139	0.409969816
17	34	9	178	0.102667462	47	41	12	157	0.159976293
18	26	14	144	0.513248826	48	43	12	88	0.470201261
19	27	7	115	0.036445866	49	34	14	155	0.171068748
20	32	12	175	0.76855878	50	25	13	127	0.080966932
21	39	15	169	0.687295763	51	37	15	105	0.661896866
22	34	13	164	0.464086254	52	44	11	138	0.665976529
23	41	11	140	0.743605475	53	29	11	148	0.503747507
24	33	10	74	0.547576058	54	28	7	80	0.446037397
25	39	13	82	0.359809661	55	37	14	119	0.16558753
26	44	12	135	0.405659276	56	45	7	114	0.546648081
27	42	15	126	0.607806714	57	25	15	90	0.570496729
28	42	8	104	0.135733924	58	38	8	99	0.252273179
29	33	7	151	0.220021216	59	43	10	173	0.814712302
30	42	15	63	0.46109265	60	32	11	95	0.209671239

4.3.2. Construction of the SSA-CNN-LSTM Model

In the CNN-LSTM model, SSA adjusts various parameters to enhance performance. The network structure has three main components: CNN for extracting spatial features, LSTM for learning time series information, and SSA for optimizing parameters (e.g., number of convolutional kernels and LSTM units). Input data pass through the CNN to extract spatial features, then into the LSTM to learn time-series information. SSA optimizes parameters through iterative searching and updating. Ultimately, this optimized model can be used for data prediction tasks.

The application of the SSA to optimize the CNN-LSTM model involves the following steps:

(1) Data preprocessing: data labeling, dataset division, data normalization, and data format conversion;
(2) SSA population initialization: setting the initial size of the sparrow population (n), the maximum number of iterations (N), the proportion of discoverers (PD), the number of sparrows perceiving danger (SD), the safety value (ST), and the alert value (R2);
(3) Computing fitness, updating discoverer positions, updating joiner positions, updating scrounger positions, and updating the optimal individual position;
(4) Feeding data into the CNN network, passing data through CNN layers, batch normalization layers, activation function layers, and average pooling layers;
(5) Entering data into the LSTM neural network after going through LSTM layers to the fully connected layer and SoftMax layer;
(6) Outputting the results.

Although the SSA algorithm has global search capabilities and adaptability, it is sensitive to the selection of initial parameters and requires tuning for specific problems. When optimizing the CNN-LSTM model, carefully selecting appropriate optimization objectives and fitness functions ensures the SSA algorithm significantly enhances performance. The parameters optimized by SSA mainly include the number of neurons in the LSTM layer, convolution kernel size, number of convolutional layers, number of neurons in the fully

connected layer, and learning rate. The schematic diagram of the SSA-CNN-LSTM mo
is shown in Figure 12.

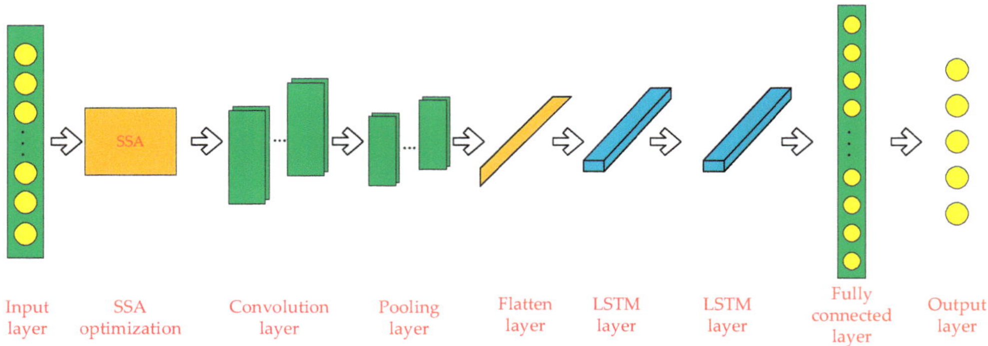

Figure 12. Schematic diagram of the SSA-CNN-LSTM model.

(1) Number of neurons in the LSTM layer: The LSTM layer is an important par
the CNN-LSTM model, and the choice of the number of neurons directly aff
the performance of the model. Through the SSA algorithm, the optimal num
of neurons in the LSTM layer can be dynamically searched, thereby improving
model's prediction accuracy.
(2) Convolution kernel size: The convolution operation in the CNN-LSTM model is v
important for extracting spatial features of the data, and the choice of convolution
nel size will affect feature extraction. The SSA can optimize the size of the convolut
kernel to further improve the accuracy of the model.
(3) Number of convolutional layers: The deeper the convolutional layers in the CN
LSTM model, the greater the computational cost required during forward propagat
but the accuracy of the model will also increase correspondingly. By optimizing
number of convolutional layers, the SSA algorithm can enhance model accuracy wit
an acceptable range of computational costs.
(4) Number of neurons in the fully connected layer: The number of neurons in the fu
connected layer determines the output dimension and prediction ability of the mo
By optimizing the number of neurons in the fully connected layer, the SSA can impr
the accuracy and generalization ability of the model.
(5) Learning rate: The CNN-LSTM model needs to specify a learning rate during train
Too high or too low a learning rate will affect the convergence and accuracy of
model. The SSA can search for the optimal learning rate to improve the efficie
and accuracy of model training. Combining the above module construction steps,
flowchart of the SSA-CNN-LSTM model is shown in Figure 13.

4.3.3. Benchmark Methods and Evaluation Criteria

In the comparative study of models, this paper adopts the following two mach
learning models:

(1) LSTM Model

LSTM is a type of Recurrent Neural Network (RNN) with two gates and one
state, allowing it to store data longer than standard RNNs. The gates regulate informat
flow by specifying activation functions within the LSTM cell. The LSTM model in
paper consists of four consecutive pairs of LSTM and Dropout layers. The Dropout lay
reduce dependency on previous data, and the final output is obtained through a fu
connected layer.

(2) CNN-LSTM Model

The CNN-LSTM model can achieve multi-feature classification prediction. The CNN is used for feature (fusion) extraction, and then the extracted features are mapped as sequence vectors input into the LSTM.

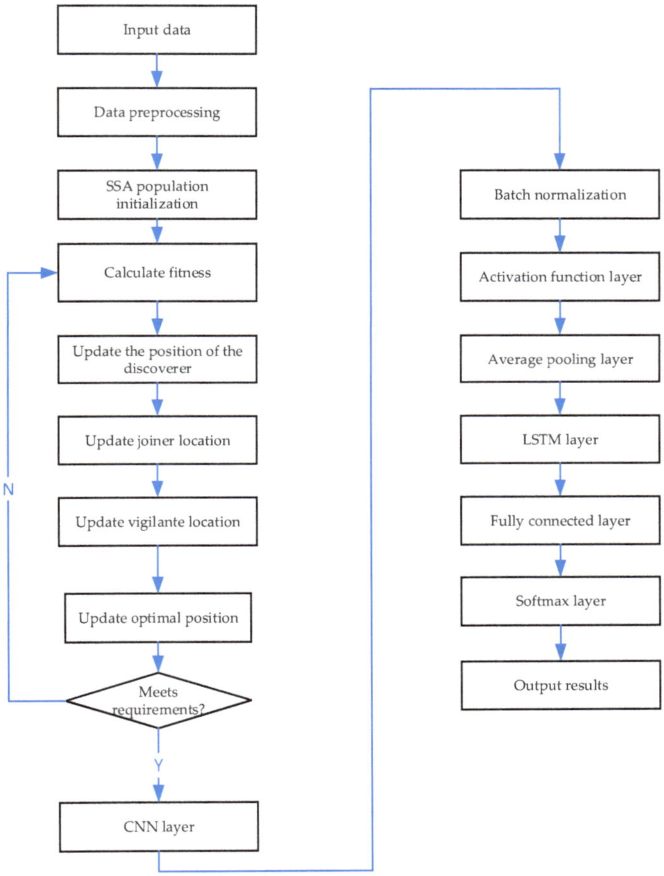

Figure 13. SSA-CNN-LSTM model flowchart.

To evaluate the effectiveness and applicability of the prediction models on the dataset, three common evaluation metrics are used to quantify the precision and errors of predictions, namely MAE, MAPE, and RMSE.

The calculation for MAE is shown below; it represents the average of the absolute errors and can reflect the actual situation of prediction errors.

$$MAE = \frac{1}{z}\sum_{i=1}^{z}|y_i - \hat{y}_i| \quad (13)$$

The calculation for MAPE is as follows, representing the fitting effect and precision of the prediction model.

$$MAPE = \frac{1}{z}\sum_{i=1}^{z}\frac{|y_i - \hat{y}_i|}{y_i} \quad (14)$$

RMSE measures the deviation between the actual and the predicted values, wit calculation shown as below.

$$RMSE = \sqrt{\frac{1}{z}\sum_{i=1}^{z} (y_i - \hat{y}_i)^2}$$

where z represents the number of samples, y_i is the actual value, and $\hat{y}_i$ is the predi value. These three indicators represent different aspects of the predictive performance, the smaller their values, the better the performance.

4.3.4. Experimental Parameter Settings

In this section, the SSA-CNN-LSTM model is employed for predicting mixing evenness. The model optimizes CNN-LSTM using the SSA, with parameters including number of iterations (M), population size pop, the proportion of discoverers (p_perc fitness function (f_n), and the dimension of parameters to be optimized (dim).

(1) Number of iterations (M): This parameter determines the computation time accuracy of the algorithm. Typically, the more iterations, the longer the computa time and the higher the accuracy. Here, it is set to 50.
(2) Population size (pop): This refers to the number of sparrows used for computat with no explicit rule for its size. Typically, it is based on a specific analysis of problem at hand. For optimizing general problems, setting 50 sparrows is sufficier solve most issues. For particularly difficult or specified problems, it might be neces to increase the population size. Generally, the smaller the population size, the easi is to fall into local optima, while a larger population size results in slower converge Here, the population size is set to 50.
(3) Proportion of discoverers (p_percent): The percentage of discoverers in the popula set here to 20%.
(4) The percentage of sparrows that are aware of danger is set to 10%.
(5) Safety threshold (ST) is set to 0.8.
(6) Fitness function (f_n): Determines the fitness of a sparrow's position.
(7) Parameter dimension (dim): Indicates the number of parameters to be optimi Here, optimization is applied to three parameters of the LSTM layer, which are optimal number of neurons, the optimal initial learning rate, and the optima regularization coefficient, making the dimension 3.

4.3.5. Comparative Analysis

Five sixths of the data from the chosen sample database were used as training d with the remaining serving as the test set. The predictive performance and prediction er of various algorithms are shown in Figures 14 and 15, respectively. It can be obser from the figures that SSA-CNN-LSTM has the smallest prediction error, not exceed 2%, followed by CNN-LSTM, with LSTM showing the poorest predictive performance the experiment, the prediction indicators were inlet diameter, outlet diameter, and ar of attack. The SSA optimized three parameters for CNN-LSTM: the optimal numbe neurons was set at 75, the optimal initial learning rate at 0.006258, and the optimal regularization coefficient at 0.0004562. By comparing the actual values and predic values, it can be seen that SSA-CNN-LSTM fits better than various baseline models.

The prediction results of the above three algorithms were statistically analyzed, an shown in Table 5 and Figure 16, SSA-CNN-LSTM achieved the best results according to evaluation metrics RMSE, MAE, and MAPE. MAPE was reduced from 12.5% to 7%, ach ing an accuracy improvement of 44 percentage points. RMSE and MAE also decrea indicating that the model's predictive performance had improved following optimizat

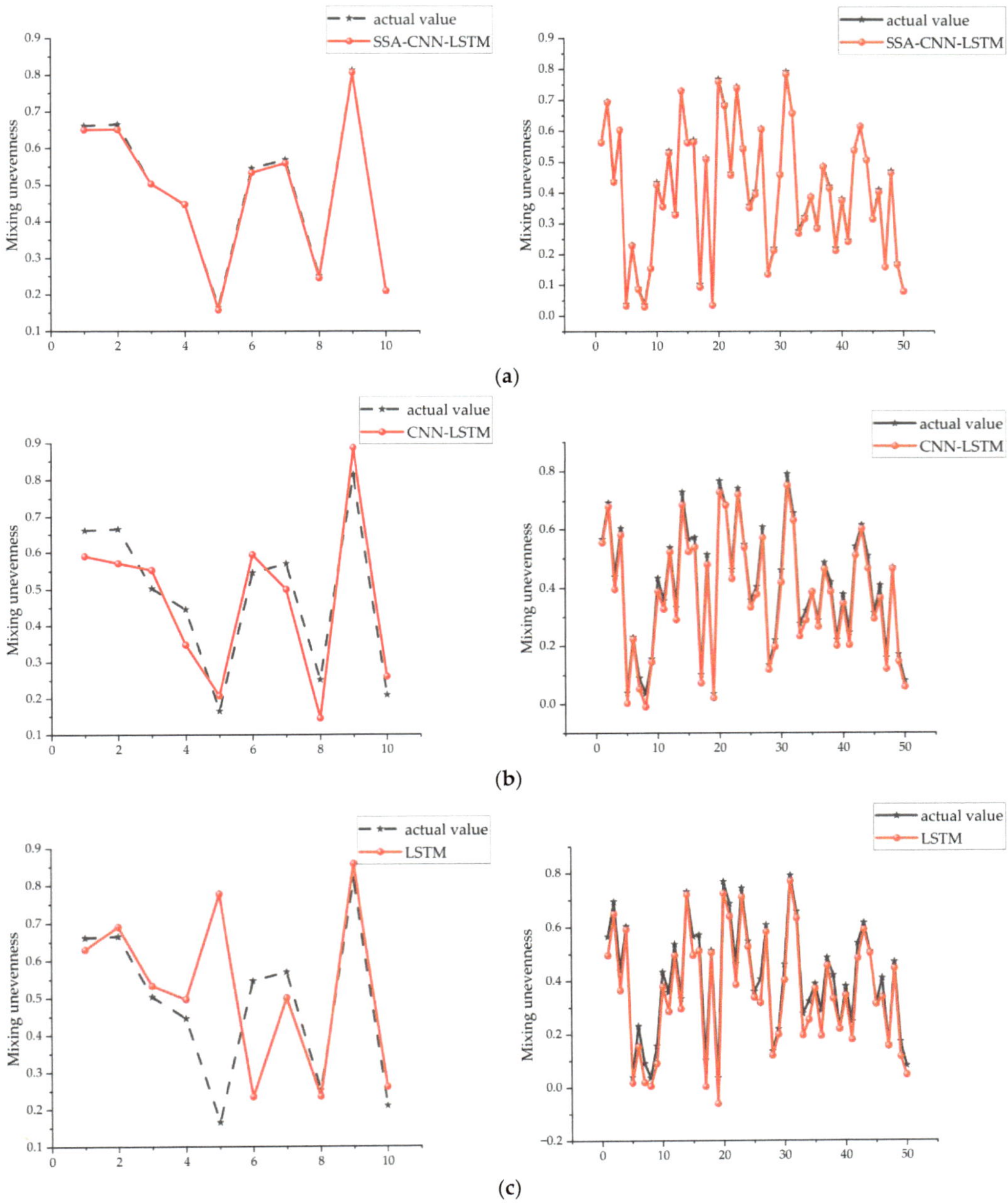

Figure 14. (**a**). SSA-CNN-LSTM prediction results. (**b**). CNN-LSTM prediction results. (**c**). LSTM prediction results.

Table 5. Predictive indicator values of various algorithms.

Method	MAE	MAPE	RMSE
LSTM	0.125	0.502	0.221
CNN-LSTM	0.071	0.18	0.074
SSA-CNN-LSTM	0.07	0.0165	0.0083

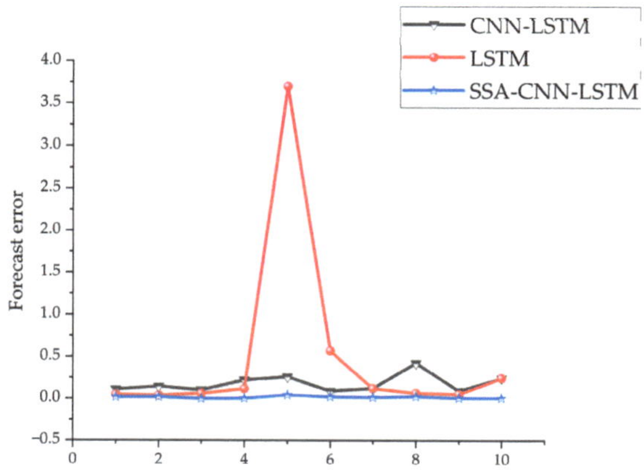

Figure 15. Prediction errors of various algorithms.

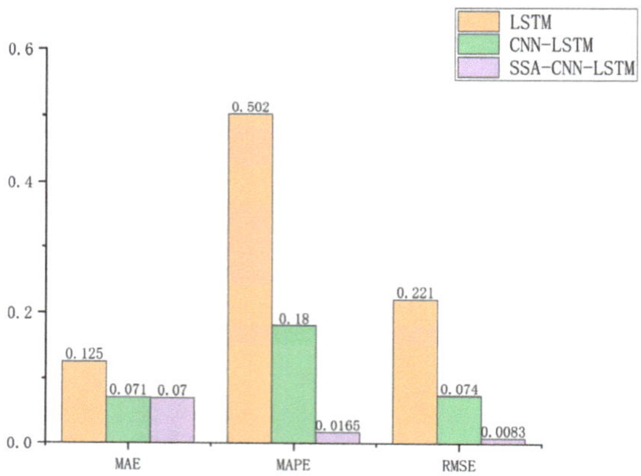

Figure 16. Comparison chart of evaluation indicators.

4.4. Joint Optimization Design of SSA-CNN-LSTM and the WOA

The WOA is a relatively novel optimization algorithm proposed in recent years 2016, Mirjalili et al. [30] from Griffith University in Australia introduced the WOA, inspi by the unique predatory behavior of humpback whales observed in the ocean. As emerging swarm intelligence optimization algorithm, the WOA is characterized by few parameters, simple structure, and high flexibility. Specifically, the advantages as follows:

(1) The algorithm's optimization mechanism is primarily controlled by a random num (Pa), a threshold parameter (A), and a random number (C), resulting in fewer con parameters.
(2) The WOA is divided into three main optimization mechanisms, with a relativ simple formula model that is easy to implement.
(3) The WOA offers high flexibility, as it can switch between optimization mechanis using the random number (Pa) and the threshold parameter (A).

As a result, the WOA can be widely applied in various engineering fields [31]. example, Li et al. [32] designed a WOA-based optimization algorithm for evaluating bipo transistor models. In the field of intelligent healthcare, Hassan et al. [33] proposed a hyb algorithm based on the WOA for diagnosing equipment faults. He et al. [34] developed improved WOA for adaptively searching for optimal parameters in stochastic resonance systems. El-Fergany et al. [35] used the WOA to enhance the accuracy of fuel cell models.

The overall optimization process of the WOA mainly consists of three parts: Encircling Prey, the Bubble-Net Attacking Technique, and Searching for Prey. Through the cooperation of these three processes, the optimization process can achieve good global search capability and accurate and fast local convergence ability. Thus, while ensuring fast convergence of the algorithm, it can also avoid falling into local optimal solutions, ensuring that the obtained optimal solution is the global optimum, fully leveraging the best performance of the optimization algorithm.

The optimization process of the WOA is as follows: First, initialize the WOA parameter values and input the structural parameters to be optimized. At the beginning of each iteration, check if the maximum number of iterations has been reached. If it has, return the current best structural parameter values and their corresponding fitness values. If the maximum number of iterations has not been reached, update the WOA parameters (A, a, C, and Pa).

The search strategy is then determined based on the random number (p): If Pa < 0.5, further check the value of $|A|$. If $|A| < 1$, select a random agent and update its position using the range search mechanism; otherwise, update the search agent's position using the spiral search strategy. Next, calculate the new fitness value for each search agent and update the structural parameters based on these fitness values. This process repeats until the maximum number of iterations is achieved. Finally, when the maximum number of iterations is reached, the algorithm returns the best structural parameter values and their corresponding fitness values.

Based on the above algorithm steps, the flowchart of the WOA is illustrated as shown in Figure 17.

Based on the information provided, the combined optimization algorithm of SSA-CNN-LSTM and the WOA can rapidly optimize structural parameters. Utilizing the determined structural model, an SSA-CNN-LSTM prediction model is built from the sample library to establish a functional relationship between structural parameters and their corresponding fitness values. This enables the input of parameter values to yield fitness values as outputs. After determining the optimization range of parameter values, the parameters of the optimization algorithm are initialized. The SSA-CNN-LSTM model determines the fitness value of each individual in each generation, facilitating selection and iteration, ultimately obtaining the best parameter values and outputting the optimal solution with its corresponding fitness value. Compared to traditional optimization methods, the combined algorithm offers higher efficiency, achieving superior optimization accuracy and better results in a shorter time. The combined optimization process of SSA-CNN-LSTM and the WOA is illustrated in Figure 18.

The prediction output of the trained model is used as the individual fitness value, eliminating individuals with inferior fitness values in favor of superior ones. The WOA is utilized to globally optimize the search for extrema in the prediction model. To demonstrate the superior capabilities of the WOA, it is compared with traditional genetic algorithms and particle swarm optimization.

The Particle Swarm Optimization (PSO) algorithm initializes a swarm of particles within the feasible solution space, characterizing each particle by its position, velocity, and fitness. As the particles move, they update their positions using the individual best position—Pbest—the position with the optimal fitness value experienced by the particle—and the global best position—Gbest—the position with the optimal fitness value found by all particles in the swarm. Pbest and Gbest positions are continuously updated by comparing and adjusting fitness values [36].

The Genetic Algorithm (GA) [37] follows Darwin's theory of survival of the fittest and simulates biological gene inheritance. The GA calculates a fitness function to evaluate the quality of individuals in the population, classifying them based on these values. Inferior individuals are eliminated, while superior ones survive. This iterative process ensures that the best individuals are passed on to subsequent generations, eventually leading population containing the optimal individuals [38,39].

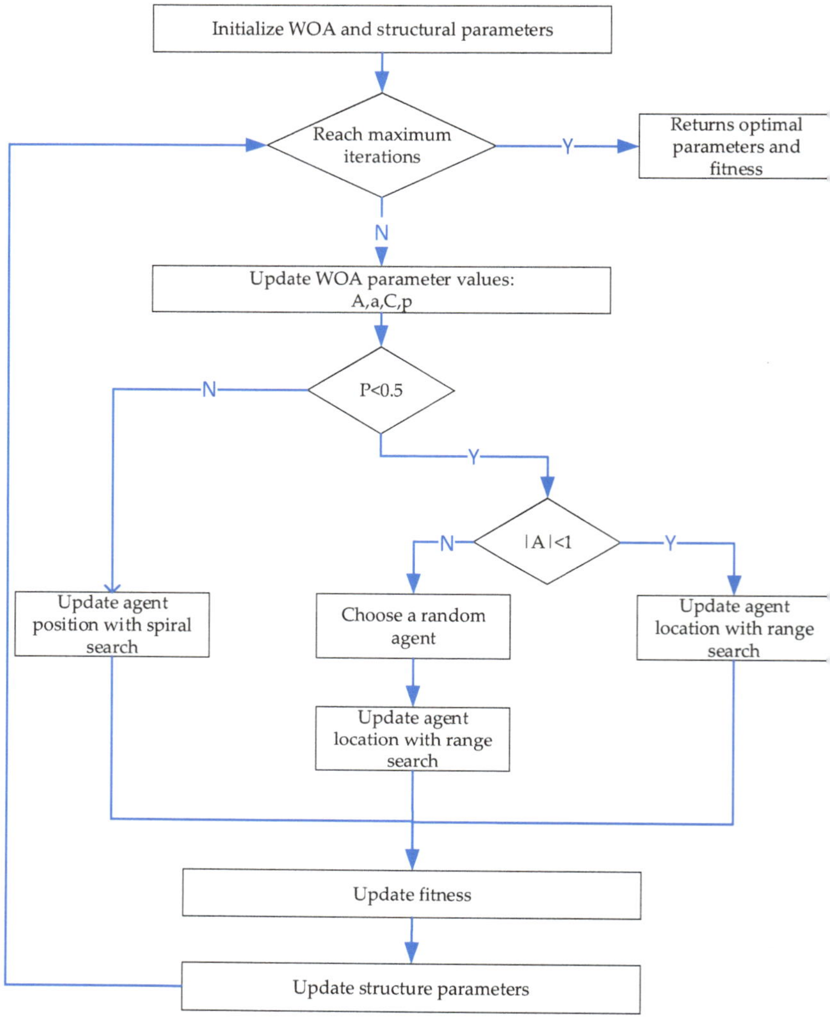

Figure 17. WOA process.

The optimization performance is highlighted by comparing optimization times the optimization effects of the best fitness values. The results are shown in Figure 19. Fr the figure, it can be seen that the GA requires 303 iterations to obtain the optimal param values, PSO also requires 41 iterations to achieve the optimal parameter values, w the WOA only needs 17 steps to obtain the best optimization solution. At the same ti compared with the GA and PSO, the WOA yields better optimization results. Accordin the previous text, the lower the mixing unevenness achieved, the better the performa Here, combined with Figures 4–12, it can be observed that the best solutions obtained the GA, PSO, and WOA are 0.039, 0.037, and 0.03, respectively. Through comparison, it be seen that the WOA has a faster optimization speed and can achieve better optimiza results with fewer iterations. After optimization by the WOA, the best structural parame obtained are: inlet diameter of 7, outlet diameter of 31.6, and impingement angle of 11

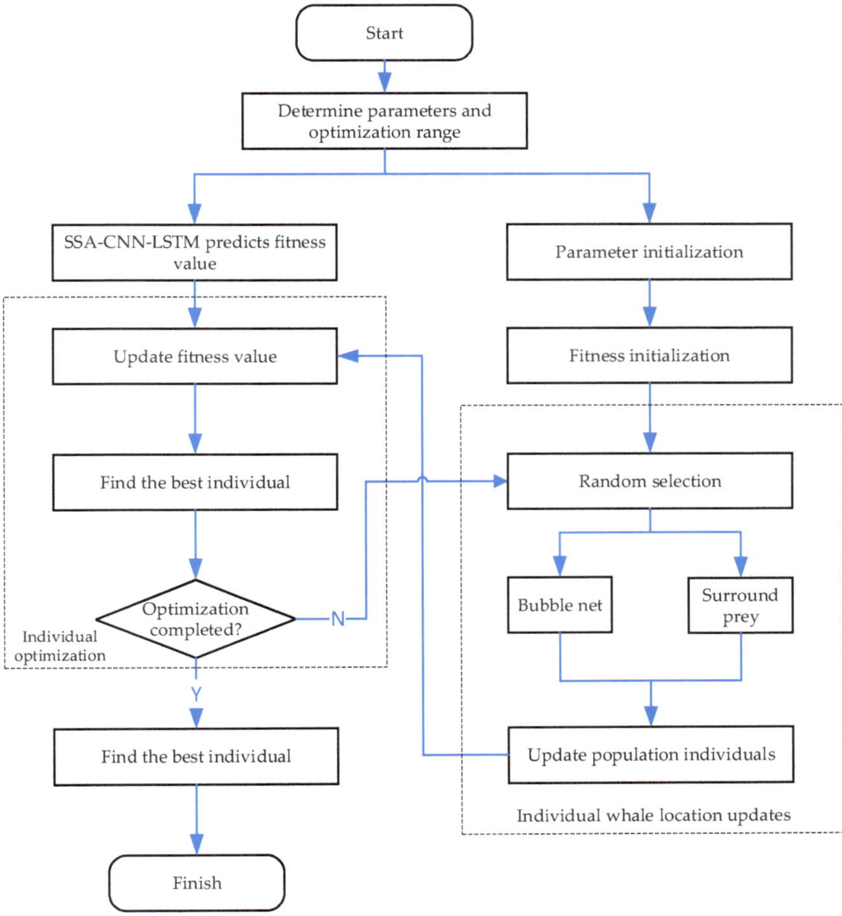

Figure 18. SSA-CNN-LSTM-WOA joint optimization process.

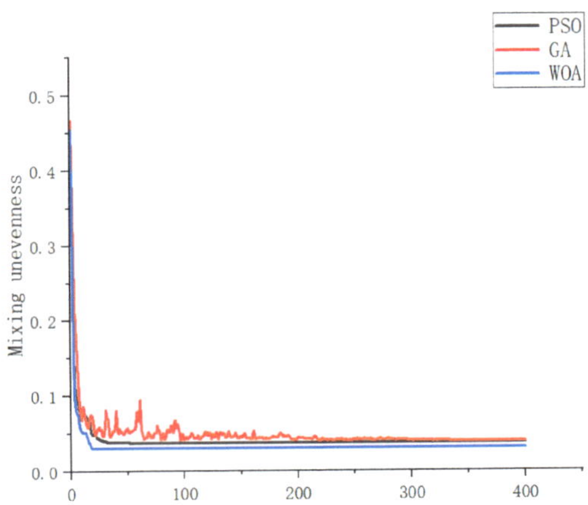

Figure 19. The fitness curve.

5. CFD Validation Results

After optimization using SSA-CNN-LSTM-WOA, we compared the density di bution at the mixer outlet and the flow conditions within the mixer before and after optimization. This comparison was conducted to verify the effectiveness of the optimizat

5.1. The Density Distribution at the Outlet after Optimization

Based on the WOA optimization results, three parameters were adjusted in 3D draw software. The obtained structural parameters will be subjected to simulation calculati The optimized outlet cross-section density cloud image and corresponding density d were obtained and calculate the mixing unevenness, which was found to be 0.031. Figur displays the density cloud image post-optimization. There is a significant differenc the density distribution before and after optimization; initially, it was more concentra whereas, post-optimization, the density distribution is noticeably more uniform compa to the original model. As illustrated in Table 6, the pre-optimization mixing unevenr was 0.11. Following WOA optimization, this value decreased to 0.031, a reduction of 7 Comparing this with the predicted mixing unevenness in MATLAB, the error margi 3.1%, demonstrating the stability of the SSA-CNN-LSTM prediction. Additionally, mixing effect of the mixer has been enhanced to some extent after the WOA optimizat

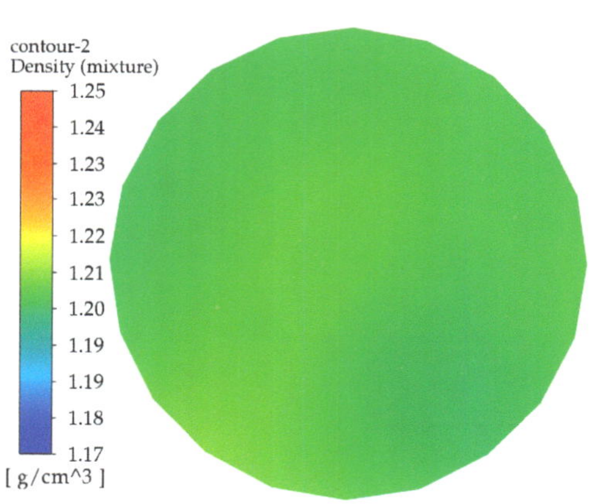

Figure 20. Density distribution of the exit cross-section after WOA optimization.

Table 6. Optimization result.

Structure	D_1 [mm]	D_2 [mm]	θ [°]	S
Original Structure	8	35	180	0.11
Joint Algorithm Optimization	7	31.6	119.5	0.031
Reduction in Mixture Non-Uniformity				70%
Prediction error				3.1%

5.2. Velocity Distribution before and after Optimization

Figure 21 shows the velocity vector diagram before optimization. As seen in figure, due to the excessive impingement angle, the liquid's velocity decreases shar after collision, almost no vortices are generated, and the high-speed region is concentra The velocity gradient is significant, and the liquid almost exits the mixer without hitt the walls, resulting in poor mixing. Figure 22 shows the velocity vector diagram a optimization using the joint optimization algorithm. The figure illustrates that the t materials collide at a certain angle, resulting in less velocity loss compared to the origi mixer, and generating a certain number of vortices, which improves the mixing eff Additionally, Figure 22 indicates that the velocity changes smoothly at the intersecti with the velocity vectors being more uniform and the velocity gradient being smaller. Based on the analysis of the velocity changes and vortex structures, it can be concluded that the flow characteristics within the optimized mixer are more conducive to mixing compared to the original mixer.

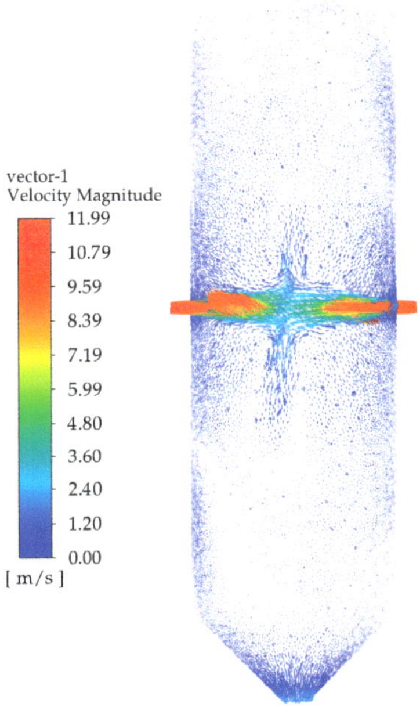

Figure 21. Velocity vector diagram before optimization.

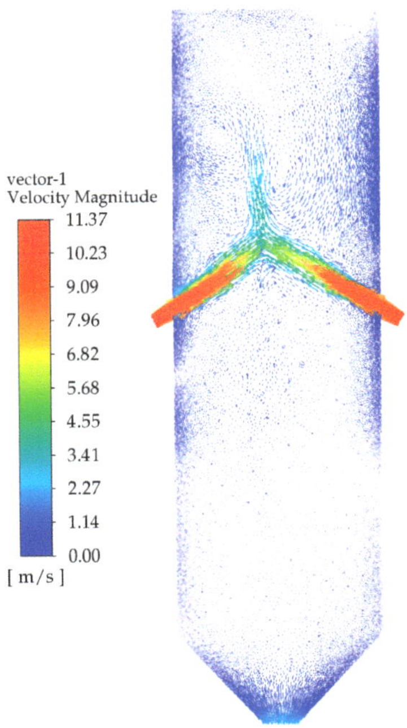

Figure 22. Velocity streamline diagram after optimization.

6. Conclusions

In this study, the analysis of the current high-pressure mixier is executed anc optimal design framework is proposed. Firstly, the most important parameters that tinctly influence mixing unevenness are selected by using the full factorial DOE. Afterw we constructed an SSA-CNN-LSTM prediction model and used the WOA algorithr determined the best parameter set. The verification is achieved in the CFD simulation the detailed conclusions are as follows:

(1) The structural parameters that obviously influence mixing unevenness are mixer ou diameter, impingement angle, inlet diameter, and impingement pressure, ranke order of decreasing importance.
(2) An SSA-CNN-LSTM prediction model achieves a prediction accuracy of up to 9 which is higher than the LSTM neural network and CNN-LSTM models. Me while, the SSA demonstrated its superior optimization and convergence capabili to determine the optimal parameters faster and accurately.
(3) A joint optimization framework based on SSA-CNN-LSTM and the WOA is propc to obtain the best parameter set. Based on the optimized mixer parameter values, mixing unevenness decreases by 70%, significantly enhancing the mixer's performa

Author Contributions: Writing—original draft preparation, G.Y.; methodology, G.H.; software, validation, Y.L. and J.L.; formal analysis, G.H.; investigation, G.Y. and G.H.; resources, G.H. and data curation, J.L.; writing—review and editing, J.L.; visualization, G.Y.; supervision, G.H. and project administration, G.H. and X.T.; funding acquisition: G.H. and X.T. All authors have read agreed to the published version of the manuscript.

Funding: This work was supported by The National Defense Supporting Project of China (J 2022-137), Panzhihua Key Laboratory of Advanced Manufacturing Technology Open Fund Prc 'Research on Reliability Design of High Temperature and High Pressure Special Equipment Ba on Digital Twin Technology' (No. 2022XJZD01) and Research and Health Assessment Progn Technology for Complex Equipment Systems (SUSE652A004).

Data Availability Statement: Data are contained within the article.

Conflicts of Interest: The authors declare no conflict of interest.

Abbreviations

DOE	Design of Experiments
SSA	Sparrow Search Algorithm
CNN	Convolutional Neural Network
LSTM	Long Short-Term Memory Network
WOA	Whale Optimization Algorithm
VOF	Volume of Fluid
CFD	Computational Fluid Dynamics
D_1	Inlet Diameter
D_2	Outlet Diameter
θ	Impinging Angle
P	Impinging Pressure

References

1. Li, E.; Yu, X.; Fan, L.; Xu, Y.; Liu, W. Development of PI Films with Active Groups. *Insul. Mater.* **2011**, *44*, 12–15.
2. Deng, S.; Wang, J.; Yang, W. Simulation and Analysis of High-Pressure High-Speed Collision Process in Fluid Inside Mixing H of Reinforced Reaction Injection Molding Machine. *China Plast.* **2022**, *36*, 130–136.
3. Pei, K.; Li, R.; Zhu, Z. Effects of geometry and process conditions on mixing behavior of a multijet mixer. *Ind. Eng. Chem.* **2014**, *53*, 10700–10706. [CrossRef]
4. Zhan, X.; Jing, D. Influence of geometric parameters on the fluidic and mixing characteristics of T-shaped micromixer. *Micro Technol.-Micro Nano Syst.-Inf. Storage Process. Syst.* **2020**, *26*, 2989–2996. [CrossRef]
5. Yuan, X.; Liao, Y. Analysis of Material Flow Field in the Mixer of Polyurethane Foaming Machine. *J. Wuhan Inst. Technol.* **2006** 67–69.

Trautmann, P.; Piesche, M. Experimental investigations on the mixing behavior of impingement mixers for polyurethane production. *Chem. Eng. Technol.* **2001**, *24*, 1193–1197. [CrossRef]
Guo, M. *Research on Key Technologies of Polymer Road Repair Foaming Machine*; Wuhan Institute of Technology: Wuhan, China, 2011.
Schneider, F.W. High Pressure Impingement Mixing Apparatus. U.S. Patent 4440500, 3 April 1984.
Bayer, A.G. Apparatus for the Continuous Preparation of a Liquid Reaction Mixture from Two Fluid Reactants. U.S. Patent 5093084, 3 March 1992.
Murphy, S.A.; Meszena, Z.G.; Johnson, A.F. Experimental and modelling studies with mixers for potential use with fast polymerisations in tubular reactors. *Polym. React. Eng.* **2001**, *9*, 227–247. [CrossRef]
van Horn, I.B. Venturi Mixer. U.S. Patent 3507626, 21 April 1970.
Pan, Y. *Research on Performance Optimization and Internal Flow Characteristics of Centrifugal Pumps Based on Particle Swarm Optimization Algorithm*; Jiangsu University: Zhenjiang, China, 2022.
Jiang, D.; Zhou, P.; Wang, L.; Wu, D.; Mou, J. Analysis of the Impeller Optimization Design for Dual-Blade Sewage Pumps Based on Response Surface Methodology. *J. Hydroelectr. Eng.* **2021**, *40*, 72–82.
Luo, M.; Zuo, Z.; Li, H.; Li, W.; Chen, H. Multi-objective Optimization Design Method for Centrifugal Compressor Impellers Based on BP Neural Network. *J. Aerosp. Power* **2016**, *31*, 2424–2431.
Jiang, M.; Mei, Y.; Luo, Y.; Yu, S.; Hu, D. Structural Parameter Optimization of the Mixing Head of Polyurethane Foaming Machine Based on GA-BP-GA. *Eng. Plast. Appl.* **2023**, *12*, 92–99.
Li, L.; Guo, Z.; Liu, Y.; Li, P. Numerical Simulation and Structural Optimization of Supersonic Spray Nozzles. *J. Dalian Jiaotong Univ.* **2023**, *3*, 58–62. [CrossRef]
Liu, G.; Zhao, T.; Wang, C.; Yang, Z.; Yang, X.; Li, S. Optimization of Structure and Operating Parameters of Y-Type Micromixers in Two-Phase Pulsatile Mixing. *J. Jilin Univ. (Eng. Technol. Ed.)* **2015**, *45*, 1155–1161.
Guo, D. *Parameter Optimization and Flow Field Characteristics Study of a New Type of Impinging Stream Reactor*; Taiyuan University of Technology: Taiyuan, China, 2020.
Dai, B. *Modular Design and Research of Polyurethane Foaming Equipment*; Soochow University: Suzhou, China, 2020.
Zhang, J.; Niu, J.; Dong, X.; Feng, Y. Numerical Simulation of the Flow Field in an Impinging Stream Reactor and the Optimization of Its Mixing Performance. *Chin. J. Process Eng.* **2022**, *22*, 1244–1252.
Zhang, J.; Shang, P.; Yan, J.; Feng, Y. Mixing Effects of a Horizontal Opposed Bidirectional Impinging Stream Mixer. *Chin. J. Process Eng.* **2017**, *17*, 959–964.
Palanisamy, B.; Paul, B. Continuous flow synthesis of ceria nano particles using static T-mixers. *Chem. Eng. Sci.* **2012**, *78*, 46–52. [CrossRef]
Niu, J. *Numerical Simulation Analysis of the Flow Field in an Impinging Stream Reactor and Research on the Preparation of Calcium Hydroxide Nano Powder*; Shenyang University of Chemical Technology: Shenyang, China, 2022.
Zhengyang, Z.; Zhuo, C.; Hongjie, Y.; Liu, F.; Liu, L.; Cui, Z.; Shen, D. Numerical simulation of gas—Liquid multi-phase flows in oxygen enriched bottom-blown furnace. *Chin. J. Nonferrous Met.* **2012**, *22*, 1826–1834.
Wenli, L.; Yanbing, Y.; Dongming, Z.; Feng, M. Water modeling on double porous plugs bottom blowing in a 150t ladle. *Jiangxi Metall.* **2022**, *42*, 1–9. [CrossRef]
Wang, D.X.; Liu, Y.; Zhang, M.; Li, X.-L. PIv measurements on physical models of bottom blown oxygen smelting furnace. *Can. Metall. Copp. Q.* **2017**, *56*, 221–231. [CrossRef]
Zhang, G.; Zhu, R. Modeling and on-Hne optimization control memod of oxygen bottom-blowing copper smelting process. *China Nonferrous Metall.* **2020**, *49*, 62–65.
Xue, W.; Liang, Y. Efficient hydraulic and thermal simulation model of the multi-phase natural gas production system with variable speed compressors. *Appl. Therm. Eng.* **2024**, *242*, 122411. [CrossRef]
Ji, N.; Zhang, W.; Yu, Y.; He, Y.; Hou, Y. Multi-Objective Optimization of Injection Molding Based on Optimal Latin Hypercube Sampling Method and NSGA–II Algorithm. *Eng. Plast. Appl.* **2020**, *48*, 72–77.
Mirjalili, S.; Lewis, A. The whale optimization algorithm. *Adv. Eng. Softw.* **2016**, *95*, 51–67. [CrossRef]
Tu, J.; Chen, H.; Liu, J.; Heidari, A.A.; Zhang, X.; Wang, M.; Ruby, R.; Pham, Q.V. Evolutionary biogeography-based whale optimization methods with communication structure: Towards measuring the balance. *Knowl. Based Syst.* **2021**, *21*, 212–240. [CrossRef]
Li, L.L.; Sun, I.; Tseng, M.L.; Zhi-Gang, L. Extreme learning machine optimized by whale optimization algorithm using insulated gate bipolar transistor module aging degree evaluation. *Expert Syst. Appl.* **2019**, *127*, 58–67. [CrossRef]
Hassan, M.K.; El Desouky, A.I.; Elghamrawy, S.M.; Sarhan, A.M. A hybrid real-time remote monitoring framework with NB-WOA algorithm for patients with chronic Diseases. *Future Gener. Comput. Syst.* **2019**, *93*, 77–95. [CrossRef]
He, B.; Huang, Y.; Wang, D.; Yan, B.; Dong, D. A parameter-adaptive stochastic resonance based on whale optimization algorithm for weak signal detection for rotating machinery. *Meas. J. Int. Meas. Confed.* **2019**, *136*, 658–667. [CrossRef]
El-Fergany, A.A.; Hasanien, H.M.; Agwa, A.M. Semi-empirical PEM fuel cells model using whale optimization algorithm. *Energy Convers. Manag.* **2019**, *201*, 112–127. [CrossRef]
Run, X.; Zuo, Y.; Huo, D. Optimal design of truss structure based on pso algorithm. *Steel Constr.* **2012**, *27*, 37–39.
Michalewicz, Z.; Janikow, C.Z.; Krawczyk, J.B. A modified genetic algorithm for optimal control problems. *Comput. Math. Appl.* **1992**, *23*, 83–94. [CrossRef]

38. Wang, G.; Chen, Y.; Li, Q.; Lu, C.; Chen, G.; Huang, L. Ultrasonic computerized tomography imaging method with combinat optimization algorithm for concrete pile foundation. *IEEE Access* **2019**, *7*, 132395–132405. [CrossRef]
39. Wang, L.; Zheng, C. Identification method of aerodynamic damping based on the genetic algorithm and random decren technique. *J. Vib. Shock.* **2023**, *42*, 16–23.

actuators

MDPI

cle

›search on Air Cushion Flow Field and Bearing Characteristics Single Row Hole Air Cushion Belt Conveyor

u Wang [1,2], Lei Wu [1,2], Ziming Kou [1,2,*] and Liqiang Tian [1,2]

1 School of Mechanical Engineering, Taiyuan University of Technology, Taiyuan 030024, China; wangyahu0027@link.tyut.edu.cn (Y.W.); wulei0029@link.tyut.edu.cn (L.W.); tlq18638008283@163.com (L.T.)
2 Shanxi Provincial Engineering Laboratory for Mine Fluid Control, Taiyuan 030024, China
* Correspondence: zmkou@163.com

Abstract: In order to obtain information regarding changes in the air cushion flow field and load-bearing characteristics of a single row hole air cushion belt conveyor, a structural model of the air cushion of the belt conveyor was established, with the single row hole air cushion belt conveyor as the research object. Firstly, according to the theory of fluid lubrication, a mathematical model of the air cushion was established. Then, the effects of air cushion thickness, pore velocity, and belt velocity on the pressure and bearing characteristics of the air cushion flow field were studied using FLUENT software. In addition, the equivalent stress, displacement, and pressure curves between the air cushion flow field and the conveyor belt and the material were analyzed using the two-way fluid–solid coupling method. Finally, the experimental platform of a single row hole air cushion belt conveyor was built, and the flow field and bearing characteristics of the belt were verified through experiments. The results show that reducing the thickness of the air cushion and increasing the pore flow rate can improve the pressure and bearing characteristics of the air cushion, while speed has little effect at lower belt speeds.

Keywords: air cushion belt conveyor; air cushion pressure; load-bearing characteristics; flow field characteristics; fluid–solid coupling method

ion: Wang, Y.; Wu, L.; Kou, Z.; L. Research on Air Cushion Field and Bearing acteristics of Single Row Hole :ushion Belt Conveyor. *Actuators* 13, 424. ://doi.org/10.3390/ 100424

lemic Editors: Hui Tang, Ven and Feng Ren

ived: 1 September 2024
sed: 14 October 2024
pted: 16 October 2024
shed: 21 October 2024

1. Introduction

The working principle of the air-cushion belt conveyor is that the fan sends a certain flow and pressure of air into the air chamber. The gas flows out through the air holes onto the disk groove, generating an air film with a certain pressure between the disk groove and the conveyor belt, thus forming an air cushion to support the conveyor belt and the materials on it. The air cushion conveyor utilizes air cushions instead of traditional idlers, converting the rolling friction between the conveyor belt and idlers into fluid friction between the conveyor belt and the disk groove, with air as the medium [1]. Therefore, the formation of a uniform and stable air cushion state is key to the design and operation of air cushion belt conveyors.

In terms of gas film thickness, Wu et al. introduced a method for assessing the thickness of the air film in an air cushion belt conveyor. They developed accompanying software to automatically generate a curve depicting the variations in air film thickness [2]. Zhu et al. constructed a mathematical model for the air film and researched on the thickness of the air film in air cushion belt conveyors, utilizing gap flow theory and experimental methods [3]. Pang summarized and introduced four calculation methods for the air film thickness of an air cushion belt conveyor based on the theory of incompressible viscous fluid flow [4]. Xiong et al. analyzed the thermal elastic deformation characteristics of air-lubricated thrust bearings based on fluid–structure coupling. They found that the thermal warping deformation of the rotor at low speeds plays a dominant role in increasing the thickness of the air film [5]. Hou et al. proposed a parallel hybrid floating height prediction model that can predict the floating height of the strip in an air-cushioned furnace ur low floating conditions [6–8].

Researching air cushion pressure, Ji et al. constructed a mathematical model of th cushion pressure field based on theoretical research, carrying out a numerical simulatio the air cushion pressure field [9]. Pang et al. devised a formula to calculate the air cush pressure at different groove positions and its theoretical distribution curve. They u FLUENT6.2 software to simulate the starting flow field [10]. Huang analyzed two kind air cushion pressure calculation methods for the purpose of reasonable fan selection Zhang et al. employed the finite element method to compute the air cushion pressure o air cushion belt conveyor and performed experimental validation [12]. Yang et al. stu the pressure distribution in the air cushion chamber of an air-cushion-supported catama under low and high-speed conditions [13]. Jing, et al. employed the boundary elem method to compute and examine the motion response of a barge platform assisted by a z pressure air cushion when subjected to both regular and irregular waves [14]. Ereme et al. proposed a dynamic mathematical model of a balloon-type air cushion vehicle can predict all the main characteristics of ship motion, such as trajectory, force, mom and air cushion pressure [15]. Xu et al. conducted a numerical study on the press response characteristics of the cushion system of an air cushion vehicle [16]. Wu e studied the contact interaction between a finger and an object when wearing air-cushio gloves and concluded that the internal air pressure can adjust the contact characteris of the fingertip object [17]. Finally, Lu et al. analyzed the influence of cushion press and airflow on the resistance performance of an air-cushion-supported catamaran from perspective of total resistance [18].

In terms of air cushion bearing capacity, Li simplified the gas film bearing prob of an air cushion belt conveyor to a two-dimensional gas lubrication problem represer by the Navier–Stokes equation and calculated the numerical relationship of the des parameters such as gas film bearing capacity using the finite element method [19]. Mo e took an air-floating guide rail in the form of a single row of orifices as the research ob and obtained the changing trend of the air-floating guide rail's bearing capacity along w the gas film's thickness [20]. Li et al. studied the influence of structural parameters the bearing capacity of spiral groove hydrodynamic thrust gas bearings under varia operating conditions [21]. Zhang et al. analyzed the bearing characteristics of micro sp groove thrust air bearings in an ammonia–air mixed environment [22]. Fan et al. stuc the characteristics of a gas film flow field and the change in bearing capacity near orifice of an aerostatic journal bearing under eccentricity conditions and different supply pressures [23]. Guo et al. studied the film-forming behavior of an air-cushio belt conveyor under a stable load [24], while Zhang et al. analyzed the loading capacit gas film and verified the stability of its bearing capacity through experiments [25]. Q et al. studied an air bearing and obtained the load-bearing capacity and its variation extruded film air bearing. Reducing the gap between the bearing and the rotor, increas the amplitude of the bearing sleeve, and increasing the contact area between the bear and the rotor can improve the load-bearing capacity of the extruded film air bearing [2

Researching air film stiffness, Yu et al. conducted numerical simulation and anal on the air film stiffness of non-contact dynamic sealing spiral grooves. The air film stiffr decreased with the increase of air film thickness, and the curve of air film stiffness w thickness tended to be stable with the rise of air film thickness [27]. Ding et al. analy the testing and stability of the air film stiffness of spiral groove dry gas seals. They tes three prototypes with different helix angles in the experiment and found that when helix angle was 74°, the air film stiffness was the highest, and the air film stability w the best [28,29]. Hashimoto et al. proposed an optimal design method to minimize friction torque and maximize the gas film stiffness of spiral and bone-suction gas film th bearings [30]. Ibrahim et al. obtained the optimized dynamic stiffness by changing allowable film thickness [31]. Lin et al. analyzed the individual stiffness and vibra error sources of a hydrostatic guideway. They obtained the micro-errors of the gas f stiffness and individual stiffness of four different cross-section bearings caused by random air vibration [32]. Moreover, Chen et al. analyzed the vibration of a gas-bearing table under the micro-scale effect. According to the Reynolds equation and the airflow characteristics, considering different microscopic factors, the stiffness (K) and damping (C) of the gas film were obtained [33].

Many scholars at home and abroad have done much work on the air cushion belt conveyor, mainly focusing on numerical simulation. However, there are few studies on the flow field characteristics and bearing characteristics of the air cushion belt conveyor through the two-way fluid–solid coupling analysis method. The change in the pressure of the air cushion field will cause the conveyor belt to deform, and the thickness of the air cushion will also change, resulting in movement and displacement in the fluid area. The two-way fluid–solid coupling analysis can better analyze the air cushion's flow field and bearing characteristics than the one-way analysis. Based on predecessors, this paper studies air cushions' flow field and bearing characteristics using theoretical derivation, numerical simulation, and experimental verification. The air cushion flow field's pressure and velocity distribution law are obtained, and the bearing capacity of the air cushion is analyzed, which provides a basis for the optimal design of the air cushion belt conveyor.

2. Theoretical Analysis of Air Cushion Flow Field

2.1. Theoretical Distribution of Air Cushion Pressure

Figure 1 shows a cross-sectional diagram of the air cushion belt conveyor. The air cushion pressure is equal to the pressure carried by the air cushion belt conveyor, which is a necessary condition for its operation. The pressure P_1 is generated by the weight of the conveyor belt, the additional bending deformation force P_2 is required for the conveyor belt to bend into a disk-shaped groove, and the pressure P_3 is generated by the weight of the material.

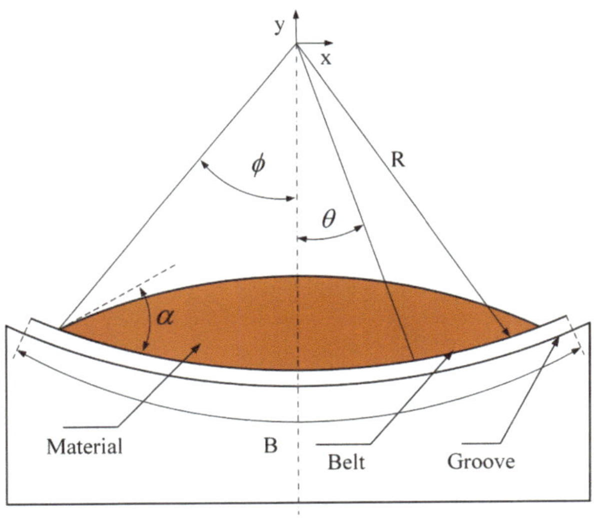

Figure 1. Air cushion belt conveyor cross-sectional diagram.

The calculation equation for the air cushion pressure P is as follows [34]:

$$P = P_1 + P_2 + P_3 \tag{1}$$

$$P_1 = G_B(2\cos\theta - \cos\phi)/B \tag{2}$$

$$P_2 = \frac{(1-\cos\theta)\theta EI}{2(1-\sin\frac{\theta}{2})} \tag{3}$$

$$P_3 = R\rho g\left\{\cos\theta - \cos\phi - \frac{\sin\phi}{\sin\alpha}\cos\alpha + \frac{\sin\phi}{\sin\alpha}\cos[\arcsin(\frac{\sin\alpha}{\sin\phi}\sin\theta)]\right\}$$

where EI is the transverse stiffness of the conveyor belt, N/m; B is the width of the conve belt, mm; θ is the groove position angle, rad; G_B is the weight of conveyor belt per u length, N/m; ρ is material density, kg/m^3; g is the acceleration of gravity, m/s^2; ϕ is maximum disk groove position angle, rad; R is the radius of the groove, mm; and α is material accumulation angle.

2.2. Gas Flow Model near the Pores

The theoretical thickness of the air cushion on the belt conveyor is theoretically 3–5 r and the variation range is relatively small. Therefore, the gas flow of the air hole attachm can be regarded as the radial gap source flow between the parallel disks [35], as show Figure 2.

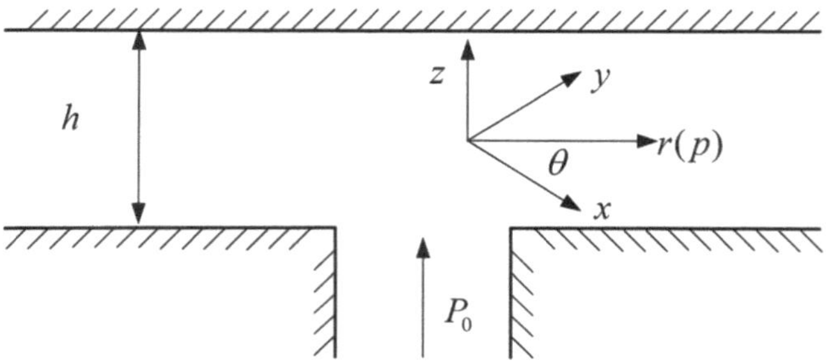

Figure 2. Schematic diagram of pore radial gap source flow.

The parallel disk model is symmetrical to the z axis. Due to the relatively sm thickness of the air cushion, it is possible to consider $u_\theta = 0$, $u_z = 0$, $u_r = u$. Navier–Stokes equation represented by a cylindrical coordinate system can be simpli as follows:

$$f_r - \frac{1}{\rho_1}\frac{\partial P}{\partial r} + v(\frac{\partial^2 u}{\partial r^2} + \frac{1}{r}\frac{\partial u}{\partial r} + \frac{\partial^2 u}{\partial z^2} - \frac{u}{r^2}) = \frac{\partial u}{\partial t} + u\frac{\partial u}{\partial r}$$

$$f_z - \frac{1}{\rho_1}\frac{\partial P}{\partial z} = 0$$

The continuity equation is simplified to

$$\frac{u}{r} + \frac{\partial u}{\partial r} = 0$$

The pressure and velocity distribution equations of the flow field near the hole car obtained as follows:

$$P = P_0 - \frac{\Delta P \ln(\frac{r_2}{r})}{\ln(\frac{r_2}{r_1})}$$

$$u = \frac{1}{2\mu}\frac{dp}{dr}(z^2 - hz)$$

where f_r is the radial mass force, m/s^2; ρ_1 is air density, kg/m^3; P is the air cushion press Pa; P_0 is the inlet pressure, Pa; ΔP is the total pressure difference, Pa; r_1 is the radius the disk, mm; r_2 is the pore radius, mm; v is air viscosity, m^2/s; u is the speed, m/s; f the mass force in the z direction, m/s^2; μ is the dynamic viscosity of air, Pa·s; and h is thickness of the air cushion, mm.

2.3. Cross-Section Gas Outflow Model

The material on the conveyor belt is distributed on the conveyor belt in the middle thick and thin areas on both sides, which also leads to the gradual decrease in air cushion pressure from the middle to the edges along the bandwidth direction. Similarly, the thickness of the air cushion is in the direction of the bandwidth, the middle thick edge is relatively thin, and the radius of the groove is much larger than the thickness of the air cushion. Therefore, the angle-wedge-shaped gap formed by the conveyor belt and the disk groove can be used, that is, the air cushion transverse gap flow model, as shown in Figure 3.

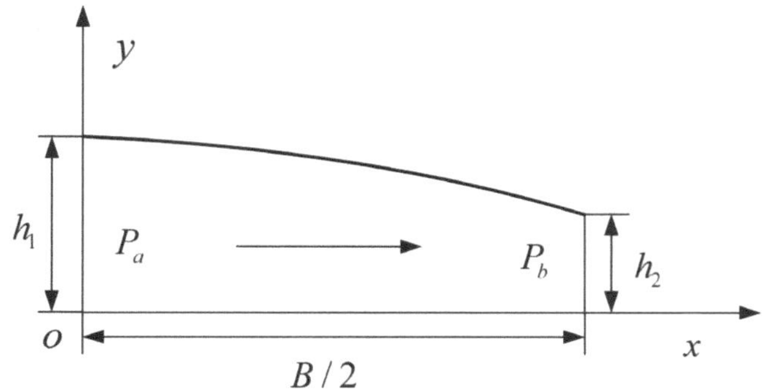

Figure 3. Wedge gap flow model.

For the above reasons, the following can be considered: $u_x = u$, $u_y = u_z = 0$, mass force $f_x = f_y = 0$, $f_z = -g$.

The pressure distribution and temperature of the cross-section of the air cushion remain basically unchanged. The air cushion is regarded as incompressible flow, and the density is constant. At this time, the N–S equation can be simplified as follows:

$$-\frac{1}{\rho}\frac{\partial P}{\partial x} + v\left(\frac{\partial^2 u}{\partial x^2} + \frac{\partial^2 u}{\partial y^2} + \frac{\partial^2 u}{\partial z^2}\right) = u\frac{\partial u}{\partial x} \quad (10)$$

$$-g - \frac{1}{\rho}\frac{\partial P}{\partial y} = 0 \quad (11)$$

$$-\frac{1}{\rho}\frac{\partial P}{\partial z} = 0 \quad (12)$$

The continuous equation can be simplified to

$$\frac{\partial u}{\partial x} = 0 \quad (13)$$

Continuing to simplify the above equations, the velocity and pressure distribution of the air cushion along the cross-section can be obtained:

$$u = \frac{y^2 - hy}{2\mu}\frac{dP}{dx} \quad (14)$$

$$P_x = P_a + \frac{{h_2}^2({h^2}_1 - h^2)}{h^2(h_2^2 - h_1^2)}(P_a - P_b) \quad (15)$$

where P_a is the pressure at the center of the air cushion, Pa; P_b is the pressure at the outlet of the air cushion, Pa; h_1 is the thickness at the center of the air cushion, mm; and h_2 is the thickness at the outlet of the air cushion, mm.

2.4. Air Cushion Stiffness Analysis

The stiffness of the air cushion can reflect its carrying capacity: the greater the stiffr the greater the carrying capacity. When the load changes, the air cushion has a stror anti-interference ability and stronger stability. Analogous to spring stiffness, air cusł stiffness can be defined as follows:

$$K = -\frac{dW}{dh}$$

where W is the air cushion bearing capacity, N, K is the stiffness of air film, N/mm.

3. Numerical Simulation of Air Cushion Field

The simulation hardware device used was an Intel (R) Xeon (R) Gold 6133 CP 2.5 GHz 2.49 GHz (two-core processor) equipped with 96 GB RAM and a Win10 64 operating system workstation. The simulation environment was Workbench 2021R1.

3.1. Model Foundation

The radius of the groove is 1007 mm, and the width of the conveyor belt is 1000 The thickness of the outlet on the left and right sides of the air cushion model is h_{min}, the thickness of the middle of the air cushion is h_{max}. Based on the upper surface of disc groove, the arcs are made at the three points determined by h_{min} and h_{max} val respectively, as the position of the lower surface of the conveyor belt. The cross-sect of the air cushion thickness is made with the upper surface of the disc groove and lower surface of the conveyor belt as the edge, and then stretched along the z direct to form a complete air cushion model. The front and rear sides of the air cushion are as symmetrical surfaces, and the left and right sides are gas outlets. The model of the cushion is shown in Figure 4.

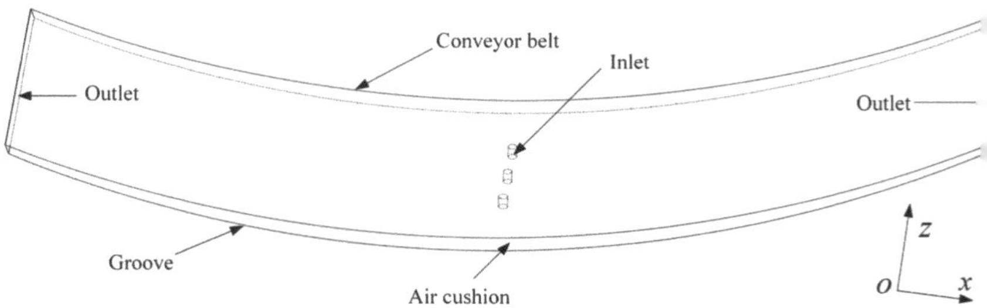

Figure 4. Air cushion model.

3.2. Influence of Air Cushion Thickness on the Pressure of Air Cushion Flow Field and Bearin Capacity of Air Cushion

Considering the symmetry of the geometric model, a half air cushion model is u for the analysis. The mesh is polyhedral mesh divided by Fluent mesh, and the minim orthogonal mass is 0.35. Due to the relatively small thickness of the air cushion, boundary layer is set to two layers. Due to the relatively low pressure and velocity of air cushion, air is considered an incompressible fluid, and the influence of air gravit ignored. The static pressure at the inlet of the air cushion pressure is set to 6.8 kPa, outlet pressure on both sides of the air cushion is set to 0 kPa, and we assume that the fl is air at temperature of 300 K. The density is constant as the default of the software, density is 1.225 kg/m^3, and the viscosity is 1.7894×10^{-5} kg/m·s. The three-dimensic dual-precision model is adopted. The turbulence model uses the standard $k - \varepsilon$ model. coupled solution method is used. At the same time, the residual curve is selected as monitoring window to observe the iterative process of the solution process.

Because this paper analyzes the single row hole air cushion belt conveyor and si the air hole spacing is not very large, the pressure between the two pores is relativ stable, and the mutual influence is not very large. However, the pressure loss of the cushion belt conveyor in the cross-section is larger than that in the running direction, so, the following focuses on the analysis of a curve pressure distribution along the bandwidth near the air cushion hole and on the cross-section.

Under the condition that the radius of the groove is constant, the diameter of the hole is 8 mm, the distance between the holes is 100 mm, and the thickness of the outlet on both sides is 0.5 mm. Figure 5a–d illustrates the wall pressure distribution corresponding to the central hole with air cushion thicknesses of 3 mm, 4 mm, 5 mm, and 6 mm. From the pressure cloud map, it can be concluded that the pressure around the central hole decreases approximately monotonically from the inside out along the radius direction of the central hole. Due to the existence of the gas eddy current effect, the pressure value of some areas in the center of adjacent pores is lower than that of adjacent areas. The smaller the thickness of the air cushion, the larger the pressure near the center of the pore, and the more uniform the distribution of the larger pressure value. The maximum pressure appears in the center of the air hole, and the pressure peaks are 6718, 6697, 6705, and 6785 Pa, respectively. The maximum value of the wall pressure does not increase with the decrease in air cushion thickness, but the distribution of a larger pressure value is more uniform, which also reflects the complexity of the pressure generation in the air cushion flow field.

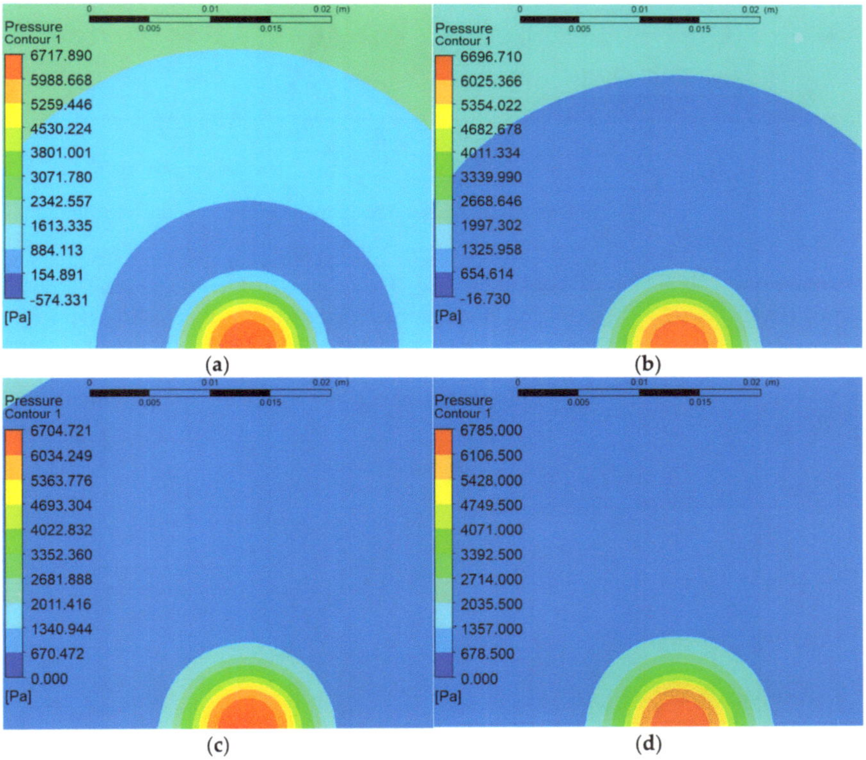

Figure 5. Wall pressure distribution corresponding to the center hole of different air cushion thicknesses. (**a**) Air cushion thickness of 3 mm. (**b**) Air cushion thickness of 4 mm. (**c**) Air cushion thickness of 5 mm. (**d**) Air cushion thickness of 6 mm.

The variation curves of the air cushion pressure with different air cushion thicknesses are shown in Figure 6. The maximum pore pressure is located in the center of the pore. Due to the generation of eddy currents near the pores, gas flow pulsations and pressure losses occur, resulting in certain fluctuations in the pressure values near the pores. The larger the thickness of the air cushion, the more intense the pressure fluctuations near the pores. The

pressure distribution curves of the different air cushion thicknesses are basically the sa
and the pressure at the same position of the air cushion decreases with an increase in
air cushion thickness. The pressure of the air cushion is symmetrical around the cen
hole, and the pressure of the cross-section of the air cushion exhibits an approxima
flat-top distribution.

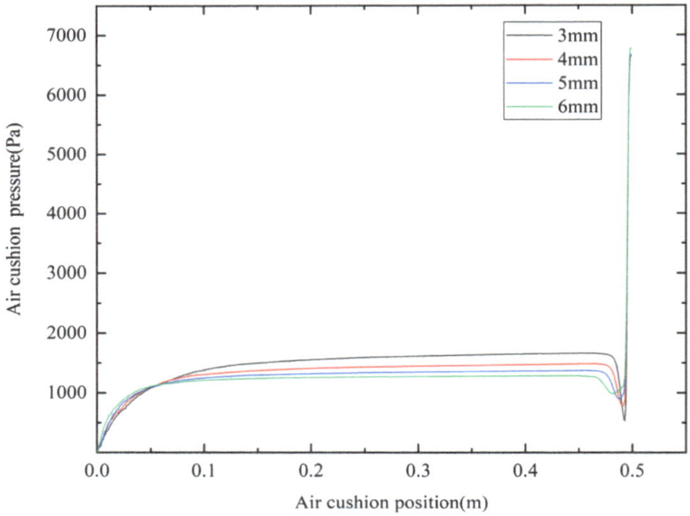

Figure 6. Air cushion pressure distribution of different air cushion thicknesses.

The thickness interval of the air cushion is 0.5 mm, and the bearing capacity of the
cushion for seven kinds of air cushion thicknesses is obtained. The relationship betw
the thickness of the air cushion and the bearing capacity is shown in Figure 7. It can
seen from the figure that as the thickness of the air cushion increases, its bearing capa
gradually decreases. Using interpolation and fitting methods, we can deduce that
thickness of the air cushion is inversely proportional to its load-bearing capacity.

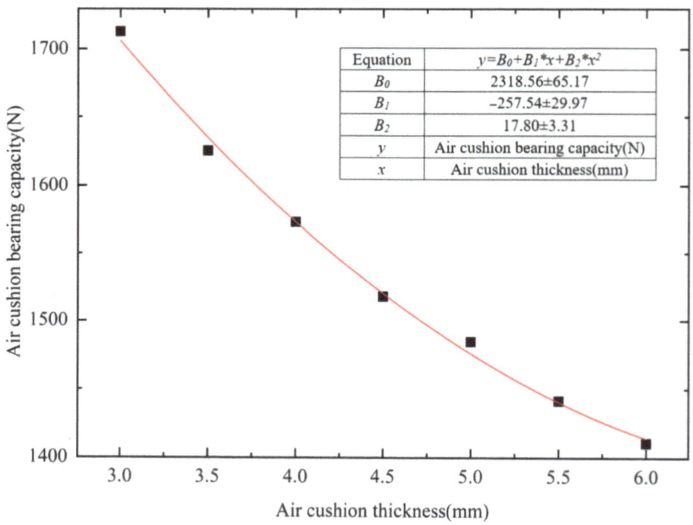

Figure 7. Relationship between air cushion thickness and air cushion bearing capacity.

According to Equation (16), the stiffness of the air cushion can be obtained by deriving the fitted air cushion bearing equation. Equation (17) shows that the air cushion stiffness has a linear relationship with the thickness of the air cushion and decreases with a decrease in thickness.

$$K = -(B_1 + 2B_2 * x) \tag{17}$$

3.3. Influence of Air Hole Velocity on Air Cushion Flow Field Pressure and Bearing Capacity

Under the condition that the radius of the groove is constant, the diameter of the hole is 8 mm, the spacing of the holes is 100 mm, the thickness of the air cushion is 3 mm, and the thickness of the outlets on both sides is 0.5 mm. Figure 8a–d shows the wall pressure distribution corresponding to the center hole of the hole velocity of 70 m/s, 80 m/s, 90 m/s, and 100 m/s. As it can be seen from the pressure cloud diagram, the maximum pressure appears at the center of the hole, and the pressure peaks are 5944, 7714, 9727, and 11,932 Pa, respectively. The pressure peak increases gradually with an increase in the flow velocity.

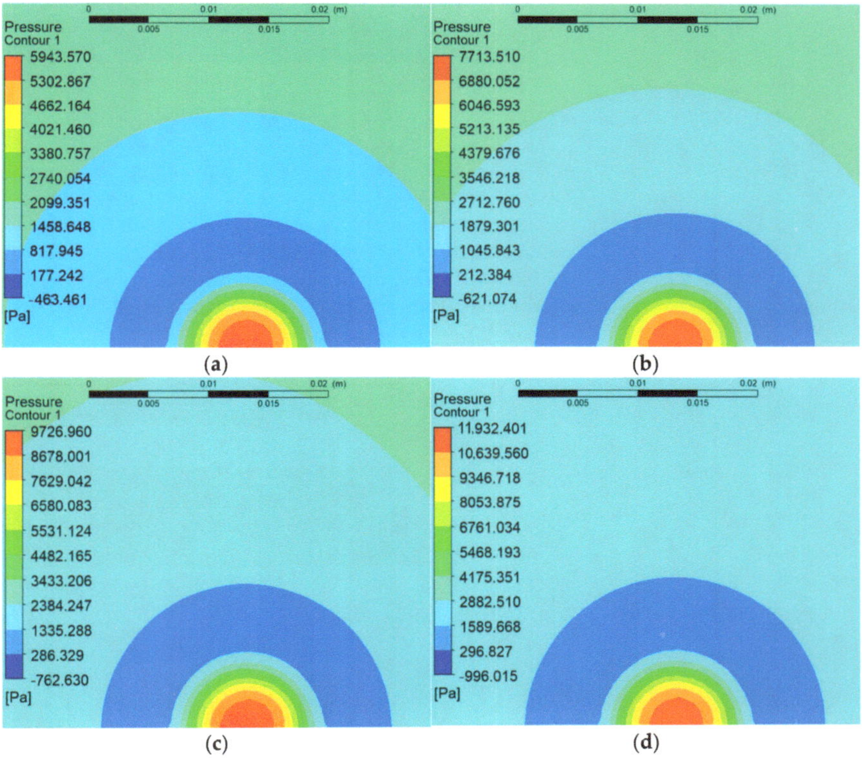

Figure 8. Wall pressure distribution corresponding to central holes of different pore velocities. (**a**) Air holel velocity of 70 m/s. (**b**) Air holel velocity of 80 m/s. (**c**) Air holel velocity of 90 m/s. (**d**) Air holel velocity of 100 m/s.

The variation curves of air cushion pressure at different pore velocities are shown in Figure 9. The maximum pore pressure is located near the center of the pore. The distribution curves of air cushion pressure at different pore velocities are basically the same, and the pressure at the same position of the air cushion increases with an increase in pore velocity.

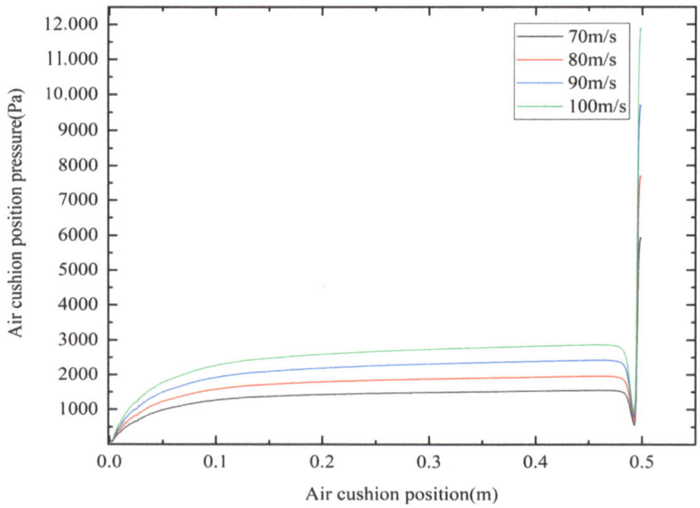

Figure 9. Air cushion pressure distribution at different pore velocities.

The pore velocity interval is 10 m/s, and the air cushion bearing capacity of se kinds of pore velocity is obtained. The relationship between air cushion velocity and cushion bearing capacity is shown in Figure 10. From the diagram, it can be seen that w an increase in the air cushion flow rate, which is equivalent to an increase in the air sup volume of the fan, the air cushion bearing capacity also gradually increases. Using interpolation fitting method, it can be concluded that the air cushion thickness has a lir relationship with the air cushion bearing capacity.

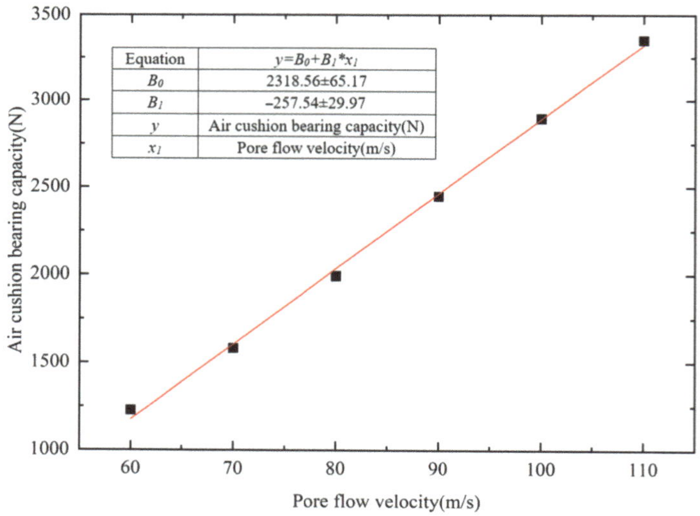

Figure 10. Relationship between pore flow velocity and air cushion bearing capacity.

3.4. *Influence of Belt Speed on Air Cushion Flow Field Pressure and Bearing Capacity*

Under the condition that the radius of the groove is constant, the diameter of the h is 8 mm, the hole spacing is 100 mm, the thickness of the air cushion is 3 mm, the thickr of the outlet on both sides is 0.5 mm, and the inlet pressure is 6800 Pa. Figure 11a–d shc the wall pressure distribution corresponding to the central hole with velocities of 1 m

3 m/s, 5 m/s, and 7 m/s. The speed of the conveyor belt is to set the upper surface of the air cushion, that is, the surface in contact with the conveyor belt, as the moving wall surface and the other wall surfaces are set unchanged. Then, the running speed of the conveyor belt is simulated by setting different speeds. The pressure cloud map reveals that the maximum pressure appears at the center of the air hole, and the peak pressures are 6678, 6677, 6677, and 6675 Pa, respectively. Compared with the data in Figure 5a, the dynamic pressure generated by the low belt speed does not have much of an effect on the pressure corresponding to the wall surface of the center hole and only slightly decreases.

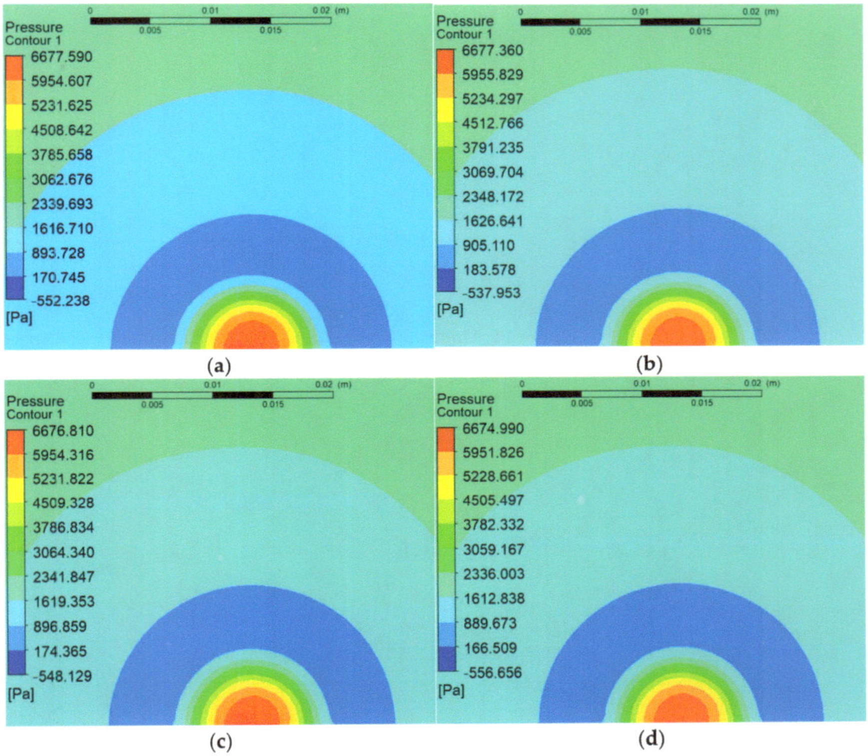

Figure 11. Wall pressure distribution corresponding to the center hole of different belt speeds. (**a**) Belt speed of 1 m/s. (**b**) Belt speed of 3 m/s. (**c**) Belt speed of 5 m/s. (**d**) Belt speed of 7 m/s.

Figure 12 shows the curves of the air cushion pressure at different belt speeds. The maximum pore pressure is located near the center of the pore. When the belt speed is low, the pressure distribution curves basically coincide; thus, it can be concluded that a lower belt speed will not have much influence on the air cushion pressure.

At a speed interval of 1 m/s, the bearing capacity of seven kinds of air cushions with speed is obtained. Figure 13 shows the relationship between the belt speed and the bearing capacity of the air cushion. The diagram reveals that the difference in the bearing capacity of the air cushion at different belt speeds is minimal, indicating that the air cushion conveyor is particularly suitable for high-speed operation.

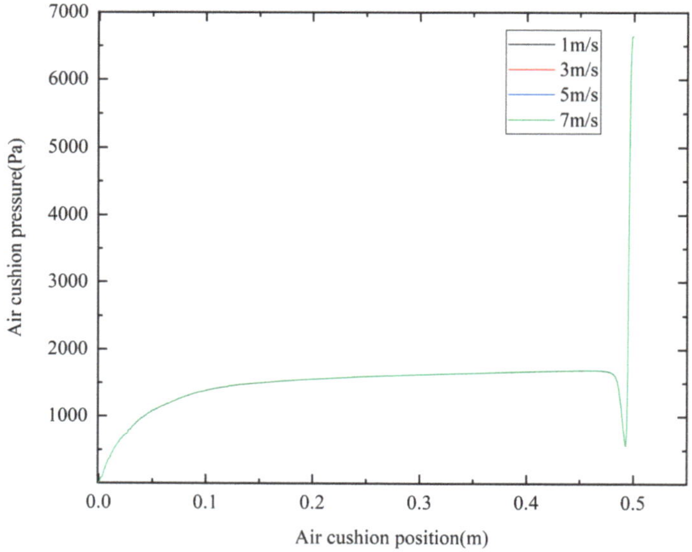

Figure 12. Air cushion pressure distribution at different belt speeds.

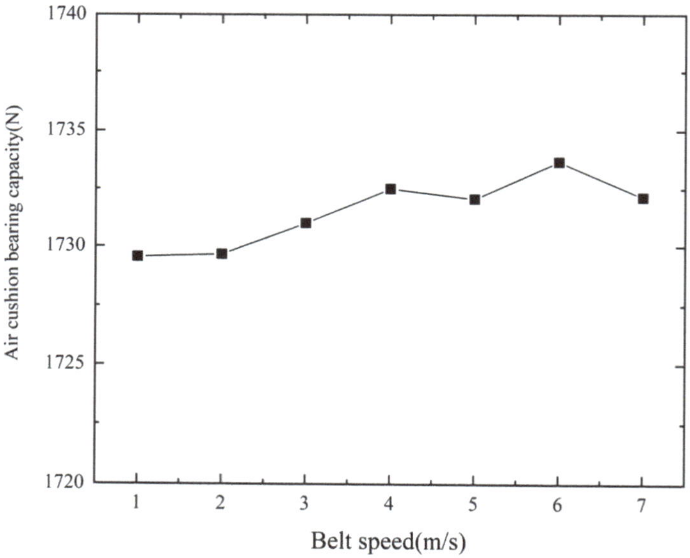

Figure 13. Air cushion bearing capacity at different belt speeds.

3.5. Fluid–Solid Coupling Analysis of Air Cushion Flow Field, Conveyor Belt, and Material

The conveyor belt and material are supported by a very thin layer of air cushion for air cushion belt conveyors. The pressure exerted by the air cushion causes deformat in the conveyor belt, and the altered shape of the conveyor belt, in turn, influences thickness of the air cushion, leading to displacement and motion within the fluid dom This interactive dynamic should be analyzed through bidirectional fluid–structure c pling. This mutually influential relationship should be analyzed using two-way fluid–s coupling. This approach considers the simulation of fluid and solid characteristics at same time so that the results are closer to the real situation. In the coupling simulati in order to provide the flow channel of the airflow and the boundary conditions of flow field, the initial thickness of the air cushion flow field is set to 1 mm, and the inlet pressure is 6800 Pa. Gravity loads are applied on conveyor belts and materials, as shown in Figure 14. The lower surface of the conveyor belt is set to the Fluid Solid Interface, and symmetrical constraints are added at the specified positions of the three surfaces.

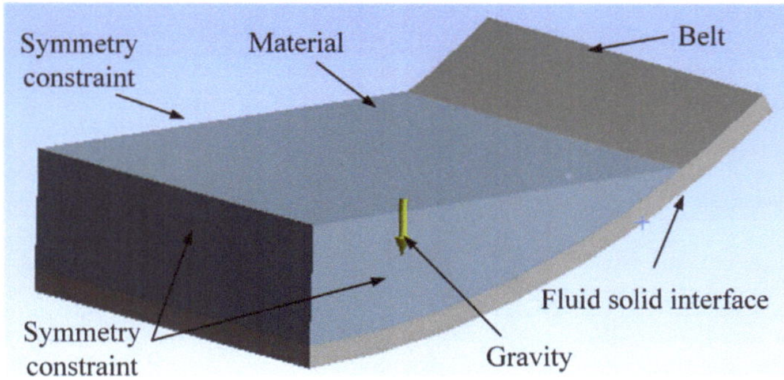

Figure 14. Constraints and loads.

The data transfer between the coupling surface in the fluid analysis and the fluid–solid interface in the solid analysis is set. The minimum number of coupling iterations is two, the maximum number of iterations is five, the time step is 0.001 s, and the simulation time is 0.1 s. The density, elastic modulus, and Poisson's ratio of the conveyor belt and the material are set as shown in Table 1.

Table 1. Solid domain material properties.

Material	Density (kg/m^3)	Elastic Modulus (MPa)	Poisson Ratio
Conveyor belt	1545	2810	0.49
Coal	1300	3500	0.32

Figure 15 is the equivalent stress cloud diagram of the conveyor belt and the material at the time of 0.1 s. It can be seen that the stress in the bearing area of the conveyor belt is very small due to the gravity of the material. However, the stress at the connection between the bearing area and the non-bearing area of the conveyor belt is large. This is because there is no material in the non-bearing area, and the amount of suspension in the conveyor belt is slightly larger under the action of air cushion pressure, resulting in a larger stress on the upper surface.

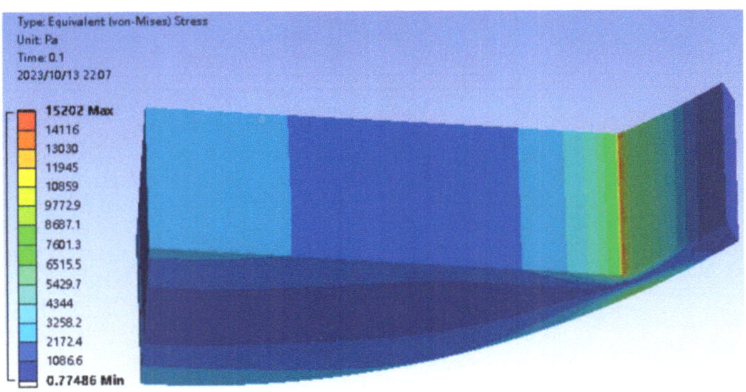

Figure 15. Equivalent stress cloud diagram at 0.1 s.

The maximum and minimum equivalent stress curves of the conveyor belt du
the fluid–solid coupling process are shown in Figure 16. The maximum equivalent st
changes considerably at 0–0.04 s, and the maximum equivalent stress is stable near 15,20
after 0.04 s. The minimum equivalent stress gradually stabilizes at 0.06 s and then stabi
near 0.78 Pa.

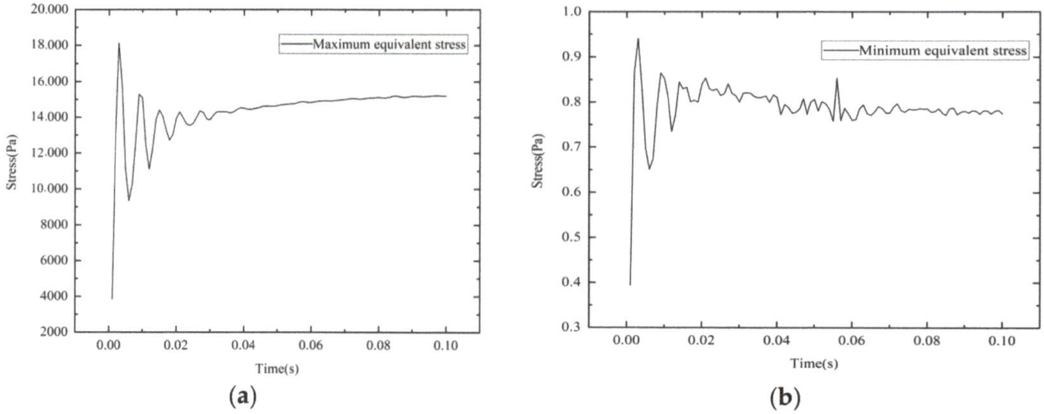

Figure 16. Equivalent stress change curves. (**a**) Maximum equivalent stress change curve. (**b**) Minin
equivalent stress change curve.

Figure 17 is the displacement curve of the conveyor belt. The variation trend of
maximum and minimum displacement curves is basically similar. From the beginnin
the conveyor belt, there is a small increase, and it gradually stabilizes at 0.1 s. From
size of the stable value, the difference between the maximum and minimum displacen
values is not large.

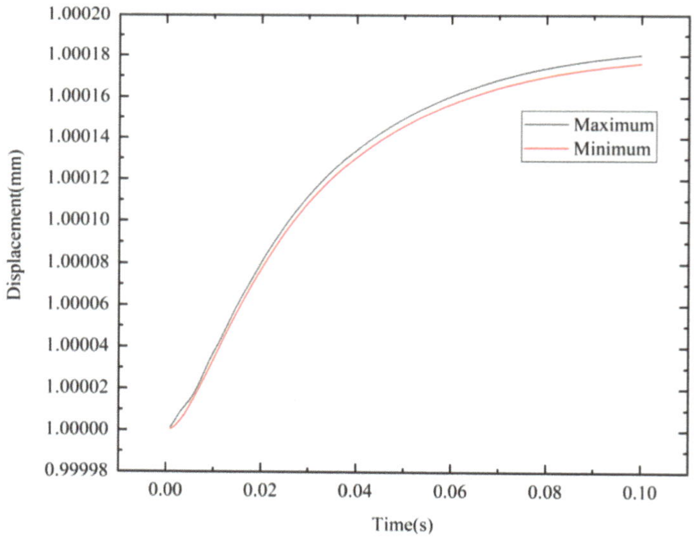

Figure 17. Displacement change curve.

Figure 18 shows the pressure distribution of the air cushion at 0.1 s. The figure sh
that the pressure near the air inlet is maximum. It then gradually decreases from the ce
of the conveyor belt towards the sides, eventually reaching zero. This behavior aligns w

the characteristic that the air cushion pressure depends on the load. After loading, the conveyor belt undergoes elastic deformation, and the air cushion pressure obtained from the fluid–solid coupling simulation is basically consistent with the theoretical calculation.

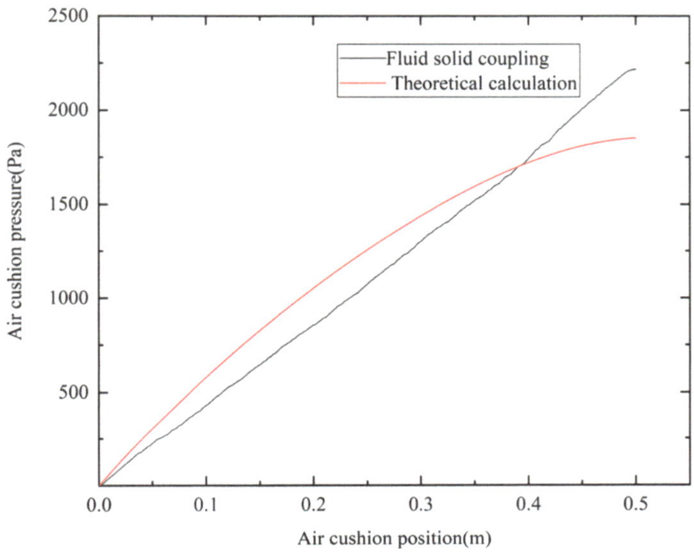

Figure 18. Air cushion pressure distribution at 0.1 s.

4. Experimental Platform Construction and Experimental Verification

The experimental platform is shown in Figure 19, with a length of 3 m. The mechanical structure mainly comprises of a disk groove and a gas chamber. The micro-differential pressure transmitter model is GH-130/Y-B8-mA, the range is 0–10 kPa, 24 V DC input, 4–20 mA DC output, and the stability is ±0.5%. The laser displacement sensor model is BL-100 NZ, its measurement range is 65~135 mm, the accuracy is 0.075 mm, the power supply voltage is 24 V, and the reaction time is 10 ms. The micro-differential pressure transmitter measures the pressure value at the lower hole of the conveyor belt. The laser displacement sensor is arranged above the conveyor belt to measure the displacement in the vertical direction of the conveyor belt, and the gas film thickness is calculated. The core of the control system is the PLC. The PLC transmits the control signal to the inverter to control the fan speed, which can change the air cushion pressure and air cushion thickness. Labview2018 software was used to design the system data acquisition software. The data acquisition module can monitor the pressure in real-time and alert when the pressure value exceeds the limit value.

Figure 20 shows the cross-section of material accumulation. The experimental material is coal, the density is 1300 kg/m^3, and the loading mass is 150 kg. The conveyor belt is a wire rope core conveyor belt of type ST-2000, with a bandwidth of 1000 mm, a thickness of 22 mm, and a mass per unit length of 34 kg.

The air cushion pressure was measured in the experiment, and the average was calculated from three sets of measurements. From Table 2, the theoretical values of the central hole air cushion pressure exhibit a slight discrepancy when compared to the measured values, with the theoretical values consistently higher than the experimental measurements. Due to the influence of the elasticity of the conveyor belt, the force condition changes when the suspension condition occurs, which increases the curvature of the conveyor belt and results in additional pressure on the air cushion. However, in the actual situation, this elastic strain will be restored, reducing additional stress. The theoretical calculation does not consider the additional stress caused by the bending of the conveyor belt, and pressure value will be smaller than the measured value.

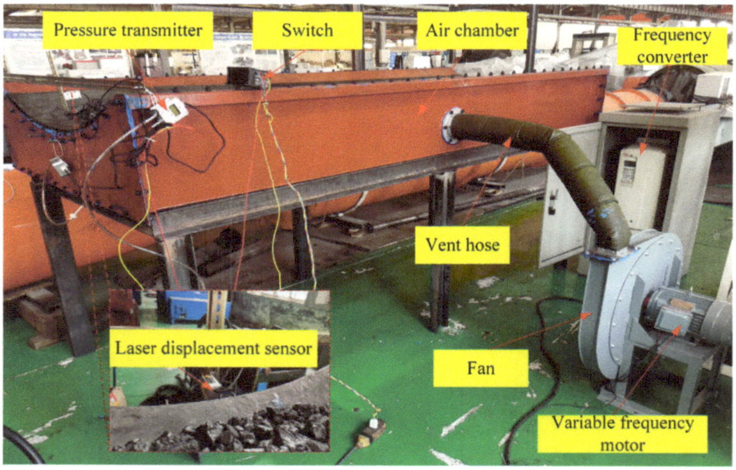

Figure 19. Experimental platform.

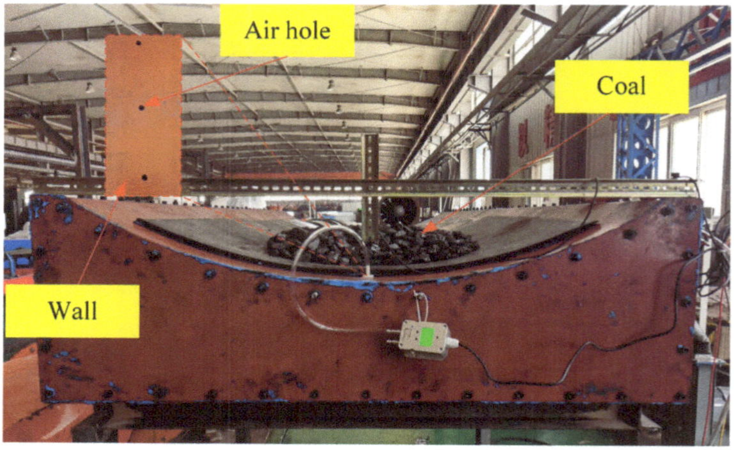

Figure 20. Material accumulation cross-section.

Table 2. Comparison of air cushion pressure results.

Method	Air Cushion Position (m)		
	$x_1 = 0.5$	$x_2 = 0.5$	$x_3 = 0.5$
Experimental measurement	2200.56 Pa	2199.68 Pa	2201.32 Pa
Theoretical calculation	1851.78 Pa	1851.78 Pa	1851.78 Pa

Figure 21 shows the thickness distribution of the air cushion. The thickness measu during the experiment is not much different from the theoretical calculation; the mid value is large, and the edge value is small. However, the edge change of the conveyor measured during the experiment is slightly different from the theoretical calculation, a the change range at the edge is faster. The reason for this may be that the displacemen the edge part of the conveyor belt is relatively unstable due to the absence of load.

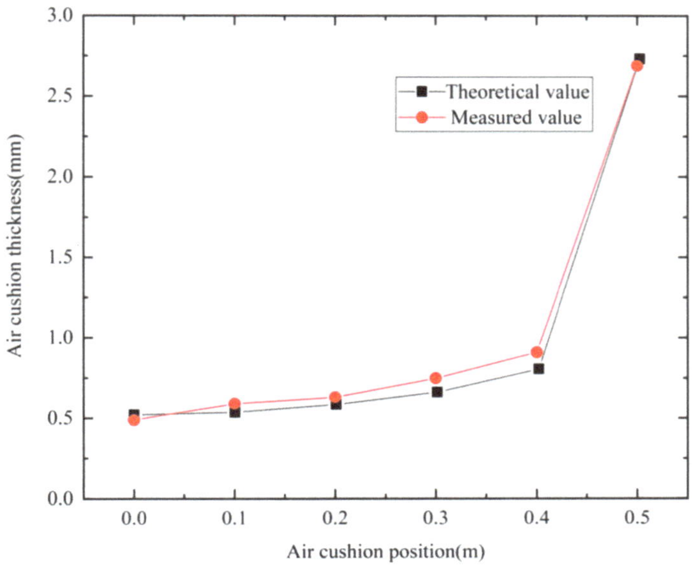

Figure 21. Air cushion thickness distribution.

5. Conclusions and Perspectives

On the basis of constructing the ideal parameter model, this paper assumes that the load of the conveyor belt is evenly distributed to simplify the research on the air cushion flow field and the bearing characteristics of the single-row hole air cushion belt conveyor, and reveals the influence of air cushion thickness, air hole velocity, and belt speed on the air cushion performance. Some conclusions are drawn as follows:

1. In this study, the air cushion structure model of the air cushion belt conveyor is first constructed, and the theoretical formula of the air cushion pressure and the velocity and pressure distribution of the gas inflow and outflow model is obtained through mechanical analysis. These theoretical results provide a solid theoretical basis for the subsequent numerical simulation and experimental analysis of air cushions.
2. Under certain conditions, this study reveals the influence of air cushion thickness, pore velocity, and belt velocity on the performance of air cushions. Specifically, reducing the thickness of the air cushion can improve its pressure and bearing capacity. The pore velocity is linearly and positively correlated with its bearing capacity. At the same time, the study also found that at a lower belt speed, the change of belt speed has little effect on the air cushion performance and bearing capacity, which indicates that the air cushion conveyor is very suitable for high belt speed operation.
3. In this study, the two-way fluid–solid coupling method is used to further analyze the maximum equivalent stress, displacement curve, and air cushion pressure distribution of the conveyor belt. Through numerical simulation and experimental verification, we obtained the relationship between air cushion pressure and air cushion thickness and found that the theoretical calculation was basically consistent with the experimentally measured values, which verified the accuracy and reliability of the research results.

Due to the uneven thickness of the air cushion during the operation of the air cushion conveyor, the research on the flow field characteristics and bearing characteristics of the air cushion conveyor is still very challenging. This paper can be regarded as a preliminary attempt to study the air cushion flow field and bearing characteristics of single row hole air cushion belt conveyors. The conclusion is helpful to the system design of single row hole air cushion belt conveyors. It should also be noted that the influence of gas eddy current effect on the flow field and bearing characteristics of the air cushion is not considered in the theoretical analysis of the air cushion in this paper, which is also important for the des of the air cushion belt conveyor. Therefore, further research work should be carried ou the interaction of gas eddies on the air cushion flow field and load-bearing characteris

Author Contributions: Conceptualization, Y.W.; methodology, Y.W.; investigation, Y.W., L.W. L.T.; writing—original draft, Y.W.; writing—review and editing, Z.K.; supervision, Z.K.; pr administration, Z.K. All authors have read and agreed to the published version of the manuscri

Funding: This work was supported by National Natural Science Foundation of China under g 52174147 (project title: Research on Intelligent Control and Fault Diagnosis of Autonomous Inspec Robot for Belt Conveyor in Harsh Environment); Key Technology Research and Industrializatio Intelligent Lower Transportation Expandable Belt Conveyor Equipment (Taishan Industry Lea Talent Program).

Data Availability Statement: The data presented in this study are available on request from corresponding author.

Conflicts of Interest: The authors declare no conflicts of interest.

References

1. Xue, H.; Fu, T.; Liu, Y. Dynamic Analysis of Air Cushion Flow Field of Multi-row Hole Air Cushion Belt Conveyor. *Hois Conveying Mach.* **2009**, *4*, 64–67.
2. Wu, F.; Jia, Z.H.; Wang, Y. Calculation of Air Film Thickness of Air Cushion Belt Conveyor. *J. Taiyuan Univ. Sci. Technol.* **200** 85–87.
3. Zhu, X.L.; Rui, Y.N.; Zhu, H.M. Study on air film thickness of air cushion belt conveyor. *J. Mech. Strength* **2017**, *39*, 723–726.
4. Pang, M.J. Calculation of gas film thickness of air cushion belt conveyor. *Colliery Mech. Electr. Technol.* **2006**, *2*, 46–49.
5. Xiong, C.; Xu, B.; Yu, H.; Huang, Z.; Chen, Z. A thermo-elastic-hydrodynamic model for air foil thrust bearings conside thermal seizure and failure analyses. *Tribol. Int.* **2023**, *183*, 108373. [CrossRef]
6. Hou, S.; Liu, J.; Lv, W. Flotation Height Prediction under Stable and Vibration States in Air Cushion Furnace Based on H Division Method. *Math. Probl. Eng.* **2019**, *2019*, 5713602. [CrossRef]
7. Hou, S.; Han, X.; Zhang, X.; Bai, M.; Hua, F. Hard-Division and Multi-Model Based Floating Height Prediction for Air Cus Furnace with Hybrid Nozzles. *IEEE Access* **2020**, *8*, 194685–194699. [CrossRef]
8. Hou, S.; Zhang, X.; Dai, W.; Han, X.; Hua, F. Multi-Model- and Soft-Transition-Based Height Soft Sensor for an Air Cus Furnace. *Sensors* **2020**, *20*, 926. [CrossRef] [PubMed]
9. Ji, X.F.; Jin, Y.; Han, W.J. Numerical Simulation Study on Air Cushion Pressure Field of Air Cushion Belt Conveyor. *J. Chang. U Sci. Technol. (Nat. Sci. Ed.)* **2009**, *32*, 102–104.
10. Pang, M.J.; Zhang, S.L. Air cushion pressure field research and numerical simulation of air cushion belt conveyor. *Coal M Mach.* **2006**, *27*, 38–40.
11. Huang, Y. Calculation of air-cushion pressure of air-cushion belt conveyor. *Coal Mine Mach.* **1997**, *5*, 10–12.
12. Zhang, L.Y.; Wang, X.M.; Sun, Y. Theoretical and Experimental Study on Air Cushion Pressure of Air Cushion Belt Conve *J. Taiyuan Univ. Sci. Technol* **1996**, *4*, 60–64+70.
13. Yang, J.L.; Sun, H.B.; Li, X.W.; Liu, X. Flow field characteristic analysis of cushion system of partial air cushion support catam in regular waves. *Pol. Marit. Res.* **2022**, *29*, 21–32. [CrossRef]
14. Jing, F.; Xu, L.; Guo, Z.; Liu, H. A Theoretical Study on the Hydrodynamics of a Zero-Pressurized Air-Cushion-Assisted B Platform. *J. Mar. Sci. Eng.* **2020**, *8*, 664. [CrossRef]
15. Eremeyev, V.O.; Peplin, F.S.; Tumanin, A.V. Mathematical model of dynamics of Air Cushion Vehicle with ballonet type ski water. *Procedia Eng.* **2017**, *206*, 354–359. [CrossRef]
16. Xu, S.; Tang, Y.; Chen, K.; Zhang, Z.; Ma, T.; Tang, W. Numerical investigation on pressure responsiveness properties of skirt-cushion system of an air cushion vehicle. *Int. J. Nav. Archit. Ocean. Eng.* **2020**, *12*, 928–942. [CrossRef]
17. Wu, J.Z.; Wimer, B.M.; Welcome, D.E.; Dong, R.G. An analysis of contact stiffness between a finger and an object when wea an air-cushioned glove: The effects of the air pressure. *Med. Eng. Phys.* **2012**, *34*, 386–393. [CrossRef]
18. Lu, S.; Zou, J.; Zhang, Y.; Guo, Z. Experimental and Numerical Study on Motion and Resistance Characteristics of the Partial Cushion Supported Catamaran. *Water* **2019**, *11*, 1033. [CrossRef]
19. Li, Y.K. Finite element analysis of air-film carrying capacity of belt air-cushion conveyor. *Lubr. Eng.* **2000**, *5*, 14–16.
20. Mo, D.Y.; Ma, P.; Lian, H.S. Mathematical model improvement and verification analysis of bearing capacity of single-row or air-floating guideway. *Lubr. Eng.* **2019**, *44*, 88–94.
21. Li, Y.L.; Dong, Z.Q. The influence of structural parameters on the bearing capacity of spiral groove hydrodynamic thrust bearing under variable operating conditions. *Bearing* **2022**, *38*, 23–28+40.
22. Zhang, C.W.; Ge, Q.J.; Xie, Z.J. Analysis of load-bearing characteristics of micro-spiral-grooved thrust air bearing in heliu mixed environment. *J. Tribol.* **2018**, *38*, 213–219.

Fan, C.; Han, S.H.; Li, B. Analysis of gas film flow field and load-bearing characteristics of aerostatic journal bearings. *Mech. Des. Manuf.* **2021**, *1*, 45–49.
Guo, S.; Liu, J.; Li, Z.; Zhang, D.; Huang, W. Experimental research on air film formation behavior of air cushion belt conveyor with stable load. *Sci. China Technol. Sci.* **2013**, *56*, 1424–1434. [CrossRef]
Zhang, B.; Meng, W.; Zhang, H. Research on the pressure field and loading characteristics of air film of air cushion belt conveyor. *AIP Adv.* **2022**, *12*, 095123. [CrossRef]
Quan, H.; Lan, J.; Du, Y. Research on load characteristics of squeeze-film air bearings based on fluid–solid coupling. *Phys. Fluids* **2023**, *35*, 016125. [CrossRef]
Yu, S.R.; Cao, X.Y.; Ding, X.X. Numerical Simulation and Analysis of Gas Film Stiffness of Non-contact Dynamic Seal Spiral Groove. *Fluid Mach.* **2012**, *40*, 16–20.
Ding, X.X.; Zhang, H.Z.; Su, H. Gas film stiffness test and stability analysis of spiral groove dry gas seal. *Vib. Shock* **2013**, *32*, 163–168+190.
Ding, X.X.; Zhang, H.Z.; Zhang, W.Z. Vibration test and stability analysis of spiral groove dry gas seal. *J. Vib. Meas. Diagn.* **2013**, *33*, 231–235+337–338.
Hashimoto, H.; Ochiai, M.; Nanba, T. Theoretical Analysis and Optimum Design of High Speed Air Film Thrust Bearings (Application to Optimum Design Problem). *J. Adv. Mech. Des. Syst. Manuf.* **2007**, *1*, 306–318. [CrossRef]
Ibrahim, M.D.; Namba, T.; Ochiai, M.; Hashimoto, H. Optimum Design of Thrust Air Bearing for Hard Disk Drive Spindle Motor. *J. Adv. Mech. Des. Syst. Manuf.* **2010**, *4*, 70–81. [CrossRef]
Lin, J.J.; Li, D.S.; Zhang, W. Analysis on the Individual Stiffness of the Aerostatic Guide Way and Error Sources of its Vibration. *Adv. Mater. Res.* **2012**, *472–475*, 862–868. [CrossRef]
Chen, D.; Dong, L.; Pan, R.; Fan, J.; An, C. Vibration analysis of air-bearing work-stage under micro-scale effect. *Proc. Inst. Mech. Eng. Part K J. Multi-Body Dyn.* **2018**, *232*, 521–535. [CrossRef]
Peng, X.X. Quantitative research on air cushion parameters of air cushion conveyor. *Mech. Res. Appl.* **2010**, *23*, 4–6+10.
Li, J.F.; Zhang, S.L.; Zhou, H.Q. Experimental study on air cushion flow field of air cushion belt conveyor. *Coal Mine Mach.* **2009**, *30*, 71–73.

actuators

MD

Article

Experimental Analysis of Flow Separation Control by a Dielectric Barrier Discharge Plasma Actuator in Burst-in-Burst Actuation Mode

Rodrigo Viguera [1,2], Yasuo Sasaki [2] and Taku Nonomura [2,*]

1 Department of Aerospace Engineering, Tohoku University, Aoba 6-6-01, Aramaki, Aoba-ku, Sendai 980-8577, Japan; viguera.rodrigo.blas.p3@dc.tohoku.ac.jp

2 Department of Aerospace Engineering Aerodynamics and Propulsion Laboratory, Nagoya University, Furo-cho, Chikusa-ku, Nagoya 464-8603, Japan; sasaki.yasuo.g8@f.mail.nagoya-u.ac.jp

* Correspondence: nonomura@nagoya-u.jp

Abstract: This study investigated the effectiveness of a dielectric barrier discharge (DBD) pla actuator operating in burst-in-burst (BIB) mode for flow separation control on a NACA 0015 ai Time-resolved particle image velocimetry measurements were conducted at a Reynolds numbe 66,000 and 13° angle of attack. Various BIB signal configurations were tested, with actuation peri of 70 ms and 150 ms, non-actuation periods ranging from 5 ms to 50 ms, and burst frequen of 300 Hz and 600 Hz. Proper orthogonal decomposition was applied to analyze the flow f dynamics. The results showed that BIB actuation maintained flow attachment with reduced po consumption compared with continuous burst actuation. However, the effectiveness was hig sensitive to the BIB parameters, with some configurations failing to achieve consistent reattachm and becoming unstable. This study reveals complex interactions between actuation vortices separation processes, highlighting both the potential and challenges of intermittent plasma actua for efficient flow control.

Keywords: plasma actuator; flow control; separation control; burst-in-burst; vortex generator; f mechanics; particle image velocimetry; proper orthogonal decomposition; experimental

Citation: Viguera, R.; Sasaki, Y.; Nonomura, T. Experimental Analysis of Flow Separation Control by a Dielectric Barrier Discharge Plasma Actuator in Burst-in-Burst Actuation Mode. *Actuators* **2024**, *13*, 435. https://doi.org/10.3390/act13110435

Academic Editors: Hui Tang, Xin Wen and Feng Ren

Received: 19 September 2024
Revised: 21 October 2024
Accepted: 23 October 2024
Published: 29 October 2024

1. Introduction

Flow control is a crucial aspect of aerodynamics and fluid mechanics, significar impacting the performance, efficiency, and stability of various systems such as aircraft, tomobiles, and wind turbines. By effectively managing the fluid flow (such as air or wa around these systems, engineers can achieve benefits like reduced drag [1,2], increa lift [3,4], and noise reduction [5,6]. Flow control methods can be broadly categorized i passive and active types. Passive flow control devices, such as vortex generators, wing or riblets, do not require an external energy source. While these devices are simple a easy to implement, they are typically designed for specific operating conditions and n not perform optimally in different situations. Conversely, active flow control devices, s as synthetic jets, movable surfaces, and plasma actuators, require an external energy in The main advantage of active flow control devices is their ability to adapt to changing f conditions, offering greater flexibility and performance across a wider range of scenari

Corke et al. [7] described DBD plasma actuators as typically consisting of two e trodes separated by a layer of dielectric material, as shown in Figure 1. When a high volt is applied to the electrodes, the surrounding air is ionized, forming non-thermal plas The charged particles interact with the electric field generated by the electrodes and coll with air particles, transferring momentum. This interaction accelerates the surround neutral air, generating a jet that is tangential to the surface and directed from the expo electrode to the grounded one [8–10].

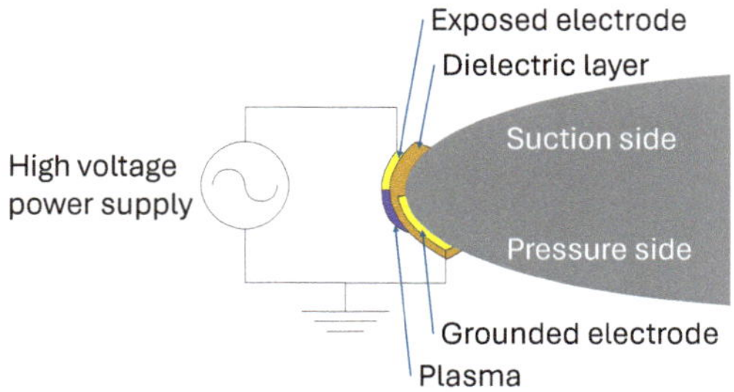

Figure 1. Diagram of plasma actuator installed on wing model.

Plasma actuators offer several advantages over traditional flow control methods. They have no moving parts, potentially reducing maintenance and increasing reliability. They can also respond almost instantaneously to control signals, which allows us to use them with high-frequency control inputs. Moreover, plasma actuators can be easily retrofitted onto existing structures, without significant modifications [11]. However, they also have disadvantages, such as significant power consumption, which can be a limiting factor for some applications [12], and a relatively small body force generation, which might not be sufficient for controlling large-scale flows without using multiple actuators.

Numerous studies have investigated the specific electrical and geometric parameters of dielectric barrier discharge (DBD) plasma actuators that influence performance [13,14]. Research indicates that increasing the applied voltage, while maintaining a small but positive electrode gap, can significantly enhance maximum ionic wind velocities [15]. The configuration of the active electrode is another major factor, where substituting a standard flat electrode with a thin wire can boost the velocity and minimize streamer formation [16]. Experimental investigations have also explored the effects of varying the number and arrangement of encapsulated electrodes within the dielectric layer [17]. Efforts have been directed towards identifying an effective multi-electrode configuration that overcomes the negative interactions between successive DBD plasma actuators [18]. The selection of dielectric material and its thickness plays a crucial role in actuator performance, with materials like Kapton demonstrating superior results at specific voltage frequencies [19].

Plasma actuators have been found to be sensitive to flow [20–28] and environmental [29–32] conditions, which can change the optimal electrical settings for the plasma actuator. Recent advancements have focused on the implementation of feedback control for flow separation using plasma actuators [33–40]. This approach aims to optimize actuator performance by dynamically adjusting to changing flow conditions, consequently enhancing the overall efficiency and effectiveness of the system.

A significant breakthrough in the field of plasma actuators was the use of burst actuation. Burst actuation involves modulating the actuation frequency, applying a high voltage in short, periodic bursts rather than continuously. While in continuous actuation a plasma actuator simply provides a steady linear momentum to the flow, a plasma actuator with a burst signal behaves like a spanwise vortex generator, creating vortices with each burst. This approach has been shown to reduce power consumption while maintaining, or even enhancing, the effectiveness of the actuators [41]. Viguera et al. [42] studied the effects of activating and deactivating control using a plasma actuator under normally separated flow conditions. The flow reattachment process when activating the plasma actuator was substantially observed to be faster than the separation process, even with burst frequencies considered sub-optimal for steady separation control. This asymmetry suggested that intermittent actuation in a burst-in-burst (BIB) pattern, in which the burst actuation is intermittently turned on and off, could further improve the efficiency of plasma actuators. Figure 2 shows an example of a new signal with a BIB pattern. This method involves adding a new layer of low-frequency modulation to a typical burst signal.

This study analyzed the effects of a leading-edge plasma actuator driven by a BIB signal on flow separation, comparing various combinations of actuation and non-actuation periods.

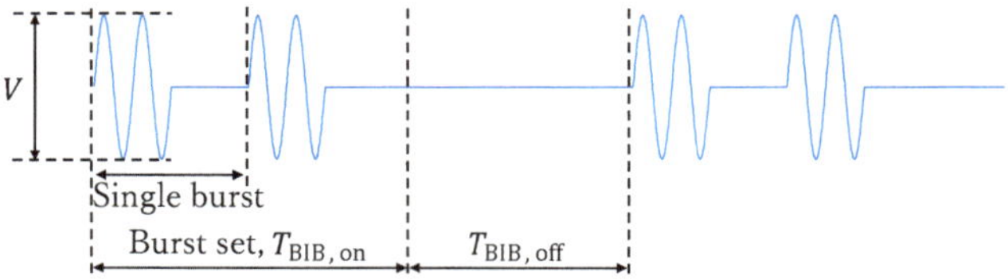

Figure 2. Example of driving signal of plasma actuator in BIB mode.

The methodology consisted of obtaining time-resolved particle image velocim (PIV) data of the flow field, centered on the recirculation region. The plasma actuator initiated along with the recording of particle images. Due to the 0.25 s required for laser to reach full operating power after the initial triggering, data from the first 0.25 each run were excluded from analysis. This was a sufficiently long time for the long BIB period signal to complete well over one full period, meaning that the reattachm transient processes from quasi-steady separated conditions should not affect the res After completing data acquisition, proper orthogonal decomposition (POD) was applie the entire set of stacked PIV data.

Here, POD is a technique that breaks down high-dimensional data into a superposi of simpler modes, which can provide a low-dimensional description of the most signific features of the flow field [43]. This analysis provides more easily interpretable insights i the data, based on the superposition of coefficients associated with leading spatial mo and is an effective tool in this scenario because the main variations in the flow field naturally defined. Figure 3 illustrating the energy ratio of POD modes shows that the f mode alone expresses 60% of the total energy and is a good indicator of the overall stat the flow field in terms of separation and attachment.

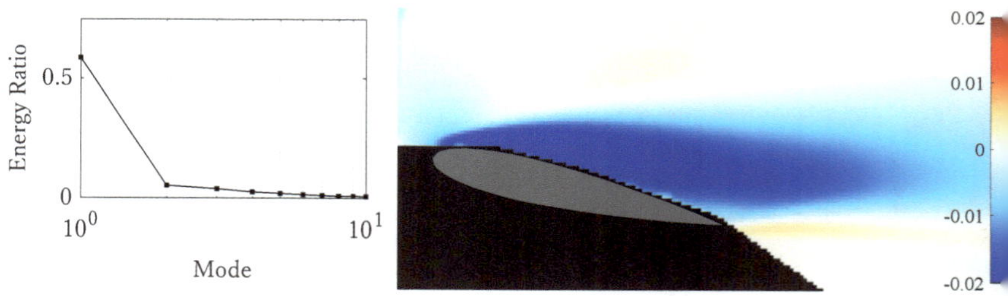

Figure 3. (**Left**): energy ratio of POD modes. (**Right**): spatial distribution of first POD mode.

The curl of the velocity fields was calculated, and the vorticity field was obtai Lastly, phase-averaged data were also analyzed, averaging the actuation and non-actua periods of each case. The time measurements were conserved across cases, rather th analyzing at the same fractions of the cycle, and flow phenomena were characterized. time required for the separation and reattachment processes was measured from the control state when the plasma actuator was turned off and on, respectively.

2. Experimental Setup and Methodology

The wind tunnel used for this work was the Tohoku University Basic Aerodyna Wind Tunnel, with a closed test section of 300 mm × 300 mm and an open circuit. The w model was a NACA 0015 airfoil with a span of 300 mm and a chord length of 100 r The wing had piezoelectric pressure sensors installed, but their data were not conside in the analysis, due to intense electromagnetic noise from the plasma actuator. The plas actuator was installed on the leading edge of the wing, with the exposed electrode on suction side, as shown in Figures 1 and 4. The electrodes of the plasma actuator were m of copper tape with a thickness of 0.07 mm. The exposed electrode had a width of 6 mm the grounded electrode was 14 mm, with a gap between them of 1 mm centered along leading edge of the wing. The layer of dielectric material was made of polyimide tape w a thickness of 0.05 mm. The overall dimensions of the plasma actuator were approximately 200 mm long spanwise and 21 mm wide chordwise. The electrodes of the plasma actuator were connected to a PSI-PA1050 (PSI, Kawagoe, Japan) high voltage amplifier with a ratio of 1:1800, and a maximum voltage of 8 kV peak to peak was delivered. The driving signal was created using a WF1948 (NF Corporation, Yokohama, Japan) function generator, with custom functions for each BIB configuration.

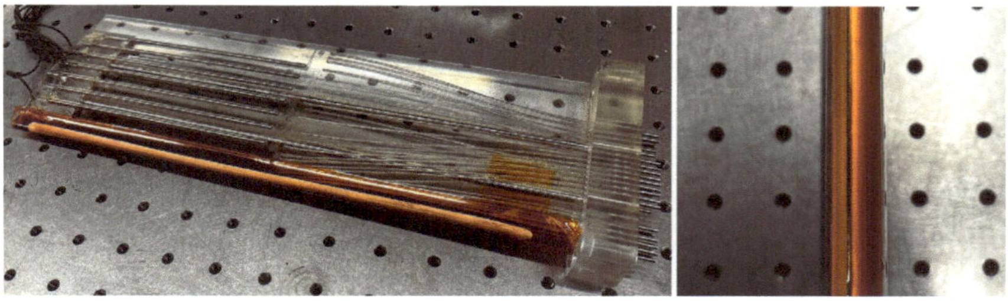

Figure 4. (**Left**): plasma actuator as installed on wing model. (**Right**): detail of gap between electrodes.

The signals sent to the amplifier of the the plasma actuator were generated with the basic structure shown in Figure 2, with the observations from our previous work in mind [42]. The reattachment process was observed to take $T_{\mathrm{on}} \approx 7$ to reach a quasi-steady condition. The separation process progressed slowly until $T_{\mathrm{off}} \approx 3$. Here, T_{on} and T_{off} nondimensional times elapsed from the start and end of actuation, respectively, nondimensionalized by $\frac{c}{U_\infty}$. Considering this, the minimum $T_{\mathrm{BIB,on}}$ (overall period of burst actuation, nondimensionalized by the same factor) was set at 7, as it was suspected that a shorter actuation period might not be able to achieve full reattachment in certain conditions. Additionally, a value of $T_{\mathrm{BIB,on}} = 15$ was also tested, so that the interruptions of the separation process could be isolated with a, in principle, sufficiently long time. Here, a sufficiently long time $T_{\mathrm{BIB,on}} = 15$ seemed to guarantee a fully quasi-steady attached condition before the actuation was stopped. The periods of non-actuation $T_{\mathrm{BIB,off}}$ were determined as a range between 0.5 and 5 in increments of 0.5, so that the effects of interrupting the separation process at some of its more relevant stages could be verified for this application. Additionally, two different burst frequencies were selected for the actuation mode inside each $T_{\mathrm{BIB,on}}$ period, 300 and 600 Hz, or nondimensional frequencies F^+ equal to 3 and 6, respectively. The burst ratio for the burst actuation was 0.1, and the frequency of the sinusoidal wave was 30 kHz.

The experiments were conducted at a freestream velocity of 10 m/s, corresponding to a Reynolds number of 66,000, and the angle of attack was 13°, conditions at which there was consistent boundary layer separation at the leading edge of the wing.

The PIV data acquisition was performed with an LDY-303PIV (Litron Lasers, Rugby, England) Nd:YLF laser illuminating dioctyl sebacate (Tokyo Chemical Industry, Tokyo, Japan) particles, and a Phantom v2640 (Vision Research, Wayne, NJ, USA) high speed camera synchronized with the laser captured images at a frequency of 10 kHz, which translates to a PIV sampling frequency of 5 kHz. The image data were then processed using Dynamic Studio 6.11 (Dantec Dynamics, Skovlunde, Denmark). After background removal, the velocity field was obtained with the program's adaptive PIV algorithm with an 8×8 px^2 interrogation window.

Pressure data were also obtained in two ways: steady and unsteady pressure sensors. The wing had 29 ports for steady and 9 for unsteady pressure measurements, although many of them were unavailable, due to either being physically blocked by the plasma actuator (the ports are concentrated near the leading edge, with fewer spread over the chord length) or individual pressure transducers not working properly. The steady pressure sensors had the disadvantage of a low sampling frequency, meaning they were not adequate for studying the quick pressure changes of an experiment of this nature, but they formed a backup of the

PIV data. The unsteady pressure sensors, on the other hand, were extremely sensitive to electromagnetic noise from the plasma actuator, because of their close proximity. For reason, and the fact that breaks in actuation were short given the control signals used, mos their data were unusable. A simplified diagram of the equipment connections is show Figure 5.

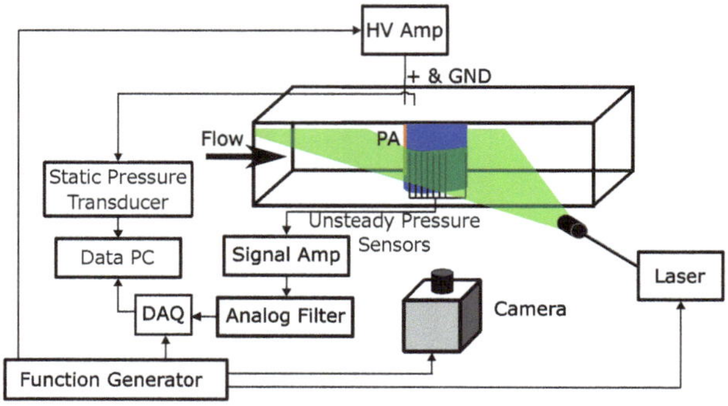

Figure 5. Schematic diagram of experimental setup.

3. Results and Discussion

The results were first divided into three sets of conditions, depending on their T_{BI} This was because each set showed different tendencies and effects, mainly caused by difference in the distinctive stages of the separation process at the time when it was interrup

3.1. $T_{BIB,off}$ 0.5 to 1.5

First, the cases with $T_{BIB,off}$ shorter than 2 were analyzed. These cases were cha terized by a stable reattachment during each actuation. Figure 6 shows phase-avera U velocity fields of four different conditions at the end of their respective periods of actuation. During the non-actuation periods, a separation bubble grew to nearly 20% of chord length at $T_{BIB,off} = 0.5$, around 30% at $T_{BIB,off} = 1$, and 35% to 45% at $T_{BIB,off} =$ The burst frequency $F^+ = 3$ induced larger vortices into the flow than $F^+ = 6$, which m the separation process slightly faster during $T_{BIB,off}$, leading to larger separation bubb by 5% to 10%.

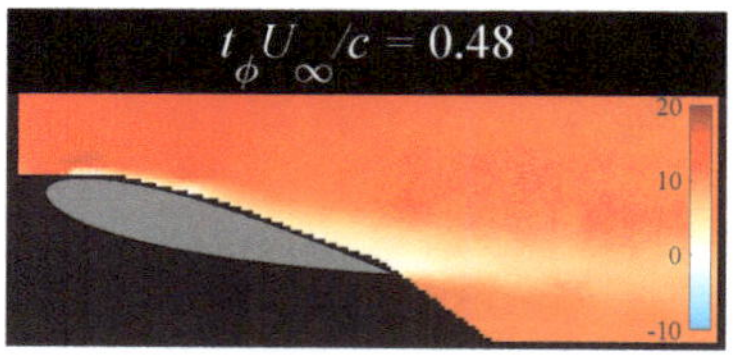

(**a**) $F^+ = 3$ and $T_{BIB,off} = 0.5$

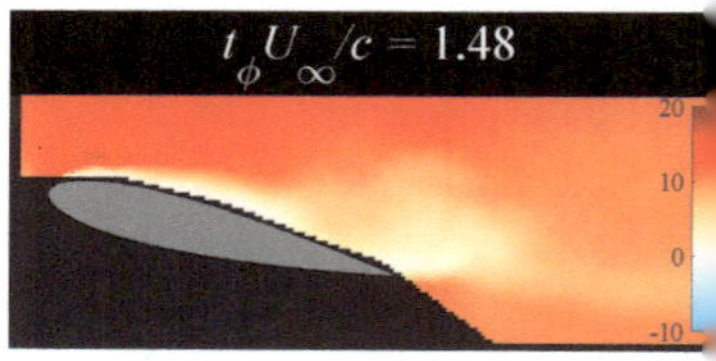

(**b**) $F^+ = 3$ and $T_{BIB,off} = 1.5$

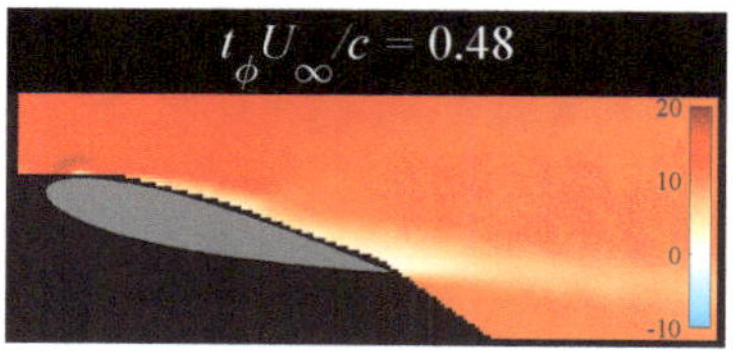

(**c**) $F^+ = 6$ and $T_{BIB,off} = 0.5$

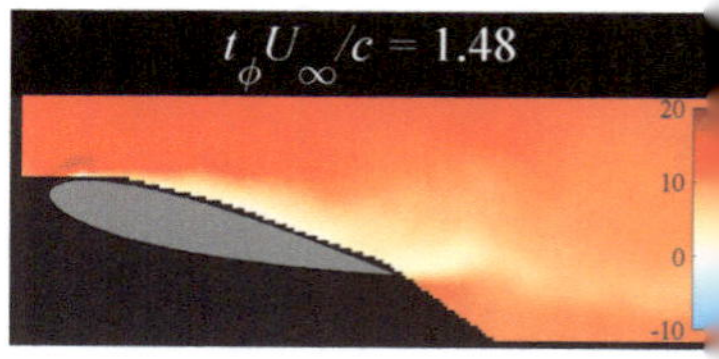

(**d**) $F^+ = 6$ and $T_{BIB,off} = 1.5$

Figure 6. Phase-averaged U velocity component at end of the non-actuation period for the case $T_{BIB,on} = 7$.

Figure 7 shows the vorticities of the flow field phase-averaged for the actuat periods, with the phase measured in nondimensional time from the start of actuat

Additionally, plots of the coefficient of the first POD mode for the corresponding cases are shown, and the coefficient is compared with the time-averaged coefficient of the first POD mode for a continuous actuation reference case ($z \approx -116$) and that of a fully separated (non-actuation) reference case ($z \approx 487$).

A small, but stronger than normal, clockwise vortex of plasma actuators flowed continuously from the leading edge as the actuation recovered in the case of $T_{\mathrm{BIB,off}} = 0.5$. These vortices are shown in Figure 7a near x/c values of 0.1 and 0.2. The reattachment vortices became even larger with greater vorticity at their core in the case of $T_{\mathrm{BIB,off}} = 1, 1.5$, as shown in Figure 7c near x/c values of 0.2 and 0.3 for $T_{\mathrm{BIB,off}} = 1$ and in Figure 7e near x/c values of 0.1 and 0.3 for $T_{\mathrm{BIB,off}} = 1.5$. These vortices were observed to immediately break down into a large set of eddies, which caused the peaks in the first POD mode coefficient for the different cases, as observed in Figure 7d,f. The vortex structures spread slightly further for the $F^+ = 3$ cases, likely due to its actuation vortices being naturally larger than those for $F^+ = 6$. It should be noted that these reattachment vortices rotate in the clockwise direction, which decreases lift.

In addition, the separation bubble was never completely eliminated, but it was contained at a small size, with only vortices influenced by the actuation shed from it.

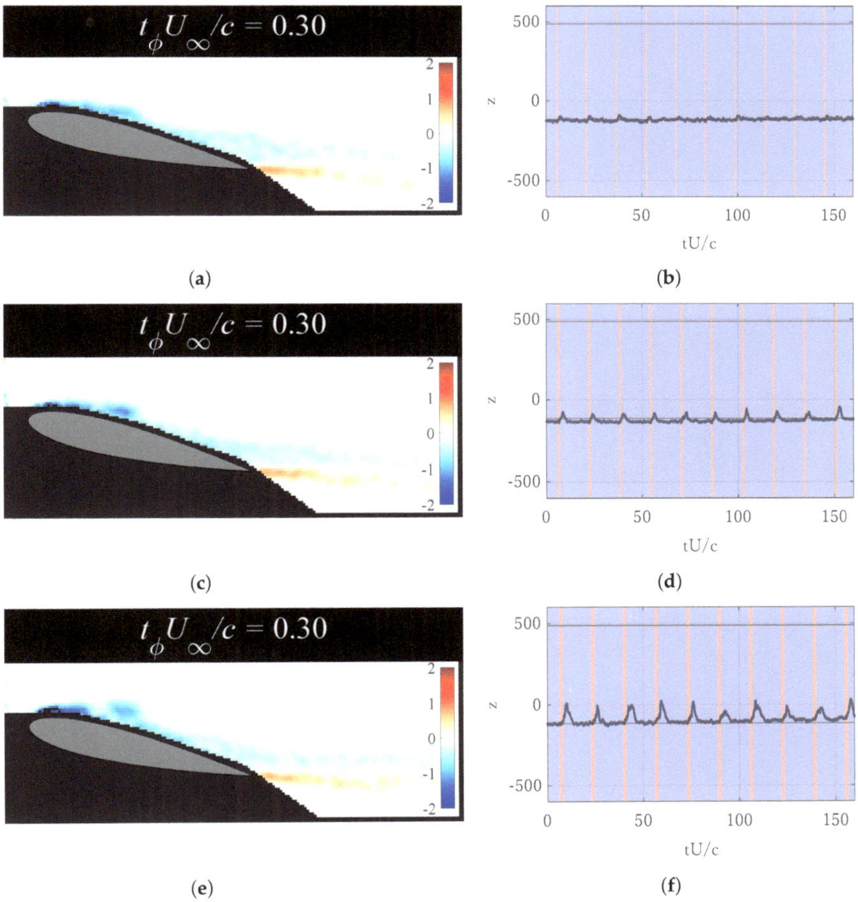

Figure 7. Vorticity and time history first POD mode coefficient for the cases of $F^+ = 6$ and $T_{\mathrm{BIB,on}} = 15$. In subfigures (**b**,**d**,**f**), blue and red shaded areas indicate periods of actuation and no actuation, respectively, and the top and bottom horizontal thin lines show the averaged first POD mode coefficient values at the fully separated and fully attached conditions, respectively. (**a**) Phase-averaged

vorticity of reattachment vortices with $T_{BIB,off} = 0.5$. (**b**) Time history of first POD mode coeffic for $T_{BIB,off}$ of 0.5. (**c**) Phase-averaged vorticity of reattachment vortices with $T_{BIB,off} = 1$. (**d**) T history of first POD mode coefficient for $T_{BIB,off}$ of 1. (**e**) Phase-averaged vorticity of reattachm vortices with $T_{BIB,off} = 1.5$. (**f**) Time history of first POD mode coefficient for $T_{BIB,off}$ of 1.5.

3.2. $T_{BIB,off}$ 2.0 to 3.5

The second set of cases had a $T_{BIB,off}$ that ranged from 2 to 3.5. The $T_{BIB,off}$ peri of these cases, particularly 3 and 3.5, fell inside the range where the separation bub became large enough to cover the entire chord length of the wing when the actuation ended, while the flow was quasi-steady. These cases showed a significant deterioratio the reattachment. Some cases did not complete reattachment before the plasma actua was switched off, even with $T_{BIB,on} = 15$. Figure 8 shows the first POD mode coeffici of characteristic cases with $T_{BIB,off}$ values of 2 and 2.5. Here, the effectiveness of the c of $F^+ = 3$, $T_{BIB,on} = 7$ and $T_{BIB,off} = 2.5$ (Figure 8a) was better than that of F^+ = $T_{BIB,on} = 15$, and $T_{BIB,off} = 2$ (Figure 8b), which had a longer actuation time and sho non-actuation time. At the same time, the cases with $F^+ = 3$, $T_{BIB,off} = 2.5$, and T_{BI} values of 7 and 15 showed significantly better reattachment characteristics than the c of $F^+ = 6$, $T_{BIB,on} = 7$, and $T_{BIB,off} = 2.5$, even though $F^+ = 6$ was considered to perfo better for the shorter $T_{BIB,off}$ periods.

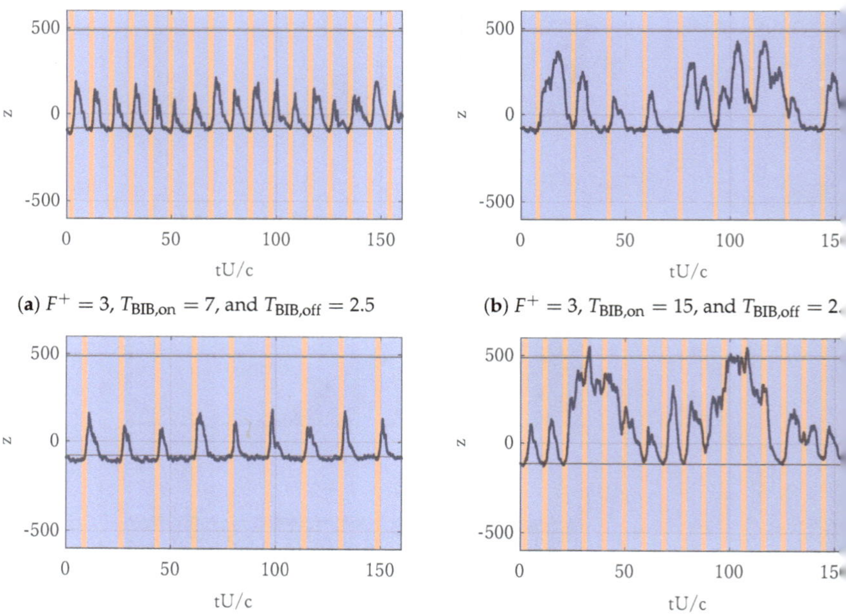

(**a**) $F^+ = 3$, $T_{BIB,on} = 7$, and $T_{BIB,off} = 2.5$ (**b**) $F^+ = 3$, $T_{BIB,on} = 15$, and $T_{BIB,off} = 2.$

(**c**) $F^+ = 3$, $T_{BIB,on} = 15$, and $T_{BIB,off} = 2.5$ (**d**) $F^+ = 6$, $T_{BIB,on} = 7$, and $T_{BIB,off} = 2.$

Figure 8. Time histories of first POD mode coefficient for $T_{BIB,off}$ of 2.0 and 2.5. Here, blue and shaded areas indicate periods of actuation and no actuation, respectively, and the top and bot horizontal thin lines show the averaged first POD mode coefficient values in the fully separated fully attached conditions, respectively.

Figure 9 shows the two worst-performing cases of the entire dataset. Figure 9 present the case with the highest average value of z, and the case that failed to achi reattachment at all actuation periods, respectively. The four worst cases happened be at $T_{BIB,off} = 3.5$, although it is difficult to establish a connection with the separat transient process from a fully attached condition, when reattachment was barely achie in these cases. When reattachment failed, large vortices were continuously shed from the leading edge, without an immediate reduction in size, as occurred when reattachment was successful. This suggests there was a complex and highly condition-sensitive interaction between the actuation vortices and the separation vortices near the leading edge.

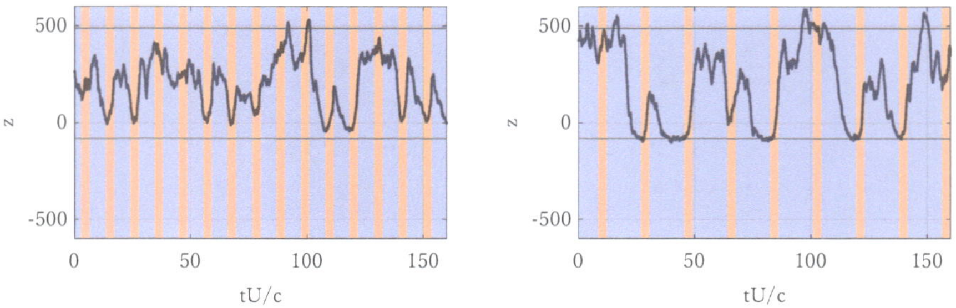

(**a**) $F^+ = 3$, $T_{\mathrm{BIB,on}} = 7$, and $T_{\mathrm{BIB,off}} = 3.5$ (**b**) $F^+ = 3$, $T_{\mathrm{BIB,on}} = 15$, and $T_{\mathrm{BIB,off}} = 3.5$

Figure 9. Time histories of first POD mode coefficient for $F^+ = 3$ and $T_{\mathrm{BIB,off}} = 3.5$. Here, blue and red shaded areas indicate periods of actuation and no actuation, respectively, and the top and bottom horizontal thin lines show the averaged first POD mode coefficient values in the fully separated and fully attached conditions, respectively.

Failure of reattachment also suggests that by the time the plasma actuators had been turned off, there was already a separated flow to varying degrees, and that the plasma actuator was not always turned on in the same separation phase. Figure 10b shows the case of $F^+ = 3$, $T_{\mathrm{BIB,on}} = 15$ and $T_{\mathrm{BIB,off}} = 3$. There is a clear peak in the z curve without any actuation at $t\frac{U_\infty}{c} \approx 80$, which indicates the shedding of the large vortex, followed by a substantial reduction in the size of the recirculation region. This kind of vortex shedding may occur at $t\frac{U_\infty}{c} > 5$ after the end of actuation from a fully attached condition, but was not observed to occur in any cases within $t\frac{U_\infty}{c} = 3$ in our previous work [42].

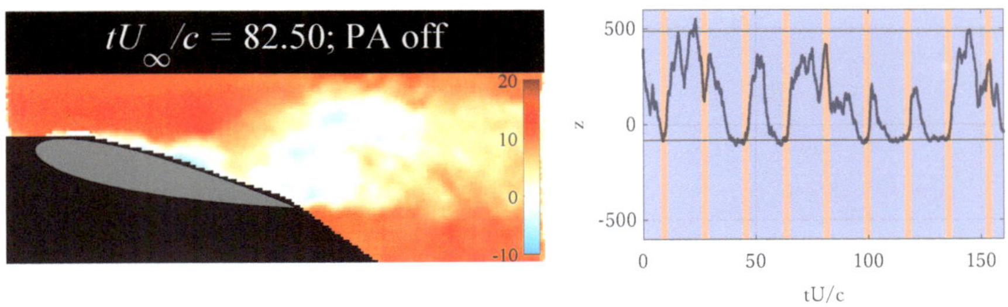

(**a**) (**b**)

Figure 10. U velocity component and time history of the first POD mode coefficient for the case of $F^+ = 3$, $T_{\mathrm{BIB,on}} = 15$, and $T_{\mathrm{BIB,off}} = 3.0$. In subfigure (**b**), blue and red shaded areas indicate periods of actuation and no actuation, respectively, and the top and bottom horizontal thin lines show the averaged first POD mode coefficient values in the fully separated and fully attached conditions, respectively. (**a**) U velocity component with large vortex shed and dissipating. (**b**) First POD mode coefficient of $F^+ = 3$, $T_{\mathrm{BIB,on}} = 15$, and $T_{\mathrm{BIB,off}} = 3.0$.

3.3. $T_{\mathrm{BIB,off}}$ *4.0 to 5.0*

Lastly, the set of cases with $T_{\mathrm{BIB,off}}$ 4.0 to 5.0, where the actuator was turned on in the vortex shedding stage of the separation process, are discussed. In these cases, a small but noticeable improvement in effectiveness over the previous set was observed. Figure 11 shows the coefficient of the first POD mode for all of the cases in this $T_{\mathrm{BIB,off}}$ range. Although the value of z at the time the plasma actuator was reactivated tended to increase significantly more than in the cases of $T_{BIB,off}$ 2.0 to 3.5, half of the cases showe relatively good control authority compared with the previous set. This suggests that flow state during $T_{BIB,off}$ had reached a degree close to a completely isolated quasi-ste state, where the plasma actuators recovered some effectiveness.

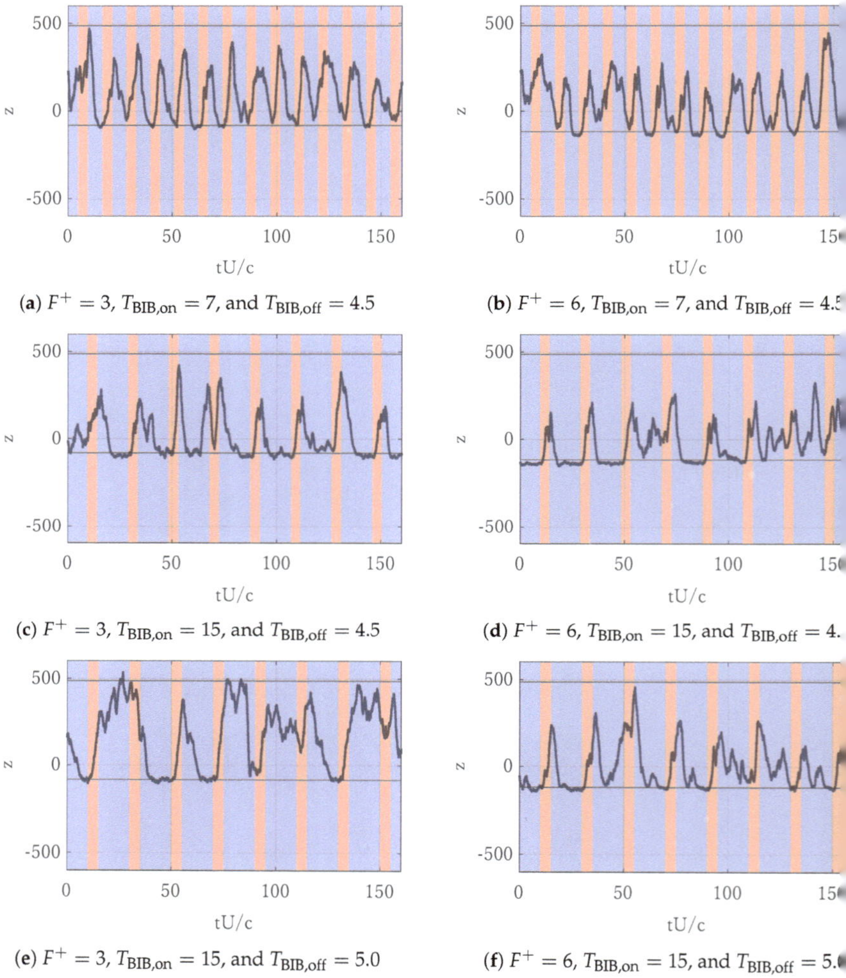

(**a**) $F^+ = 3$, $T_{BIB,on} = 7$, and $T_{BIB,off} = 4.5$ (**b**) $F^+ = 6$, $T_{BIB,on} = 7$, and $T_{BIB,off} = 4.$

(**c**) $F^+ = 3$, $T_{BIB,on} = 15$, and $T_{BIB,off} = 4.5$ (**d**) $F^+ = 6$, $T_{BIB,on} = 15$, and $T_{BIB,off} = 4.$

(**e**) $F^+ = 3$, $T_{BIB,on} = 15$, and $T_{BIB,off} = 5.0$ (**f**) $F^+ = 6$, $T_{BIB,on} = 15$, and $T_{BIB,off} = 5.$

Figure 11. Time histories of the first POD mode of the $T_{BIB,off}$ 4.5 and 5.0 cases. Here, the blue red shaded areas indicate periods of actuation and no actuation, respectively, and the top and bott horizontal thin lines show the averaged first POD mode coefficient values in the fully separated fully attached conditions, respectively.

An unexpected occurrence was observed during one of the actuation periods the case of $F^+ = 6$, $T_{BIB,on} = 15$, and $T_{BIB,off} = 4.5$ (Figure 11d), where the plas actuator failed to keep the flow attached, even after seemingly achieving full reattachm Figure 12 shows the U velocity component at the times when the flow became fu attached, and when it began separating again during actuation. While this behavior w not observed in any other actuation period or case, it raises concerns about the stabi and the repeatability of intermittent control. Further experiments are necessary to find cause of this type of failure, as key information may be hidden outside the camera fra chosen in this study.

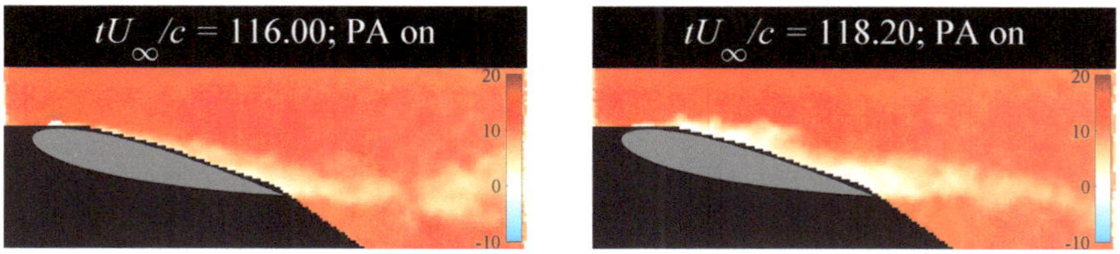

Figure 12. U velocity component of $F^+ = 6$, $T_{\text{BIB,on}} = 15$, and $T_{\text{BIB,off}} = 4.5$ before (**left**) and after (**right**) separation during actuation.

3.4. Effectiveness Comparison

Figure 13 shows the phase-averaged coefficient of the first POD mode on the actuation periods, directly comparing all of the cases and the effect of the different parameters in the different stages of actuation.

Although F^+ had a relatively small influence on the decrement rate of z in most cases and, as mentioned in Section 3.1, the separation bubble in the shorter $T_{\text{BIB,off}}$ cases, the higher burst frequency noticeably improved the quasi-steady state, which translated to a weaker separation level during $T_{\text{BIB,off}}$, and in turn smaller reattachment vortices after the plasma actuator was turned on again. These vortices had a significant impact nonetheless, and proportionately they had a larger effect on the z value of cases with successful reattachment than those that failed to reattach consistently.

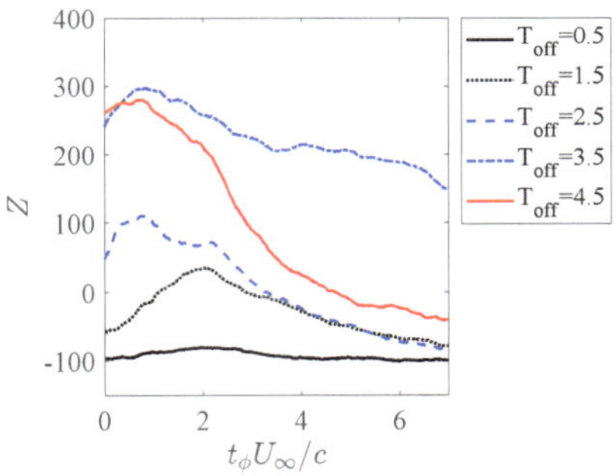

(**a**) $F^+ = 3$ and $T_{\text{BIB,on}} = 7$

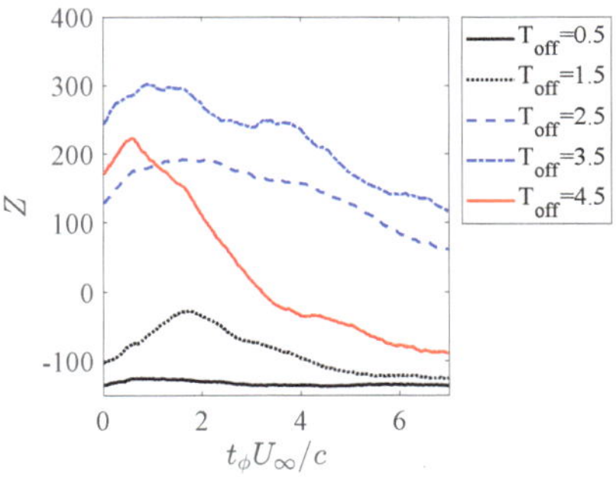

(**b**) $F^+ = 6$ and $T_{\text{BIB,on}} = 7$

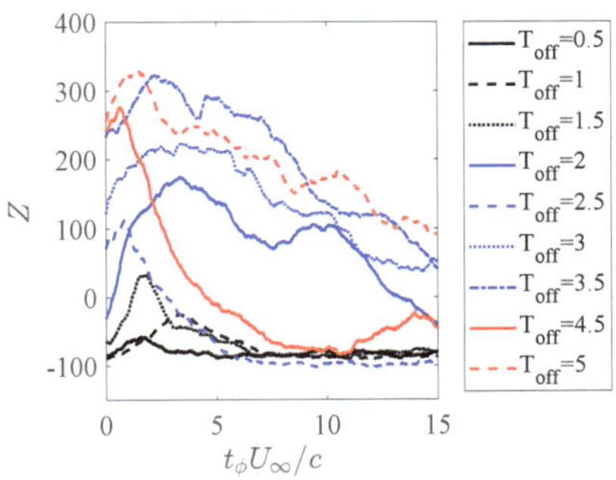

(**c**) $F^+ = 3$ and $T_{\text{BIB,on}} = 15$

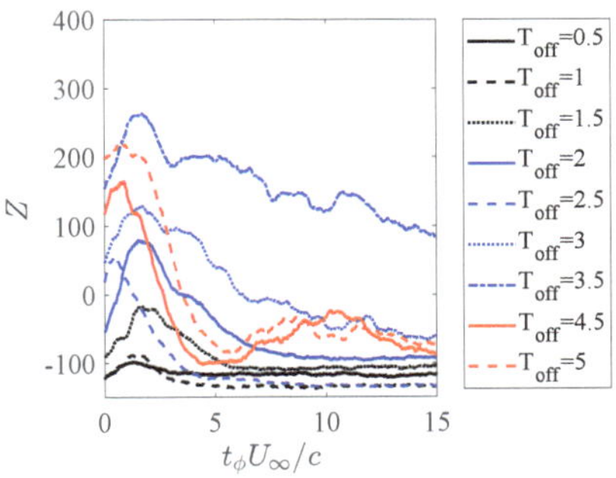

(**d**) $F^+ = 6$ and $T_{\text{BIB,on}} = 15$

Figure 13. Comparison of phase-averaged first POD mode during actuation for all cases.

Some cases, particularly those where actuation was generally effective with few or no reattachment failure events, showed a reduction in the time at which the peak of z occurred. In these cases, the decrease in the z value began earlier and the z value peak was smaller with $F^+ = 6$. Other cases, such as the ones with $T_{\text{BIB,on}} = 7$ and $T_{\text{BIB,off}} = 2.5$, and th with $T_{\text{BIB,on}} = 15$ and $T_{\text{BIB,off}} = 5$, presented drastically different behaviors at either value. This trend reinforces the idea that the effectiveness of a burst frequency is hig dependent on very specific flow conditions.

Figure 14a shows a simplified curve of the separation process, based on the res of our previous work [42], for visual reference. The cases this curve represents ha quasi-steady separation first POD mode value of approximately 300, and a quasi-ste attached value of approximately −400, explaining the differences with the values of current tests. The time was measured from the instant the plasma actuator was turned Additionally, the standard deviation is included to illustrate the differences across run their attached conditions and the evolution of the different separation processes. On other hand, Figure 14b–e show the phase-averaged first POD mode for the non-actua period of the current test.

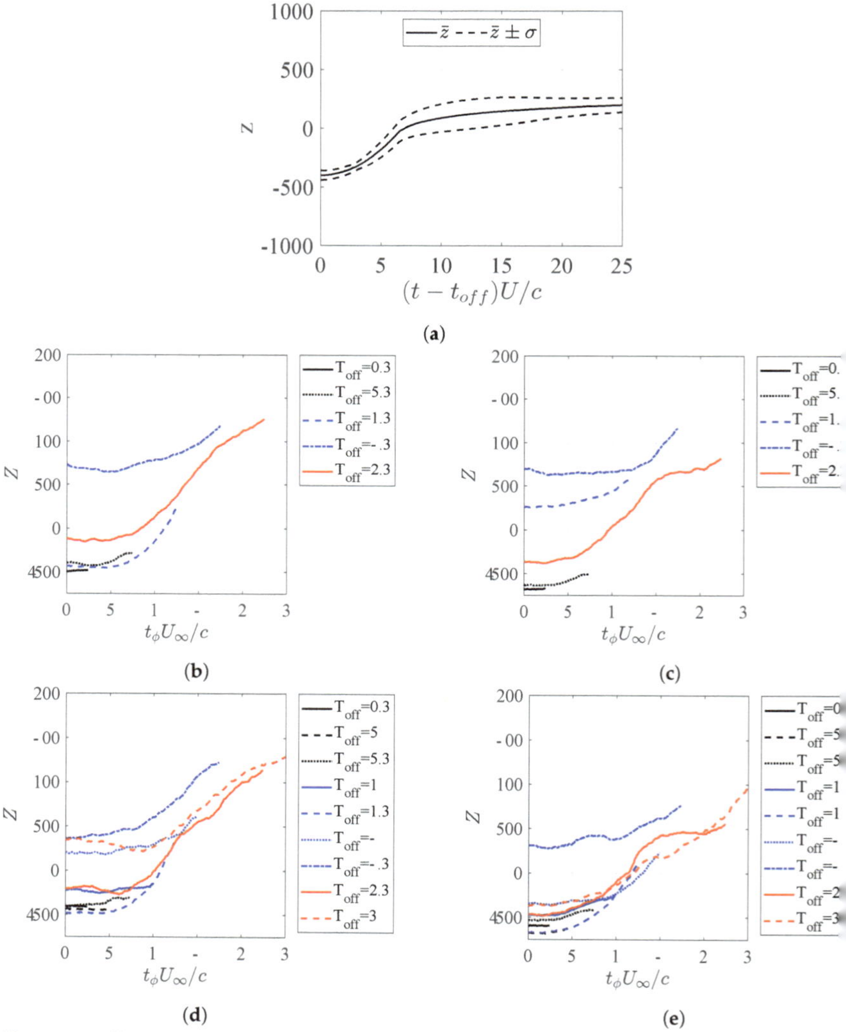

Figure 14. Comparison of phase-averaged first POD mode during non-actuation for all ca (**a**) Illustration of z after separation started [42]. (**b**) $F^+ = 3$ and $T_{\text{BIB,on}} = 7$. (**c**) $F^+ = 6$ $T_{\text{BIB,on}} = 7$. (**d**) $F^+ = 3$ and $T_{\text{BIB,on}} = 15$. (**e**) $F^+ = 6$ and $T_{\text{BIB,on}} = 15$

The first point to notice is the difference in the initial value of the different cases. It is clear that failure to achieve reattachment led to a significantly increased initial z value. These cases exhibited smaller variations in z over time compared to cases of similar duration with more effective reattachment. This is partially because its growth tends to slow down once the separation region is established over the entire chord of the wing. In addition, the flow being separated at the start of $T_{\mathrm{BIB,off}}$ means that large vortices can shed early, potentially decreasing the area of the recirculation region, which leads to smaller values of z on the latter half of the $T_{\mathrm{BIB,off}}$ period, an effect that rarely appears early in the separation process from a fully attached condition, as explained in Section 3.2. Nevertheless, these cases showed a critical failure in the BIB settings, as during this period it is necessary that the flow remains attached for as long as possible.

Comparison of the cases in which reattachment was achieved illustrates that there was a variation of 100 in the z values. This is likely explained by Figure 14a, which shows the z value history in the same operating conditions that achieved quasi-steady attached flow for all 30 runs of the ensemble. Despite differing initial values, the curves tend to converge near $t_\phi U_\infty / c = 2.5$. Figure 14a also shows this convergence as a narrowing of the standard deviation band at the same point in the separation process. After this point, the curves tend to diverge, possibly indicating different behaviors in the vortex shedding stages of the separation process. However, it is difficult to discern the impact of the failure of reattachment events in the phase-averaged data.

The shorter $T_{\mathrm{BIB,off}}$ cases showed minimal changes in the z value, which is the expected behavior for a working BIB signal.

Lastly, we discuss the difference between the first POD mode coefficients of each case and the reference value for the fully attached flow established in Section 3.1, to create a metric for the effectiveness of each case. We also evaluate the power consumption using the ratio of the actuation time of the plasma actuator. The attachment effectiveness and the power consumption are defined as follows:

$$\text{Effectiveness} = 100 \times \left(1 - \frac{\bar{z}_{\mathrm{BIB}} - \bar{z}_{\mathrm{att}}}{\bar{z}_{\mathrm{sep}} - \bar{z}_{\mathrm{att}}}\right), \tag{1}$$

$$\text{Power consumption} = 100 \times \frac{T_{\mathrm{BIB,on}}}{T_{\mathrm{BIB,off}} + T_{\mathrm{BIB,off}}}, \tag{2}$$

where $\bar{z}_{\mathrm{BIB}}$ is the mean value of the first POD mode of the BIB case during fully captured cycles, and $\bar{z}_{\mathrm{att}}$ and $\bar{z}_{\mathrm{sep}}$ are the reference values for fully attached flow and fully separated flow, respectively. The effectiveness in Equation (1) is expressed as a percentage of the effectiveness of continuous burst actuation at $F^+ = 6$. The power consumption is also expressed as the percentage of the consumption of the plasma actuator under continuous burst. Given that the burst ratio is the same, it is estimated as the fraction of time with actuation over the total BIB signal time, as shown in Equation (2).

Figure 15 shows the estimated effectiveness and power consumption for all the cases. Since the reference values for separated and attached conditions are approximate and inter-run variance is significant, effectiveness values slightly above 100% do not necessarily indicate the superiority of a particular BIB signal over continuous burst actuation. Likewise, small differences between conditions may not be statistically significant.

Figure 15 gives us two main observations: $F^+ = 6$ works better than $F^+ = 3$ at equal $T_{\mathrm{BIB,off}}$ almost every time, and more importantly for BIB as a control method, the drop in effectiveness follows the trend of the reduction in consumption. For $F^+ = 6$, the effectiveness is maintained briefly when $T_{\mathrm{BIB,off}}$ values are up to 1. However, the effectiveness rapidly declines as $T_{\mathrm{BIB,off}}$ increases beyond this point. In the range of $T_{\mathrm{BIB,off}}$ between 2 and 3.5, the actuation tends to fail to achieve reattachment, causing the effectiveness to power consumption ratio to drop significantly in most cases. This leads to a highly ineffective and inefficient actuation. A few exceptions to this were noticed, notably $T_{\mathrm{BIB,off}} = 2.5$, where three of the four cases consistently achieved reattachment. With successful reattachment, the effectiveness was at the level of the power consumption, leading to a similar efficiency (regarded as the ratio of effectiveness to power consumption, relative to the continu burst mode) as that of cases $T_{\mathrm{BIB,off}} < 2$. As $T_{\mathrm{BIB,off}}$ increases further, the effectiver appears to align once again with the decreasing trend in power consumption.

It should be noted that this effectiveness metric could be overestimating the effect ness of the plasma actuator, as the region immediately surrounding the leading edg the wing could not be observed in the images for PIV, implying that there is informat missing to comprehensively quantify the state of the flow field around the wing. Al this means that, at most, an efficiency increase of approximately 15% can be expected fr using BIB signals, although this value might decrease noticeably in actual use.

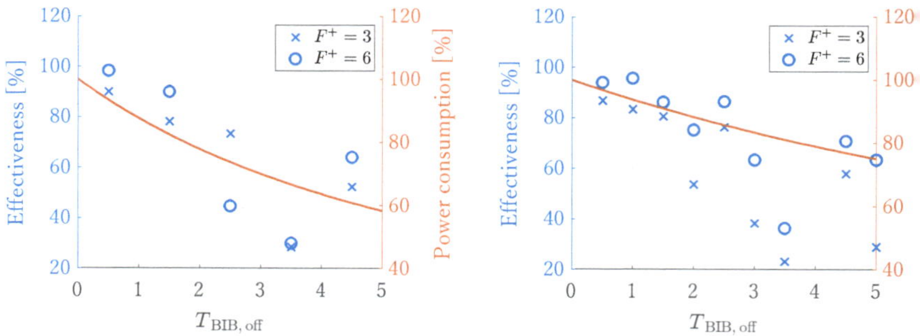

(**a**) $T_{\mathrm{BIB,on}} = 7$ (**b**) $T_{\mathrm{BIB,on}} = 15$

Figure 15. Comparison of actuation effectiveness and power consumption of all cases.

4. Conclusions

This study investigated the effectiveness of a DBD plasma actuator operating in mode for flow separation control on a NACA0015 airfoil. The results demonstrated t BIB actuation could effectively control flow separation, while reducing power consumpt by up to 15% compared to continuous burst actuation. However, the effectiveness v highly sensitive to BIB parameters, particularly $T_{\mathrm{BIB,off}}$. Shorter non-actuation peri ($T_{\mathrm{BIB,off}} \leq 1.5$) consistently achieved reattachment, while longer periods often resultec incomplete or inconsistent reattachment.

Higher burst frequencies ($F^+ = 6$) generally outperformed lower frequencies ($F^+ =$ especially in maintaining quasi-steady attached flow states. When full reattachment v consistently achieved, the separation bubble increased with $T_{\mathrm{BIB,off}}$, with correspond large reattachment vortices at the start of each actuation cycle. Complex interacti between actuation vortices and separation processes were observed with larger T_{BI} values, particularly for intermediate $T_{\mathrm{BIB,off}}$ values (2.0–3.5), where reattachment was of incomplete. This suggests a delicate balance between actuation and flow dynamics, wh requires careful optimization. At larger $T_{\mathrm{BIB,off}}$ values (4.0–5.0), the flow at the enc each $T_{\mathrm{BIB,off}}$ period more resembled a quasi-steady separated flow, which improved reattachment process, though complex vortex interactions persisted. Notably, the effect ness of the plasma actuator in BIB mode decreased at roughly the same rate as the pow consumption, except for $T_{\mathrm{BIB,off}}$ values between 2.0 and 3.5, where the effectiveness co drop significantly. This indicates that the benefits of BIB control are limited to short T_{BI} periods.

Future work should focus on refining BIB strategies to enhance stability and explor their applicability across a wider range of flow conditions and geometries. Additiona further investigation into the mechanisms of reattachment failure during actuation peri is necessary to improve the robustness of BIB control systems.

Author Contributions: Conceptualization, R.V. and T.N.; methodology, R.V. and T.N.; software, F validation, R.V. and T.N.; formal analysis, R.V. and T.N.; investigation, R.V. and T.N.; resources, and T.N.; data curation, R.V.; writing—original draft preparation, R.V., Y.S., and T.N.; writing—rev and editing, R.V., Y.S., and T.N.; visualization, R.V., Y.S., and T.N.; supervision, Y.S. and T.N.; pro administration, T.N.; funding acquisition, T.N. All authors have read and agreed to the published version of the manuscript.

Funding: This research was supported by the Japan Science and Technology Agency, FOREST Grant Number, JPMJFR202C.

Data Availability Statement: Data are contained within the article.

Acknowledgments: We are grateful to our colleagues in the Experimental Aerodynamics Laboratory at Tohoku University and the Fluid Dynamics Laboratory at Nagoya University for their assistance with the experimental setup and their valuable discussions and insights throughout this research. We also acknowledge the support of the Japan Science and Technology Agency, FOREST Grant Number JPMJFR202C, which made this research possible.

Conflicts of Interest: The authors declare no conflicts of interest.

Abbreviations

The following abbreviations are used in this manuscript:

σ	Standard deviation
BIB	Burst-in-burst
c	Wing chord
DBD	Dielectric barrier discharge
F^+	Nondimensional burst frequency
PIV	Particle image velocimetry
POD	Proper orthogonal decomposition
t	Time
$T_{\mathrm{BIB,on}}$	Nondimensional period of time with actuation
$T_{\mathrm{BIB,off}}$	Nondimensional period of time without actuation
U	Horizontal component of wind velocity
U_∞	Freestream velocity
z	First POD coefficient
$\bar{z}_{\mathrm{att}}$	Time average of z during quasi-steady controlled condition
$\bar{z}_{\mathrm{BIB}}$	Time average of z over fully captured BIB cycles
$\bar{z}_{\mathrm{sep}}$	Time average of z during fully separated condition

erences

Yagiz, B.; Kandil, O.; Pehlivanoglu, Y.V. Drag minimization using active and passive flow control techniques. *Aerosp. Sci. Technol.* **2012**, *17*, 21–31. [CrossRef]

Roy, S.; Zhao, P.; DasGupta, A.; Soni, J. Dielectric barrier discharge actuator for vehicle drag reduction at highway speeds. *Aip Adv.* **2016**, *6*, 025322 . [CrossRef]

Shehata, H.M.; Zakaria, M.Y.; Woolsey, C.A.; Hajj, M.R. Lift enhancement by a flapped trailing edge at low Reynolds number: A frequency response approach. *J. Fluids Struct.* **2022**, *110*, 103518. [CrossRef]

Vatsa, V.N.; Duda, B.M.; Lin, J.C.; Melton, L.G.P.; Lockard, D.P.; O'Connell, M.; Hannon, J., Comparative Study of Active Flow Control Strategies for Lift Enhancement of a Simplified High-Lift Configuration. In Proceedings of the AIAA Aviation 2019 Forum, Dallas, TX, USA, 17–21 June 2019. [CrossRef]

Yadegari, M.; Ommi, F.; Karimian Aliabadi, S.; Saboohi, Z. Reducing the aerodynamic noise of the axial flow fan with perforated surface. *Appl. Acoust.* **2024**, *215*, 109720. [CrossRef]

Avallone, F.; van der Velden, W.; Ragni, D. Benefits of curved serrations on broadband trailing-edge noise reduction. *J. Sound Vib.* **2017**, *400*, 167–177. [CrossRef]

Corke, T.C.; Enloe, C.L.; Wilkinson, S.P. Dielectric barrier discharge plasma actuators for flow control. *Annu. Rev. Fluid Mech.* **2010**, *42*, 505–529. [CrossRef]

Moreau, E. Airflow control by non-thermal plasma actuators. *J. Phys. D Appl. Phys.* **2007**, *40*, 605. [CrossRef]

Sundaram, P.; Sengupta, S.; Suman, V.K.; Sengupta, T.K.; Bhumkar, Y.G.; Mathpal, R.K. Flow control using single dielectric barrier discharge plasma actuator for flow over airfoil. *Phys. Fluids* **2022**, *34*, 095134. [CrossRef]

Versailles, P.; Gingras-Gosselin, V.; Vo, H.D. Impact of pressure and temperature on the performance of plasma actuators. *AIAA J.* **2010**, *48*, 859–863. [CrossRef]

Tanaka, M.; Kubo, N.; Kawabata, H. Plasma actuation for leading edge separation control on 300-kW rotor blades with chord length around 1 m at a Reynolds number around 1.6×10^6. *J. Phys. Conf. Ser.* **2020**, *1618*, 052013. [CrossRef]

12. Kriegseis, J.; Möller, B.; Grundmann, S.; Tropea, C. Capacitance and power consumption quantification of dielectric ba discharge (DBD) plasma actuators. *J. Electrost.* **2011**, *69*, 302–312. [CrossRef]
13. Seth, U.; Traoré, P.; Duran-Olivencia, F.; Moreau, E.; Vazquez, P. Parametric study of a DBD plasma actuation based on Suzen-Huang model. *J. Electrost.* **2018**, *93*, 1–9. [CrossRef]
14. Wojewodka, M.M.; White, C.; Ukai, T.; Russell, A.; Kontis, K. Pressure dependency on a nanosecond pulsed dielectric ba discharge plasma actuator. *Phys. Plasmas* **2019**, *26*, 063512. [CrossRef]
15. Forte, M.; Jolibois, J.; Pons, J.; Moreau, E.; Touchard, G.; Cazalens, M. Optimization of a dielectric barrier discharge actu by stationary and non-stationary measurements of the induced flow velocity: Application to airflow control. *Exp. Fluids* **2** *43*, 917–928. [CrossRef]
16. Moreau, E.; Debien, A.; Benard, N.; Jukes, T.; Whalley, R.; Choi, K.; Berendt, A.; Podlinski, J.; Mizeraczyk, J. Surface diele barrier discharge plasma actuators. *ERCOFTAC Bull.* **2013**, *94*, 5–10.
17. Erfani, R.; Zare-Behtash, H.; Hale, C.; Kontis, K. Development of DBD Plasma Actuators: The Double Encapsulated Electr *Acta Astronaut.* **2015**, *109*, 132–143. [CrossRef]
18. Sato, S.; Furukawa, H.; Komuro, A.; Takahashi, M.; Ohnishi, N. Successively accelerated ionic wind with integrated dielec barrier-discharge plasma actuator for low-voltage operation. *Sci. Rep.* **2019**, *9*, 5813. [CrossRef]
19. Wojewodka, M.M.; White, C.; Kontis, K. Effect of permittivity and frequency on induced velocity in ac-DBD surface and cha plasma actuators. *Sens. Actuators A Phys.* **2020**, *303*, 111831. [CrossRef]
20. Abdollahzadeh, M.; Pascoa, J.; Oliveira, P. Comparison of DBD plasma actuators flow control authority in different mode actuation. *Aerosp. Sci. Technol.* **2018**, *78*, 183–196. [CrossRef]
21. Asada, K.; Ninomiya, Y.; Oyama, A.; Fujii, K. Airfoil flow experiment on the duty cycle of DBD plasma actuator. In Proceed of the 47th AIAA Aerospace Sciences Meeting including The New Horizons Forum and Aerospace Exposition, Orlando, FL, U 5–8 January 2009; American Institute of Aeronautics and Astronautics Inc.: Starbase, TX, USA, 2009. [CrossRef]
22. Fujii, K. Three Flow Features behind the Flow Control Authority of DBD Plasma Actuator: Result of High-Fidelity Simulat and the Related Experiments. *Appl. Sci.* **2018**, *8*, 546. [CrossRef]
23. Moreau, E.; Debien, A.; Breux, J.M.; Benard, N. Control of a turbulent flow separated at mid-chord along an airfoil with D plasma actuators. *J. Electrost.* **2016**, *83*, 78–87. [CrossRef]
24. Sato, M.; Nonomura, T.; Okada, K.; Asada, K.; Aono, H.; Yakeno, A.; Abe, Y.; Fujii, K. Mechanisms for laminar separated- control using dielectric-barrier-discharge plasma actuator at low Reynolds number. *Phys. Fluids* **2015**, *27*, 117101. [CrossRef
25. Sato, M.; Aono, H.; Yakeno, A.; Nonomura, T.; Fujii, K.; Okada, K.; Asada, K. Multifactorial effects of operating condition dielectric-barrier-discharge plasma actuator on laminar-separated-flow control. *AIAA J.* **2015**, *53*, 2544–2559. [CrossRef]
26. Sekimoto, S.; Nonomura, T.; Fujii, K. Burst-mode frequency effects of dielectric barrier discharge plasma actuator for separa control. *AIAA J.* **2017**, *55*, 1385–1392. [CrossRef]
27. Sekimoto, S.; Fujii, K.; Anyoji, M.; Miyakawa, Y.; Ito, S.; Shimomura, S.; Nishida, H.; Nonomura, T.; Matsuno, T. Flow Con around NACA0015 Airfoil Using a Dielectric Barrier Discharge Plasma Actuator over a Wide Range of the Reynolds Num *Actuators* **2023**, *12*, 43. [CrossRef]
28. Visbal, M.; Gaitonde, D.; Roy, S., Control of Transitional and Turbulent Flows Using Plasma-Based Actuators. In Proceeding the 36th AIAA Fluid Dynamics Conference and Exhibit, San Francisco, CA, USA, 5–8 June 2006. [CrossRef]
29. Benard, N.; Balcon, N.; Moreau, E., Electric Wind Produced by a Surface Dielectric Barrier Discharge Operating Over a W Range of Relative Humidity. In Proceedings of the 47th AIAA Aerospace Sciences Meeting Including the New Horizons Fo and Aerospace Exposition, Orlando, FL, USA, 5–8 January 2009. [CrossRef]
30. Erfani, R.; Zare-Behtash, H.; Kontis, K. Plasma actuator: Influence of dielectric surface temperature. *Exp. Therm. Fluid Sci.* **2** *42*, 258–264. [CrossRef]
31. Lilley, A.J.; Roy, S.; Michels, L.; Roy, S. Performance recovery of plasma actuators in wet conditions. *J. Phys. D Appl. Phys.* **2** *55*, 155201. [CrossRef]
32. Wicks, M.; Thomas, F.O. Effect of Relative Humidity on Dielectric Barrier Discharge Plasma Actuator Body Force. *AIAA J.* **2** *53*, 2801–2805. [CrossRef]
33. Benard, N.; Moreau, E.; Griffin, J.; Cattafesta, Louis, N., III. Slope seeking for autonomous lift improvement by plasma sur discharge. *Exp. Fluids* **2010**, *48*, 791–808. [CrossRef]
34. Benard, N.; Cattafesta, L.; Moreau, E.; Griffin, J.; Bonnet, J. On the benefits of hysteresis effects for closed-loop separation co using plasma actuation. *Phys. Fluids* **2011**, *23*, 083601. [CrossRef]
35. Broglia, R.; Choi, K.S.; Houston, P.; Pasquale, L.; Zanchetta, P. Output feedback control of flow separation over an aerofoil u plasma actuators. *Int. J. Numer. Anal. Model.* **2018**, *15*, 864–883.
36. Pasquale, L.; Durante, D.; Broglia, R. Flow separation prevention around a NACA0012 profile through multivariable feedb controlled plasma actuators. *Comput. Fluids* **2019**, *182*, 85–107. [CrossRef]
37. Patel, M.; Sowle, Z.; Corke, T.; He, C. Autonomous Sensing and Control of Wing Stall Using a Smart Plasma Slat. *J. Aircr.* **2** *44*, 516–527. [CrossRef]
38. Ogawa, T.; Asada, K.; Sekimoto, S.; Tatsukawa, T.; Fujii, K. Dynamic Burst Actuation to Enhance the Flow Control Authorit Plasma Actuators. *Aerospace* **2021**, *8*, 396. [CrossRef]

Segawa, T.; Suzuki, D.; Fujino, T.; Jukes, T.; Matsunuma, T. Feedback control of flow separation using plasma actuator and FBG sensor. *Int. J. Aerosp. Eng.* **2016**, *2016*, 8648919. [CrossRef]
Shimomura, S.; Sekimoto, S.; Oyama, A.; Fujii, K.; Nishida, H. Closed-Loop Flow Separation Control Using the Deep Q Network over Airfoil. *AIAA J.* **2020**, *58*, 4260–4270. [CrossRef]
He, C.; Corke, T.C.; Patel, M.P. Plasma Flaps and Slats: An Application of Weakly Ionized Plasma Actuators. *J. Aircr.* **2009**, *46*, 864–873. [CrossRef]
Viguera, R.; Anzai, Y.; Sasaki, Y.; Nonomura, T. Experimental Observations of Transient Flows in Separation Control Using a Plasma Actuator. *Actuators* **2023**. *12*, 218. [CrossRef]
Chatterjee, A. An introduction to the proper orthogonal decomposition. *Curr. Sci.* **2000**, *78*, 808–817.

actuators

MDI

Article

Flow Control of Flow Boiling Experimental System by Whale Optimization Algorithm (WOA) Improved Single Neuron PID

Yan Li [1], Miao Qian [1,2], Daojing Dai [3], Weitao Wu [1,4,*], Le Liu [1], Haonan Zhou [5] and Zhong Xiang [1,2]

[1] School of Mechanical Engineering, Zhejiang Sci-Tech University, Hangzhou 310018, China; 18378692126@163.com (Y.L.); meqiaomiao@zstu.edu.cn (M.Q.); 2023210501024@mails.zstu.edu.cn (L.L.); xz@zstu.edu.cn (Z.X.)
[2] Zhejiang Provincial Innovation Center of Advanced Textile Technology, Shaoxing 312000, China
[3] Zhejiang Xindebao Machinery Co., Ltd., Wenzhou 325000, China; db5@debaochina.com
[4] School of Computer and Information Technology, Shanxi University, Taiyuan 030006, China
[5] Zhejiang Yinlun New Energy Thermal Management System Co., Ltd., Taizhou 317200, China; zhouhaonan@yinlun.cn
* Correspondence: wwt@zstu.edu.cn

Abstract: In the present study, to address the issue of flow rate instability in the fl boiling experimental system, a flow rate adaptive control system is developed usin single-neuron PID adaptive algorithm, enhanced with the whale optimization algorit (WOA) for parameter tuning. A recursive least-squares online identification metho integrated to adapt to varying operating conditions. The simulation results demonstr that in step response the WOA-improved single-neuron PID significantly mitigates overshoot, with a mere 0.31% overshoot observed, marking a reduction of 98.27% compa to the traditional PID control. The output curve of the WOA-improved single-neuron I closely aligns with the sinusoidal signal, exhibiting an average absolute error of 0.1 which is lower than that of the traditional PID (0.209) and fuzzy PID (0.296). The WC improved single-neuron PID (1.01 s) exhibited a faster return to a stable state comparec the traditional PID (2.46 s) and fuzzy PID (1.28 s). Finally, the effectiveness of the algorit is validated through practical application. The results demonstrate that, compared traditional PID and single-neuron PID algorithms, the WOA-improved single-neuron I algorithm achieves an average flow stability of 9.9848 with a standard error of 0.0914394 exhibits superior performance, including faster rise and settling times, and higher stabi

Keywords: flow boiling experimental system; WOA-improved single-neuron PID; R flow control

Academic Editors: Hui Tang, Xin Wen and Feng Ren

Received: 4 November 2024
Revised: 23 December 2024
Accepted: 25 December 2024
Published: 27 December 2024

Citation: Li, Y.; Qian, M.; Dai, D.; Wu, W.; Liu, L.; Zhou, H.; Xiang, Z. Flow Control of Flow Boiling Experimental System by Whale Optimization Algorithm (WOA) Improved Single Neuron PID. *Actuators* **2025**, *14*, 5. https://doi.org/10.3390/act14010005

1. Introduction

Heat exchangers play a crucial role in various sectors including chemical engineeri automotive, food, and many others [1–3]. they facilitate processes such as heat trans temperature regulation, and energy efficiency enhancement. Fin and tube heat exchang are one of the most common types of compact heat exchangers [4]. Many factors influe the performance of heat exchangers [5–8]. The flow distribution characteristics of gas–liq two-phase fluid within the plates under various operating conditions are crucial fact influencing the performance of heat exchangers [9]. The flow distribution characteris of heat exchangers include flow distribution, pressure loss, temperature distribution, a so on. In the study of heat exchanger performance, research methods typically inclu physical experiments and numerical simulations. Mehrpooya M et al. [10,11] employ numerical simulation methods to model and analyze various types of heat exchang investigating their heat transfer performance under different structural configurations and operating conditions. Mehdi M et al. [12]. performed a three-dimensional CFD numerical simulation of a coaxial borehole heat exchanger and determined the optimal operating conditions for the heat exchanger. Their findings offer valuable insights for optimizing heat exchanger design. Many researchers have also conducted experimental studies on the heat exchange performance of various types of heat exchangers. Li Jun et al. [8] studied the influence of flow distribution deviation and nonuniformity on heat transfer performance and outlet temperature distribution by controlling flow rate, and the results show that the heat exchanger has the best performance under uniform inlet flow. Wei chao Lao et al. [13] conducted experimental studies on the flow boiling of refrigerant R134a in a plate heat exchanger with certain temperatures, pressures, and mass flow rates while also analyzing the effects of vapor quality and mass flux on the variation of heat transfer coefficient (HTC). Burak Markal et al. [14] studied the effect of inlet temperature on flow boiling behavior of expanding micro-pin-fin type heat sinks under a certain flow rate, and it was concluded that inlet temperature is an influential parameter for thermal characteristics, as are pressure drop values. From the previous references, it can be seen that scholars pay special attention to the parameters of flow, temperature, and pressure when studying the boiling flow characteristics of heat exchangers. These parameters directly influence the heat transfer mechanism, flow regime, system stability, and boiling efficiency. A systematic study of these parameters facilitates a better understanding and optimization of the boiling process, improves heat exchange efficiency, and reduces energy waste. Tahseen et al. [15]. believe that flow is an extremely important parameter that can affect the heat transfer coefficient and pressure drop of a heat exchanger. Flow rate directly impacts heat transfer efficiency, system stability, and safety, so flow control is critical in flow boiling experimental systems. Real-time flow control can track changes in operating conditions, preventing issues such as equipment overheating and bubble dry-out. Therefore, it is very important to develop a flow boiling experimental system with adaptive flow control. The system encompasses the measurement of system temperature, pressure, and flow rate, along with the adaptive control of pipeline flow and heating power.

To control the flow, many researchers have conducted studies in this area. Yunbin Cheng [16] designed a PID controller for the water circulation system in a dynamic flow detection device. Experimental results demonstrated that the control system effectively enhanced the stability of the water circulation system's flow rate. Qixing Liu [17] proposed an expert PID control algorithm for flow control systems, and simulation results indicated that the algorithm enhanced control accuracy and dynamic performance. E. B. Priyanka et al. [18] devised a fuzzy PID controller designed to automate oil pipeline systems. Experimental results demonstrated that the controller maintained flow within a specified range while achieving optimal performance and minimal time delay.

In summary, previous studies have primarily focused on the effects of flow, temperature, and pressure on heat exchanger performance; however, several research gaps remain, which this study aims to address:

Although PID, fuzzy PID, and expert PID controllers are widely used for flow control, they struggle to maintain optimal performance under the varying operating conditions of this experimental system. Traditional PID requires manual parameter tuning, which can result in instability and inefficiency as operating conditions change. Although fuzzy PID and expert PID methods offer greater flexibility, they still demand significant expertise for tuning and are sensitive to external disturbances, limiting their practical application in dynamic flow boiling experimental systems. In addition to this, the flow boiling experimental system lacks real-time and adaptive control, as most studies focus on the flow boiling characteristics of plate heat exchangers, with limited attention to the development of adaptive control systems. Real-time adaptive control is essential for maintaining timal performance and preventing issues such as overheating or bubble drying, wh directly impact the system's efficiency and safety. Absence of Effective Control Soluti for Variable Operating Conditions: Existing flow control methods often fail to prov adequate real-time adjustments to changes in load and operating conditions. As a re maintaining consistent and reliable performance in flow boiling experimental syst remains a challenge.

Novelty of This Research: Introduction of WOA-Improved Single-Neuron PID Ad tive Control: This study introduces a novel whale optimization algorithm (WOA)-impro single-neuron PID adaptive control system. Unlike traditional PID or fuzzy PID controll this method allows for more efficient parameter tuning without the need for man adjustments, making it suitable for flow boiling systems that experience varying ope ing conditions.

Adaptive Control for Flow Boiling Experimental Systems: The proposed con system is specifically designed to enhance the performance of flow boiling experime systems. By implementing real-time adaptive control, this research ensures that flow ca adjusted automatically, thus improving the system's ability to handle dynamic conditi By addressing these gaps, this research provides a more adaptive flow control solution flow boiling experimental systems.

The following are the novelties of this paper:

A single-neuron PID adaptive control algorithm optimized by the whale optimiza algorithm (WOA) is proposed for the flow control of the flow boiling experimental tem. The algorithm optimizes the scaling factor through WOA, overcomes the defec traditional PID requiring manual parameter adjustment, and significantly improves response speed and robustness of the system.

In order to solve the problem that the parameters of the flow rates model fluctu due to the change in operating conditions, this paper combines the RLS method for onl identification. This method improves the accuracy of model identification and enables control algorithm to dynamically adapt to the changes of the flow model.

2. System Design

Our research focuses on developing an experimental system for flow boiling t incorporates adaptive control ability. The experimental system requires to measure press temperature, and flow. Given that the refrigerant must enter the heat exchange plate w a specific dryness, it is essential to regulate the flow and temperature within the syst The system's experimental temperature is controlled by the heating power, the flow rat regulated by the motor speed, and the pressure is influenced by the quality of the injec refrigerant, heating power, and flow rate.

The experimental system comprises a computer (Savior y7000, Lenovo, Beijing, Chi data acquisition cards (I-7017, Hongge Technology Co., Ltd., Shanghai, China), flow me (DMF-1, Zhongxing Bona automation equipment Co., Ltd., Xian, China), reservoir (M by oneself), temperature sensors (thermocouple and PT100) (K type thermocouple PT100, Jiumao Automation (Dalian) Co., Ltd., Dalian, China), pressure sensors (JM- Anhui Jiamin instrument Co., Ltd., Anhui, China), preheater (Make by oneself), heat rods (Make by oneself), MS20CA power controllers (MS20CA, Zibo Silicon Microelectro Technology Co., Ltd., Zibo, China), gear magnetic pump (CS.Y015, Shenzhen Weik precision Technology Co., Ltd., Shenzhen, China), product (fins) (Make by oneself), c densers (Make by oneself), and chiller (DHF-50/80,Shanghai Sinovac Instrument Co., L Shanghai, China). The schematic diagram of the system is illustrated in Figure 1. The en experimental setup constitutes a cyclic system. At the onset of the experiment, refrigera injected into the reservoir, then the magnetic pump propels the refrigerant to flow through the pipeline. The condenser transforms the refrigerant from a gas state to a gas–liquid two-phase state. Then, the refrigerant flows through the flowmeter. The flowmeter is responsible for measuring the flow of refrigerant in the pipeline. Then, the refrigerant flows through the preheater and enters the product. Finally, the refrigerant passes through the second condenser and returns to the reservoir to form a cycle.

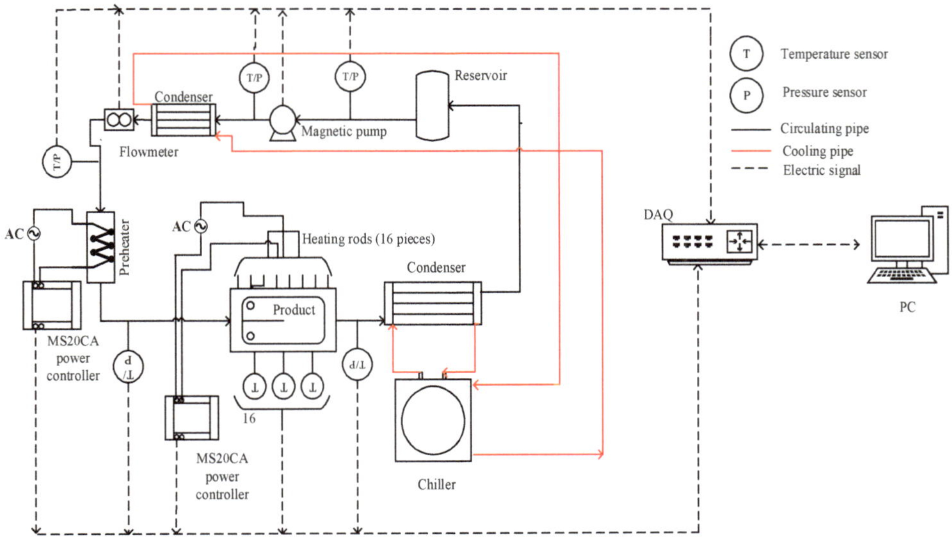

Figure 1. Flow boiling experimental system.

The chiller and the condenser form a cooling cycle, primarily tasked with reducing the temperature of the refrigerant within the experimental loop. The temperature sensor is used to measure the fluid temperature in each section of the pipeline, as well as the temperature of the experimental product. Likewise, the pressure sensor is utilized to measure the pressure of each section of the pipeline. The data acquisition card transmits the data collected by the sensors to the computer. The experimental conditions are established based on the measured temperature, pressure, and flow. With stable power output from the preheater and heating rod, the flow rate of the refrigerant at the inlet of the experimental section is regulated by adjusting the motor speed. This adjustment ensures that the refrigerant enters the product at the desired dryness level. Table 1 lists the technical specifications for key components.

Table 1. Technical specifications of key components.

Measured Parameter	Equipment/Model	Range	Accuracy
Temperature of product	K-type thermal couple	−50–800 °C	0.1 + 0.0017 |t|
Temperature of pipeline	Pt100 thermal resistance	−50–200 °C	±0.3% of full scale
Pressure	Pressure sensor/JM-801	0–1 MPa	±0.5% of full scale
Flow of pipeline	Flowmeter/DMF-1	0–25 kg/h	±0.2% of full scale
Magnetic pump speed	CS.Y015	0–4000 rpm	±0.5% of full scale

3. System Flow Rates Model Identification

In order to understand the dynamic characteristics of the system flow rates, design a more effective controller, and optimize the system performance, it is necessary to identify and simulate the control system. Typically, mathematical models can be established using theoretical analysis or experimental methods [19]. In this paper, the experime modeling method is used. During the offline identification process of the system, heating is turned off, and the system heating is considered as noise when designing flow adaptive controller.

The system flow rates model is solved through offline identification. A step sign used as the identification signal, with the servo motor controlling the system flow rate adjusting its speed. During the identification process, it was observed that various fac such as the refrigerant quality, coolant flow rate, and type of refrigerant, could resul different flow rates at the same motor speed. From the above description, the outpu the system (i.e., flow rate) is influenced not only by the current input (servo motor sp but also by the system's inherent characteristics and external factors, such as refriger quality, coolant flow rate, and refrigerant quality.

It is a common system identification method to use a step signal as the system in for model identification [20]. Sharma et al. [21] proposed using step input to iden continuous-time systems with dead time. This method can estimate the model parame more accurately. Dazi Li et al. [22]. presented a three-step closed-loop identification met based on step response and open-loop transformation. This proposed method avoids complexity of direct closed-loop identification and enhances the results and accurac identification for continuous systems with large time delays. Yongli Zhang et al. identified the experimental platform of a resonant system using the step response sys identification method, and the results indicate that the identified model fits well. previous literature indicates that a step signal can be used to identify continuous-t systems with dead time and delays. In practical applications, most systems are of this t including the system identified in this paper. Therefore, this paper employs a step sig for system identification. To determine the appropriate sampling frequency, data w collected at a higher sampling rate and analyzed using Fourier transformation, reveal that most of the system's spectral components are below 0.5 Hz. The system's traffi sampled every 1 s until it reaches a steady state, after which the data is imported into MATLAB System Identification Toolbox for offline identification. Due to the ambigu order of the system, the primary models utilized in practical applications are first-, seco and third-order models. Therefore, this paper primarily identifies these three mod compares their fit, and conducts a comprehensive analysis to select the best model.

The main operating range for this system is 800 rpm to 2500 rpm. To identify system, flow response data were collected at motor speeds of 1000 rpm, 1500 rpm, 2000 rpm, under conditions characterized by a coolant temperature of −10 °C, a cool speed duty cycle of 80%, a refrigerant mass of 1.8 kg, and an initial pressure of 370 k Table 2 shows the corresponding experimental conditions.

Table 2. Experimental condition.

Experiment	Coolant Temperature/°C	Coolant Speed Duty Cycle/%	Refrigerant Mass/kg	Initial Pressure/kPa
Case1	−10	80	1.8	370
Case2	−10	100	1.3	300

From (a) in Figure 2, it is evident that the steady-state value of the system fl increases with motor speed. The system exhibits an overshoot in response to the step sig indicating oscillatory characteristics, which suggests that the system model is at lea second-order system [24]. Furthermore, the repeated experiment with a motor spee 2000 rpm in (a) of Figure 2 demonstrates that the system operates stably.

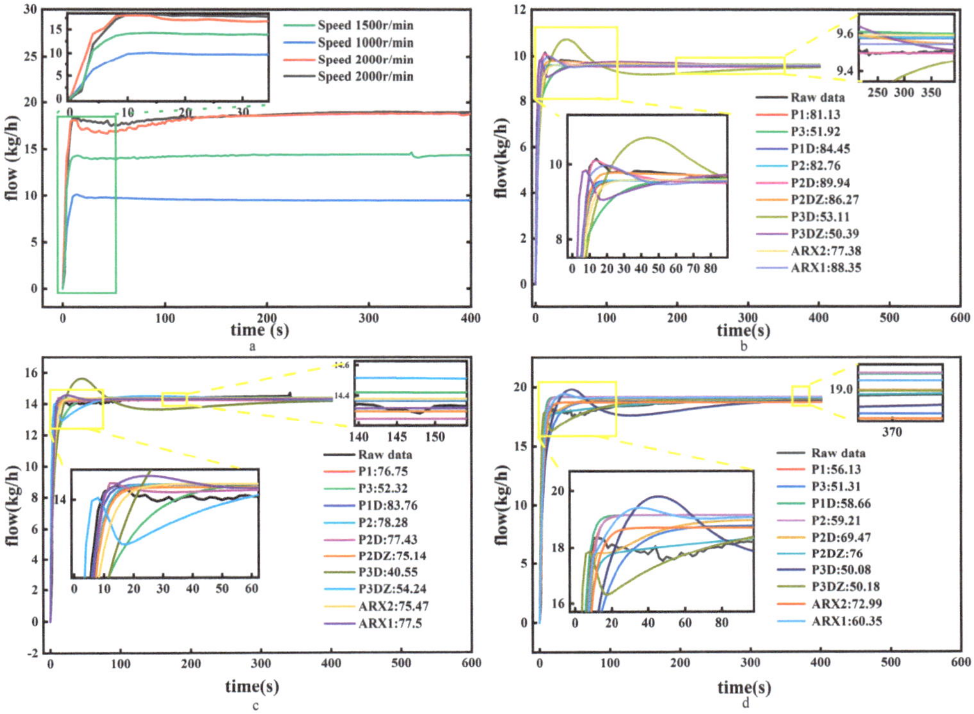

Figure 2. (**a**) The system flow step response; (**b**) fitness of identification flow model at 1000 rpm speed, (**c**) fitness of identification flow model at 1500 rpm speed, and (**d**) fitness of identification flow model at 2000 rpm speed model fitness.

The flow data identification model at a motor speed of 1000 rpm was utilized, while the flow data at 1500 rpm and 2000 rpm were employed to validate the model. The model fitness [25] was employed to evaluate the identification results. A higher model fitness value indicates better performance of the identification system, approaching the actual system.

From the results of the identification in (b) of Figure 2, it is evident that the models P1 (first-order without delay), P1D (first-order with delay), P2 (second-order without zero point and delay), P2D (second-order without zero point but delay), P2DZ (second-order with one zero point and delay), and ARX1 (linear regression equation) demonstrate higher fitness, all exceeding 80%. From the verification set presented in (c) and (d) of Figure 2, it can be observed that the model fitness of P1, P1D, P2, P2D, P2D, and ARX1 under the data at 1500 rpm remains relatively consistent, all exceeding 70%. However, the fitness of P1, P1D, P2, P2D, and ARX1 models under the data at 2000 rpm is notably reduced to less than 60%, whereas P2DZ remains above 70%. Thus, it can be concluded that the P2DZ model, which includes two poles and one zero point with delay, performs the best. The identified model is as follows:

$$G(s) = \frac{0.09654s + 0.009529}{111.4s^2 + 71.11s + 1} e^{(-2s)} \tag{1}$$

s is a complex variable, a variable of the Laplace transform. $e^{(-2s)}$ represents the dynamic behavior of a pure delay (or pure lag). The linear dynamic response of the system to the input is represented as 0.09654 s, and is related to the rate of change in the input signal. The constant term 0.009529 usually represents the static gain of the system, that is, the steady-state value of the system output when there is no dynamic change in the input.

It relate to the initial flow rates of the system. The quadratic term $111.4s^2$ indicates that system is a second-order system, usually with inertial or lagging behavior. This kin behavior is often encountered in motor control systems, especially when the system mechanical delays or inertia. The value 71.11 s is related to the damping characteris of the system. The smaller the damping, the more likely the system is to oscillate. denominator constant term is set to 1 to standardize the transfer function. $e^{(-2s)}$ indic that the input signal of the system begins to affect the system after 2 s.

In practical applications, the system transfer function in the s-domain shoul discretized [26]. The z-transform results are as follows:

$$y(k) = 0.001072u(k-1) - 0.0008663u(k-2) \\ -(1.3e-5)u(k-3) + 1.259y(k-1) \\ -0.2788y(k-2)$$

$u(k-1)$ represents the previous system input, $u(k-2)$ represents the input fr two steps ago, $u(k-3)$ represents the input from three steps ago, $y(k-1)$ represents previous system output, and $y(k-2)$ represents the output from two steps ago. From discretization formula, it is evident that the current system output is influenced by inputs of the first three. This delay indicates that the system's response to the input sig exhibits a certain historical dependence. Furthermore, the system's output also depe on the outputs from the previous moment and the two preceding moments, sugges the presence of a feedback effect. Notably, the current output is most influenced by previous output.

When factors such as the quantity of refrigerant introduced into the system cha the flow model of the system may also be affected. To verify the variation of the f rates model under different working conditions, flow data of the system were collec under the conditions of Table 2 when the input signal of the system was set to 1000 r Coolant temperature and coolant speed duty cycle are properties of the cooling system influence the pressure, flow, and temperature within the circulating system. Refrige mass is a characteristic of the circulatory system that affects both the pressure and flov the system.

From Figure 3, it is evident that the steady-state value of the system flow is influer by factors such as refrigerant quality. Changes in refrigerant mass may cause the tran function to change. This varying property complicates accurate modeling of the sys using a transfer function with fixed parameters, necessitating the use of adaptive contro online adjustment control strategies. The model identified under case 2 is as follows:

$$G_2(s) = \frac{0.79719s + 0.0056841}{1114.3s^2 + 122.58s + 1} e^{(-2s)}$$

Equation (3) represents the relationship between the input speed and the output f rate. Upon comparing Equations (1) and (3), it becomes evident that the coefficient the system model vary under different conditions. These variations suggest that a fi control strategy may not be adequate to maintain optimal performance across all opera scenarios. Therefore, an adaptive control strategy, which dynamically adjusts the con parameters based on real-time feedback, is essential.

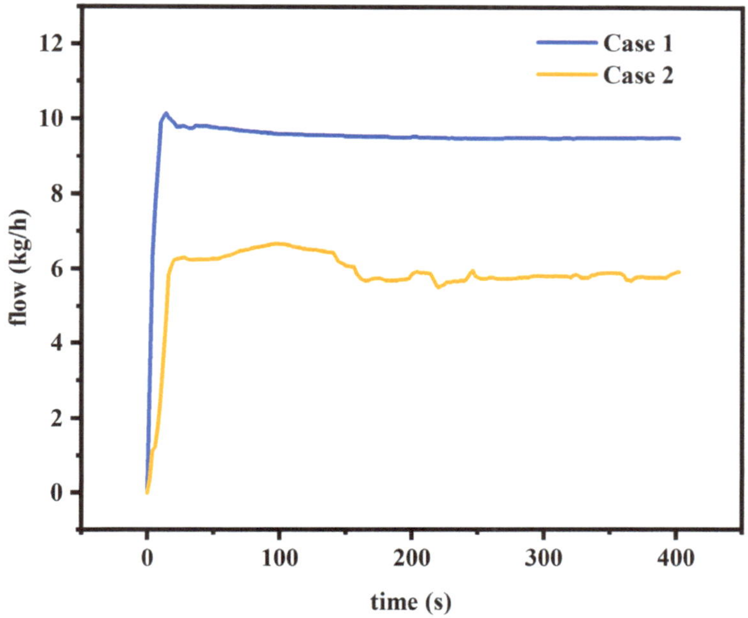

Figure 3. Flow response at motor speed of 1000 rpm under case 1 and case 2.

4. WOA Improved Single Neuron PID Control Algorithm

Due to the variation in system model parameters with changing conditions, an adaptive algorithm is necessary to control the system flow. The single-neuron PID [27] serves as an adaptive control algorithm. This paper employs an incremental PID combined with a single neuron to enhance control performance.

The control model of the single-neuron PID is depicted in Figure 4, where $r(k)$ represents the set flow (expected value) and $e(k)$ denotes the difference between the expected value and the output value of the device $w1'(k)$, $w2'(k)$, and $w3'(k)$ denote the connection weights of the single neuron, while k represents the gain of the system output variable. $\Delta u(k)$ stands for the input increment of the controlled object, $u(k)$ represents the total input of the controlled object, Z^{-1} signifies the system delay, $y(k)$ denotes the output of the controlled system, and $e(k)$ is processed to obtain $x1(k)$, $x2(k)$, and $x3(k)$. These are expressed as:

$$x1(k) = e(k) \tag{4}$$

$$x2(k) = e(k) - e(k-1) \tag{5}$$

$$x3(k) = e(k) - 2e(k-1) + e(k-2) \tag{6}$$

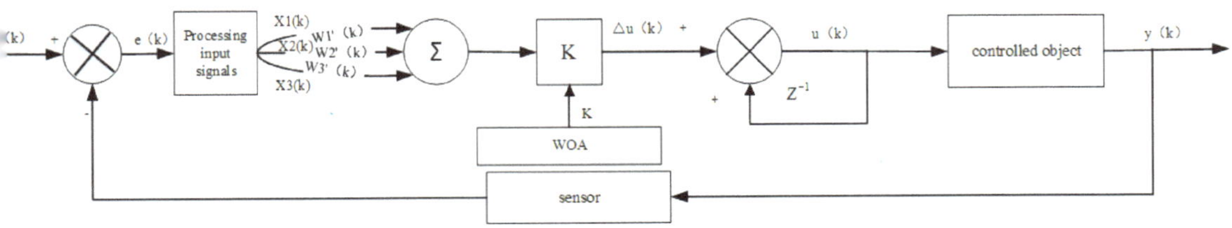

Figure 4. Single-neuron PID control block diagram.

In numerous practical applications, engineers have observed that the online adj ment of weights is primarily associated with $e(k)$ and $e(k-1)$.

Therefore, the weight formula is expressed as follows:

$$w_i(k) = w_i(k-1) + \eta e(k)u(k)(2e(k) - e(k-1))$$

At this time, the input control of the controlled system can be written as follows:

$$u(k) = u(k-1) + K\sum_{i=1}^{3} w_i'(k)x_i(k)$$

$$w_i' = \frac{w_i(k)}{\sum_i^3 |w_i(k)|}$$

In a single-neuron PID controller, the scale factor K is a crucial parameter that de mines the relationship between the neuron output and input. The scale factor K is clos linked to the dynamic characteristics, error margin, expected control response, and exte interference of the system. If the system's dynamic response is slow, a larger k value r be necessary to enhance the response speed. Conversely, for systems with a fast dyna response, a smaller k value can help prevent overly aggressive control. By adjusting th value through an optimization algorithm, the controller's performance can be significa improved, allowing the system to maintain good response characteristics under comp working conditions. This paper optimizes this value using the WOA.

The WOA [28] is a novel heuristic optimization approach inspired by the hunt tactics of humpback whales. The WOA mimics the predatory behaviors of whales, wh can be categorized into three main strategies: Surrounding Predation, Bubble Net Att and Search for Prey. The WOA algorithm used in this paper comes from the literature [Through comparing the optimization results of the WOA and the Particle Swarm Optim tion (PSO) algorithm for one-dimensional functions, it is observed that the converge speed of the WOA surpasses that of PSO. The WOA reaches the optimal value wit 5 iterations, whereas PSO requires 10 iterations. These results suggest that WOA exhi superior optimization ability compared to PSO.

In the controller designed in this article, ITAE [30] is chosen as the fitness funct of the WOA algorithm. ITAE not only indicates the magnitude of the error (con accuracy) but also reflects the speed of error. To ensure that the ITAE fitness functio the WOA algorithm places greater emphasis on system overshoot, an overshoot pen term is introduced when overshoot occurs. The updated fitness function is provide Equation (10).

$$F = \begin{cases} \int_0^\infty t|e(t)|dt, e > 0 \\ \int_0^\infty t|e(t)|dt + \exp(|e|), e \le 0 \end{cases}$$

To improve the accuracy of approximating the real system when using the wh optimization algorithm for optimization, the Recursive Least Squares [31] method is ployed to perform online identification of the flow rate model. To evaluate the accurac the Recursive Least Squares (RLS) method, simulation identification is performed on aforementioned Equation (2). The true values of a1, a2, b1, b2, and b3 are 1.259, −0.2 0.001072, −0.0008663, and −0.000013. And with approximately 500 iterations, the e mated values of a1, a2, b1, b2, and b3 are 1.258644, −0.278447, 0.001071, −0.0008663, a −0.00001362; the estimated values closely approximate the true values. It is evident t the RLS method exhibits strong identification capability. Considering the variability of flow model coefficients under different working conditions, Equation (1) is used as initial system during online identification. When the actual output of the system devi from the output of the identification model beyond a certain threshold, the RLS method is employed to conduct the online identification process and obtain new model parameters.

The combination of a single-neuron PID and the whale optimization algorithm is shown in Figure 5. The parameter changes of the system flow rates model are primarily attributed to variations in the system itself, such as the quality of the refrigerant injected into the system and the system's heating power. Based on the above analysis, before starting single-neuron PID control, only one run of RLS is needed to identify the flow model of the current experimental system and one run of the WOA optimization algorithm to find the optimal scale factor K of the current model. This scale factor will be used continuously in subsequent control. When the output error between the identification system and the actual system exceeds 10%, the RLS and WOA methods should be reactivated for optimization.

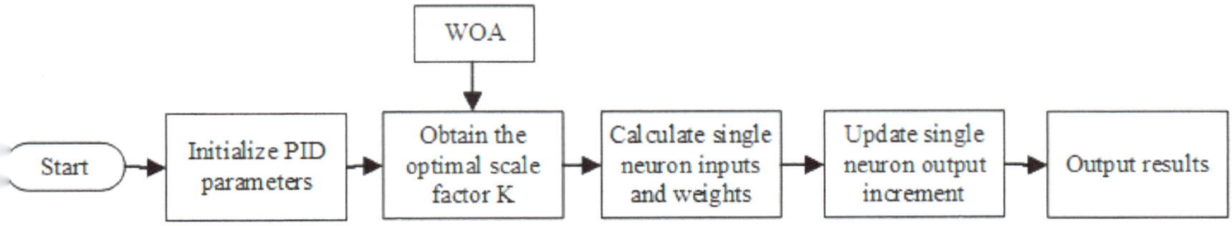

Figure 5. WOA improved single-neuron PID control block diagram.

5. Results

To assess the feasibility and positional tracking capability of the enhanced whale-optimization-based single-neuron PID control algorithm, simulations are conducted against the standard PID algorithm, the WOA-improved single-neuron PID controller, and the fuzzy PID control algorithm. All simulations are conducted on LabVIEW.

The simulation system employs the model described by Equation (2). The resulting step response and sine signal response are shown in Figures 6 and 7. In the step response, the initial step value is 10, and the step value transitions to 15 at t = 3 s. In the sine signal response, the sinusoidal signal frequency is 3, the amplitude is 2, and the offset is 2.

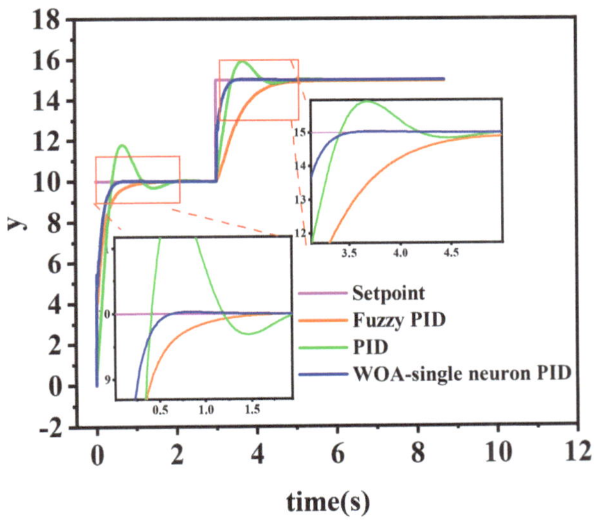

Figure 6. Step response simulation of three control algorithms.

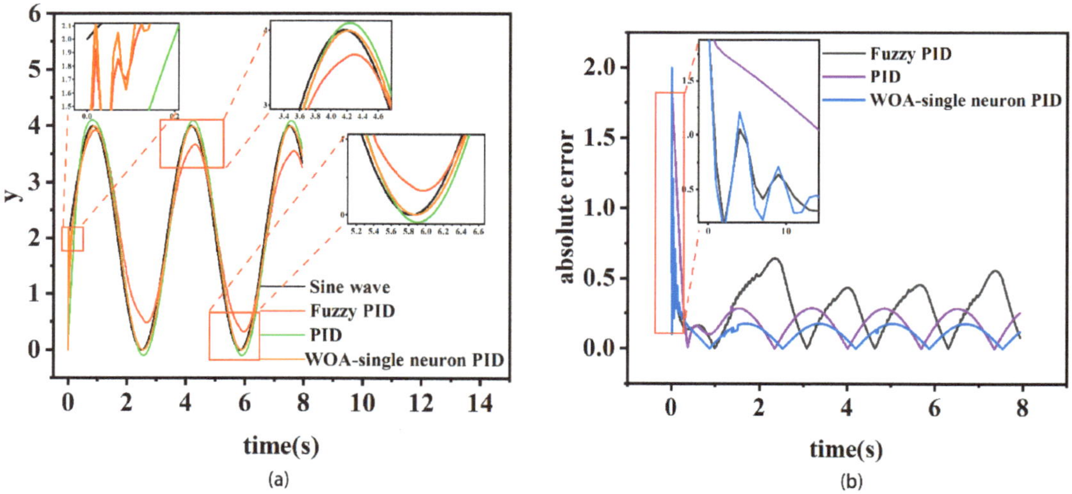

Figure 7. (**a**) Sine signal response and (**b**) absolute error of sine signal response.

From the Figure 6, it is apparent that the conventional PID control curve demonstr considerable overshoot and requires a longer time to achieve a steady state; the in overshoot reaches 18.01%, and the initial adjustment time needs 1.69 s. Although fuzzy PID has no overshoot, its adjustment time is longer, reaching 0.88 s. In contrast, WOA-improved single-neuron PID significantly mitigates the overshoot, with a mere 0. overshoot observed, marking a reduction of 98.27% compared to the traditional PID con The adjustment time is 0.45 s, presenting a reduction of 48.86% compared to that of fuzzy PID control method. In the system's steady state, when the step value changes, traditional PID control curve still exhibits overshoot, with the overshoot peaking at 6.2 the adjustment time is recorded at 1.04 s. The fuzzy PID method similarly demonstr no overshoot, with a longer adjustment time, extending to 1.42 s. In contrast, the W improved single-neuron PID showcases a minimal overshoot of only 0.22%, markir reduction of 96.46% compared to the traditional PID control. Additionally, the adjustm time is 0.32 s, representing a reduction of 69.23% compared to PID. The Y-axis repres the constants.

It can be seen from the sine signal response diagram that the traditional PID parame can achieve better results after adjustment. However, the traditional PID control cu cannot accurately fit the sinusoidal signal, and there is a phenomenon of tracking ah The fuzzy rules of fuzzy PID are taken from the literature [32]. The output curve of WOA-improved single-neuron PID is closer to the sinusoidal signal. Upon compari of the tracking absolute error curves, it is evident that the tracking absolute error of WOA-improved single-neuron PID is markedly smaller than that of the traditional and fuzzy PID methods. The average absolute error [33] for the traditional PID stand 0.209, while that for the fuzzy PID is 0.296. In contrast, the average absolute error for WOA-improved single-neuron PID is 0.120, representing a 45.58% reduction compare the traditional PID. This underscores the superior position tracking effectiveness of improved single-neuron PID based on WOA among the three algorithms.

To ascertain the resilience of the WOA-improved single-neuron PID against exte interference, an instantaneous disturbance with an amplitude of 5 is introduced to system output when it is in a stable state [34]. The step response of the three con algorithms to this external interference is illustrated in Figure 8. The Y-axis repres the constants. From Figure 8, it is evident that the WOA-improved single-neuron exhibits the highest resilience against external interference. The traditional PID requires 2.46 s to return to a stable state, while the fuzzy PID necessitates 1.28 s. In comparison, the WOA-improved single-neuron PID demonstrates the swiftest recovery to a stable state, taking only 1.01 s. This represents a 1.45 s improvement over the traditional PID and a 0.27 s enhancement over the fuzzy PID.

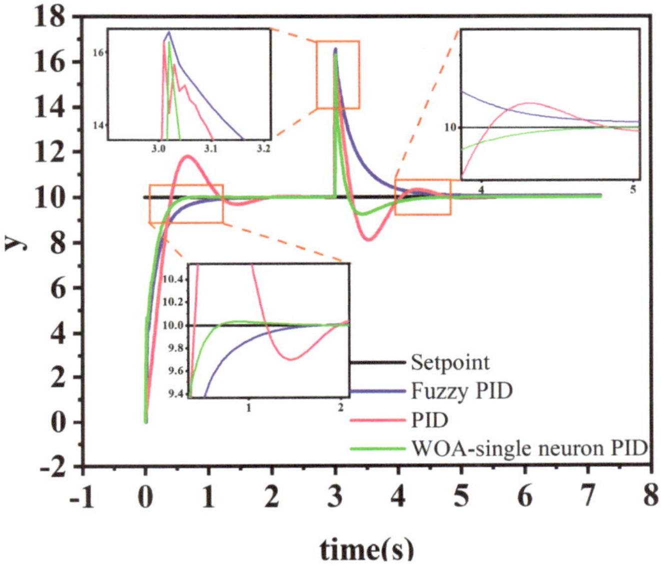

Figure 8. Anti-interference ability test of three algorithms.

The online identification simulation results of the system are presented in Figure 9. As illustrated, when the parameters of the original system model change, the model identified by the algorithm can rapidly adapt to fit the system.

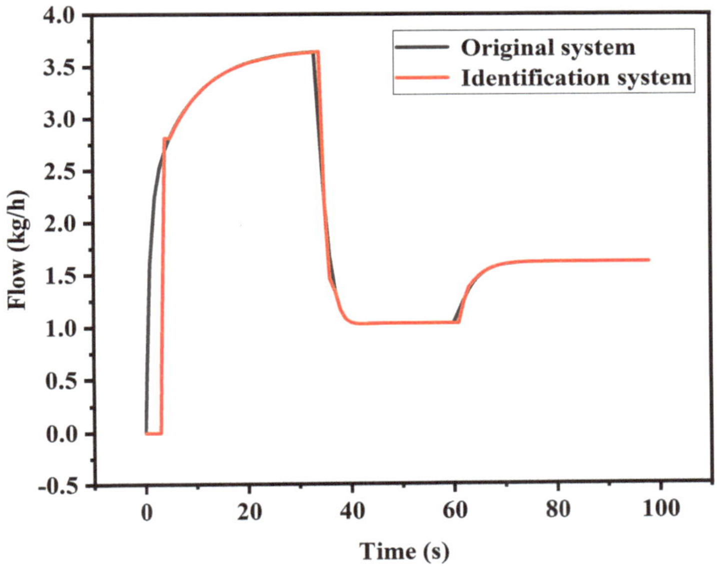

Figure 9. System online identification simulation.

In order to verify the effectiveness and reliability of the RLS method and the WC improved single-neuron PID control algorithm in the flow boiling experimental system, experiment was carried out in the flow boiling experimental system. Figure 10 is a phys diagram of the entire system.

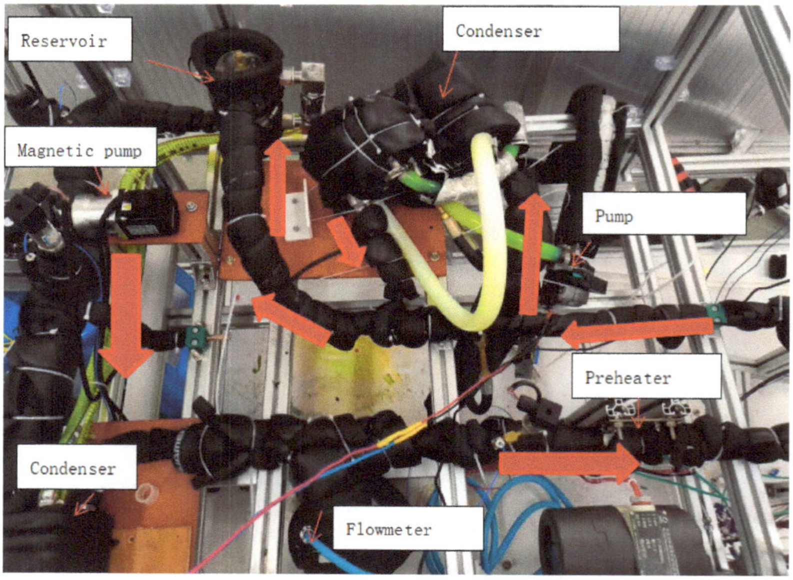

Figure 10. Physical system.

In order to verify the effectiveness and reliability of the RLS method and the WC improved single-neuron PID control algorithm in the flow boiling experimental syst the experiment was carried out in the flow boiling experimental system. Figure 10 i physical diagram of the entire system. The step response of identification system and act system are as Figure 11. When the experimental refrigerant volume changes from 1.8 k 1.7 kg and the coolant temperature changes from −10 °C to −9 °C, we conducted onl identification of the traffic model. The initialization values of a1, a2, b1, b2, and b3 are 1. −0.2788, 0.001072, −0.0008663, and −0.000013. After online identification, a1, a2, b1, and b3 are 1.084, −0.2504, 0.003644, −0.0032430, and 0.001175. The step response of identified model is presented in Figure 11a. It can be seen that when the working conditi change, the system model identified by the RLS method is in good agreement with actual system. The error at each sampling point is basically within 10%, and the maxim is 8%. The final stable value tends to be consistent, and the steady-state error between identification system and the actual system is within 5%. Therefore, it is feasible to use RLS method in this system.

To validate the efficacy of the control algorithms in the flow boiling experimen system, flow data is collected upon issuing control signals, with an acquisition inter of 1 s. The control strategies employed include PID, single-neuron PID, and the WC improved single-neuron PID. PID parameters are determined using the empirical tr and-error method. The control parameters of the single-neuron PID are derived from previous experiment.

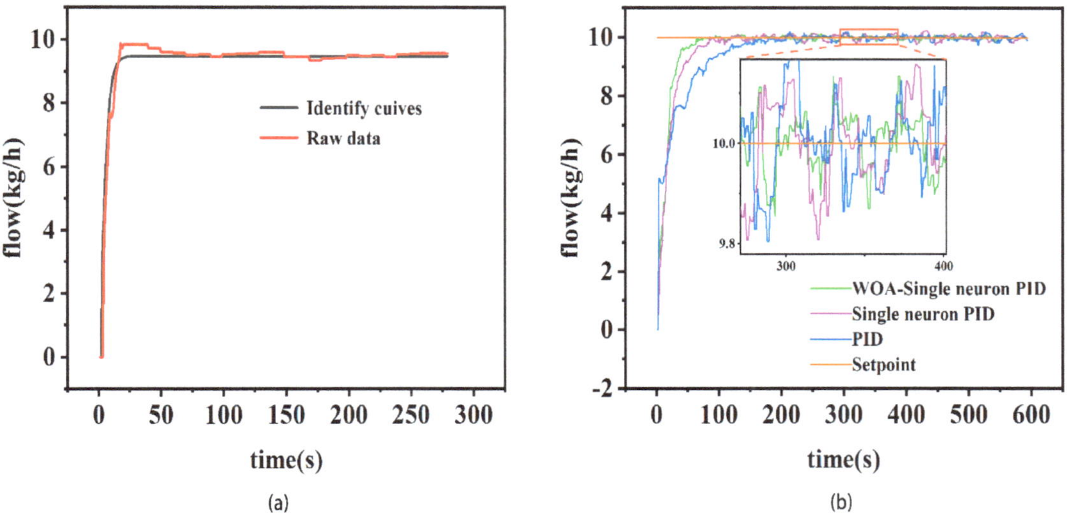

Figure 11. (**a**) Step response of identification system and actual system and (**b**) the control effect of three control algorithms on system flow.

The flow control outcomes are presented in (b) of Figure 11. Table 3 presents the performance of the three control algorithms. The table and figure demonstrates that the WOA-improved single-neuron PID algorithm outperforms traditional PID and single-neuron PID algorithms across multiple performance metrics. In terms of dynamic response, the WOA-improved single-neuron PID achieves the shortest rise time of 29 s, significantly faster than the 40 s of single-neuron PID and the 78 s of traditional PID. This indicates that the WOA-improved single-neuron PID enables the system to reach the target value more quickly, thereby reducing the time required for startup or adjustments in practical applications.

Table 3. Control performance index.

Control Algorithm	Rise Time/s	Adjusting Time/s	Steady-State Average Flow Rates
PID	78	110	9.9659
Single-neuron PID	40	63	9.9799
WOA-improved single-neuron PID	29	38	9.9848

Regarding adjustment time, the WOA-improved single-neuron PID exhibits superior performance, with an adjustment time of only 38 s, compared to 63 s for single-neuron PID and 110 s for traditional PID. This highlights the effectiveness of incorporating the WOA-improved single-neuron PID in improving the system's adaptability, enabling it to stabilize faster after disturbances or initial changes while minimizing overshoot and oscillations.

The steady-state average flow achieved by the WOA-improved single-neuron PID is 9.9848, which is closer to the target value compared to 9.9799 for single-neuron PID and 9.9659 for traditional PID. Additionally, the WOA-improved single-neuron PID achieves the lowest standard error of 0.091439, outperforming the 0.109586 of single-neuron PID and the 0.126469 of traditional PID. This reflects the ability of the WOA-improved single-neuron PID to maintain higher stability and precision during long-term operation, effectively reducing fluctuations and errors.

6. Discussion

In this paper, the WOA-improved single-neuron PID algorithm is integrated with Recursive Least Squares algorithm (RLS) to establish a robust adaptive control framew The findings of this study demonstrate the significant advantages of the WOA-impro single-neuron PID algorithm in addressing flow control challenges in flow boiling ex imental systems. The proposed approach demonstrates substantial improvements traditional PID and fuzzy PID controllers. The simulation and experimental results re significant reductions in overshoot, adjustment time, and steady-state error achieved by WOA-improved single-neuron PID algorithm. Traditional PID relies on fixed parame and manual adjustment. Similarly, fuzzy PID, though more flexible, requires dom specific expertise and struggles with parameter adaptation under varying conditions. WOA-improved single-neuron PID algorithm effectively adapts to variations in sys parameters, such as changes in refrigerant mass and coolant temperature, ensuring st and accurate performance. This adaptability is particularly advantageous in dyna operating environments. Future investigations could expand upon these findings by e uating the algorithm's performance under more diverse and extreme conditions, suc multi-variable system configurations or transient disturbances in large-scale indus systems. While these advancements highlight superior control precision, the computati demands introduced by iterative WOA optimization warrant further analysis. Fu studies could explore strategies to balance computational efficiency with real-time con performance, particularly in applications requiring high-speed response.

The proposed control framework aligns with the broader need for intelligent adaptive control systems in modern industrial applications. The demonstrated abilit integrate the WOA-improved single-neuron PID algorithm with identification techniq (RLS) suggests that such hybrid approaches could be extended to other dynamic syste including multi-phase flow systems, chemical reactors, and distributed energy man ment systems. Future research could focus on augmenting the proposed framework v predictive or model-based control methods to enhance its anticipatory capabilities. A tionally, integrating mechanisms to mitigate external disturbances or sudden param shifts could further improve system reliability and robustness.

In conclusion, the WOA-improved single-neuron PID algorithm not only addresses limitations of traditional control methods but also provides a flexible and adaptive solu with significant implications for both academic research and industrial applications.

7. Conclusions

The research focus of this paper is flow rate control in a flow boiling experime system. We obtained the flow model of the experimental system through offline identi tion and adopted the RLS identification method to address the issue of changes in mo parameters. We then compared the control effects of the WOA-improved single-neu PID controller and the other two control algorithms through simulation. Finally, the fe bility of the WOA-improved single-neuron PID controller was verified through spe experiments. The main conclusions drawn are as follows.

1. This paper identifies the flow model of the experimental system, and the identi model is a second-order system, with model parameters influenced by factors s as the quantity of refrigerant injected into the system. Therefore, in order to con the flow rates of the system, this paper adopts the WOA-improved single-neu PID algorithm.
2. For the changes in the system flow model parameters, this paper adopts RLS on identification to improve control accuracy. The online identification results dem strate that the identified system model exhibits improved fitting to the actual sys

The steady-state error between the identification system and the actual system is within 5%.

3. Based on the simulation and experimentation, it has been concluded that the WOA-improved single-neuron PID adaptive controller outperforms PID and single-neuron PID in terms of speed and accuracy in controlling the flow of the experimental platform. It eliminates the need for manual parameter setting and can automatically determine the current optimal control parameters based on the online identification of the system model. The algorithm improved flow stability, achieving a standard error of 0.0914394, lower than that of the compared methods, and a steady-state flow average of 9.9848, closely matching the set value.
4. Under external interference, the WOA-improved single-neuron PID (1.01 s) controller exhibited the fastest recovery to stability, outperforming the traditional PID (2.46 s) and fuzzy PID (1.28 s), showcasing its robustness and adaptability.
5. The control strategy proposed in this article has certain engineering application value in the field of control systems with model parameter change.

Author Contributions: Conceptualization, Y.L. and M.Q.; methodology, Y.L., M.Q., and Z.X.; software, Y.L. and M.Q.; validation, M.Q., W.W., and Y.L.; formal analysis, W.W.; investigation, D.D.; resources, H.Z.; data curation, Y.L.; writing—original draft preparation, Y.L. and L.L.; writing—review and editing, W.W. and M.Q.; visualization, Y.L.; supervision, M.Q.; project administration, D.D., H.Z., and Z.X.; funding acquisition, W.W. All authors have read and agreed to the published version of the manuscript.

Funding: This work was supported partly by the Major Scientific and Technological Project of Zhejiang Province (No. 2024C03118).

Data Availability Statement: Data are contained within the article.

Conflicts of Interest: The company was not involved in the study design, collection, analysis, interpretation of data, the writing of this article or the decision to submit it for publication. The remaining authors declare that the research was conducted in the absence of any commercial or financial relationships that could be construed as a potential conflict of interest.

Abbreviations

WOA	whale optimization algorithm
PID	proportional-integral-derivative control
RLS	recursive least squares
ITAE	integrated time and absolute error
RPM	revolutions per minute

·ferences

Michel, B.; Dufour, N.; Börtlein, C.; Zoude, C.; Prud, E.; Gremillard, L.; Clausse, M. First experimental characterization of $CaCl_2$ coated heat exchanger for thermochemical heat transformer applications in industrial waste heat recovery. *Appl. Therm. Eng.* **2023**, *227*, 120400. [CrossRef]

Luo, D.; Wu, Z.; Yan, Y.; Ji, D.; Cheng, Z.; Wang, R.; Li, Y.; Yang, X. Optimal design of a heat exchanger for automotive thermoelectric generator systems applied to a passenger car. *Appl. Therm. Eng.* **2023**, *227*, 120360. [CrossRef]

Hedayati, S.; Ansarifar, E.; Jafari, S.M. 4—Plate heat exchangers in the food industry. In *Thermal Processing of Food Products by Steam and Hot Water*; Jafari, S.M., Ed.; Woodhead Publishing: Cambridge, UK, 2023; pp. 111–128. [CrossRef]

Reshaeel, M.; Abdelsamie, M.; Ali, M. A critical review of the thermal-hydraulic performance of fin and tube heat exchangers using statistical analysis. *Int. J. Thermofluids* **2024**, *24*, 100858. [CrossRef]

Peng, Q.; Luo, Y.; Sun, X.; Li, J.; Chen, Y.; Dang, C. Study on the effects of heat transfer fluid (HTF) temperature and flow velocity on energy storage/release performance of shell and tube phase change heat exchanger. *J. Energy Storage* **2025**, *107*, 114935. [CrossRef]

6. Wang, M.; Wang, J. Flow boiling of nanofluids in microchannel heat exchangers: A critical review. *Therm. Sci. Eng. Prog.* **2** *55*, 102930. [CrossRef]
7. Simon, J.R.; Bandhauer, T.M. An experimentally validated evaporative phase change heat transfer model for low mass applications using R134a in plate heat exchangers. *Int. J. Refrig.* **2021**, *131*, 604–614. [CrossRef]
8. Li, J.; Gao, Y.; Liu, C.; Yan, X.; Zhao, H.; Xu, Y.; Meng, E. Effect of flow maldistribution on heat transfer performance temperature field of plate-fin heat exchangers. *Int. Commun. Heat Mass Transf.* **2023**, *149*, 107135. [CrossRef]
9. Zheng, W.; Jiang, Y.; Cai, W.; Li, F.; Wang, Y. Numerical investigation on the distribution characteristics of gas-liquid flow at entrance of LNG plate-fin heat exchangers. *Cryogenics* **2021**, *113*, 103227. [CrossRef]
10. Mehrpooya, M.; Ghafoorian, F.; Mohammadi Afzal, S.P.; Mirmotahari, S.R.; Ganjali, M.R. A comprehensive transient heat trar simulation of U-tube borehole heat exchanger considering porous media and subterranean water seepage. *Chem. Pap.* **2** *78*, 6315–6329. [CrossRef]
11. Mehrpooya, M.; Mirmotahari, S.R.; Ghafoorian, F.; Karimkhani, M.; Ganjali, M.R. Investigation of a packed bed energy stor system with different PCM configurations and heat transfer enhancement with fins using CFD modeling. *Chem. Pap.* **2** *78*, 2453–2467. [CrossRef]
12. Mehdi, M.; Farzad, G.; Shayan, F. 3D-Modeling of a Coaxial Borehole Heat Exchanger in Sahand Field, Northwest Considering the Porous Medium and Presence of Nanofluids. *Iran. J. Chem. Chem. Eng.* **2024**, *42*, 3898–3916. Available on https://www.magiran.com/paper/2714981 (accessed on 3 November 2024).
13. Lao, W.-c.; Fang, Y.-d.; Chen, Q.-h.; Xu, L.-j.; Yang, H.-n.; Huang, Y.-q. Experimental investigation on the boiling of R134a plate heat exchanger with mini-wavy corrugations. *Int. J. Refrig.* **2024**, *162*, 53–63. [CrossRef]
14. Markal, B.; Evcimen, A.; Aydin, O. Effect of inlet temperature on flow boiling behavior of expanding micro-pin-fin type h sinks. *Int. Commun. Heat Mass Transf.* **2023**, *149*, 107143. [CrossRef]
15. Tahseen, T.A.; Ishak, M.; Rahman, M.M. An overview on thermal and fluid flow characteristics in a plain plate finned un-finned tube banks heat exchanger. *Renew. Sustain. Energy Rev.* **2015**, *43*, 363–380. [CrossRef]
16. Cheng, Y.; Hou, X. Water. Circulation Control System of Dynamic Flow Detection Device Based on PID. In Proceedings of 2019 4th International Conference on Mechanical, Control and Computer Engineering (ICMCCE), Hohhot, China, 24–26 Octo 2019; pp. 324–328.
17. Liu, Q.X. Design of Flow Control System Based on Expert PID. In Proceedings of the 2016 International Symposium on Compu Consumer and Control (IS3C), Xi'an, China, 4–6 July 2016; pp. 1031–1034. [CrossRef]
18. Priyanka, E.B.; Maheswari, C.; Thangavel, S. Online Monitoring and Control of Flow rate in Oil Pipelines Transportation Sys by using PLC based Fuzzy-PID Controller. *Flow Meas. Instrum.* **2018**, *62*, 144–151. [CrossRef]
19. Padhee, S.; Pati, U.C.; Mahapatra, K. Modelling switched mode DC-DC converter using system identification techniques review. In Proceedings of the 2016 IEEE Students' Conference on Electrical, Electronics and Computer Science (SCEECS), Bho India, 5–6 March 2016; pp. 1–6.
20. Tan, K.C.; Li, Y. Evolutionary system identification in the time domain. *Proc. Inst. Mech. Eng. Part I J. Syst. Control. Eng.* **1** *211*, 319–323. [CrossRef]
21. Sharma, S.; Padhy, P.K. A data driven approach to identify continuous-time systems with dead-time using step input. Proceedings of the TENCON 2019—2019 IEEE Region 10 Conference (TENCON), Kerala, India, 17–20 October 2019; pp. 632–
22. Li, D.; Zeng, F.; Jin, Q.; Tan, T. A three-step identification method based on step response and open loop-closed loop transforma In Proceedings of the 2010 8th World Congress on Intelligent Control and Automation, Jinan, China, 7–9 July 2010; pp. 461–
23. Zhang, Y.; Wang, X.; Geng, L.; Li, X. Parameter Identification of Resonant System Based on Step Response Method. In Proceedi of the 2023 IEEE International Conference on Image Processing and Computer Applications (ICIPCA), Changchun, China, 11 August 2023; pp. 1076–1081.
24. Sarwito, S.; Semin Hidayaturrahman, T. Analysis of transient response first order and second order theory in pneumatic con system using feedback instrument type PCM140. In Proceedings of the 2017 International Conference on Advanced Mechatror Intelligent Manufacture, and Industrial Automation (ICAMIMIA), Surabaya, Indonesia, 12–14 October 2017; pp. 155–160.
25. Ahsan, J.; Ahsan, M.; Jamil, A.; Ali, A. Grey Box Modeling of Lateral-Directional Dynamics of a UAV through System Identificat In Proceedings of the 2016 International Conference on Frontiers of Information Technology (FIT), Islamabad, Pakistan, 19 December 2016; pp. 324–329.
26. Goncharov, V.; Sokolova, E.; Kudryavtsev, A.; Song, J. The Transformation of Continuous Transfer Function in Discrete. Proceedings of the 2015 International Conference on Computational Intelligence and Communication Networks (CICN), Jabal India, 12–14 December 2015; pp. 585–588.
27. Ma, F.; Lu, J.; Liu, L.; He, Y. Application of Improved Single Neuron Adaptive PID Control Method in the Angle Predefined L of Active Radar Seeker for Anti-radiation Missile. In Proceedings of the 2021 IEEE 4th Advanced Information Managem Communicates, Electronic and Automation Control Conference (IMCEC), Chongqing, China, 18–20 June 2021; pp. 2160–216
28. Mirjalili, S.; Lewis, A. The Whale Optimization Algorithm. *Adv. Eng. Softw.* **2016**, *95*, 51–67. [CrossRef]

Li, Z.; Chen, W.; Sun, X.; Chen, W.; Fang, Z.; Yang, G. Single neuron adaptive control and parameter optimization of lower limb rehabilitation exoskeleton. In Proceedings of the 2022 IEEE 17th Conference on Industrial Electronics and Applications (ICIEA), Chengdu, China, 16–19 December 2022; pp. 313–318.
Rao, C.S.; Santosh, S. Tuning optimal PID controllers for open loop unstable first order plus time delay systems by minimizing ITAE criterion. *IFAC-PapersOnLine* **2020**, *53*, 123–128. [CrossRef]
Gonzalez-Villagomez, J.; Gonzalez-Villagomez, E.; Rodriguez-Donate, C.; Palillero-Sandoval, O. FPGA-based reconfigurable unit for systems identification through RLS algorithm. In Proceedings of the 2023 IEEE International Autumn Meeting on Power, Electronics and Computing (ROPEC), Ixtapa, Mexico, 18–20 October 2023; pp. 1–6.
Demir, M.H.; Cetin, S.; Haggag, O.; Demir, H.G.; Worek, W.; Premer, J.; Pandelidis, D. Independent temperature and humidity control of a precooled desiccant air cooling system with proportional and fuzzy logic + proportional based controllers. *Int. Commun. Heat Mass Transf.* **2022**, *139*, 106451. [CrossRef]
Zhang, Y.; Yin, Y.; Meng, Z.; Chen, D.; Qin, W.; Wang, Q.; Dai, D. Development and testing of a grain combine harvester throughput monitoring system. *Comput. Electron. Agric.* **2022**, *200*, 107253. [CrossRef]
Cao, S.; Zhao, W.; Zhu, A. Research on intervention PID control of VAV terminal based on LabVIEW. *Case Stud. Therm. Eng.* **2023**, *45*, 103002. [CrossRef]

MDPI AG
Grosspeteranlage 5
4052 Basel
Switzerland
Tel.: +41 61 683 77 34

Actuators Editorial Office
E-mail: actuators@mdpi.com
www.mdpi.com/journal/actuators

www.ingramcontent.com/pod-product-compliance
Lightning Source LLC
Chambersburg PA
CBHW041408210926
51850CB00013B/3211

9783725834815